D1205706

MATHEMATICS
The Alphabet
of Science

MATHEMATICS
The Alphabet of Science
SECOND EDITION

MARGARET F. WILLERDING
Professor of Mathematics
San Diego State University

RUTH A. HAYWARD
Senior Engineer
General Dynamics Corporation
Electro Dynamic Division

JOHN WILEY & SONS, INC.
New York   London   Sydney   Toronto

WILLIAM MADISON RANDALL LIBRARY UNC AT WILMINGTON

PICTURE CREDITS

Chapters 1–5, 7, 8, 11–13, The Bettmann Archive, Inc.;
Chapters 6 and 9, Charles Phelps Cushing;
Chapter 10, Wide World Photos; Chapter 14, IBM.

Copyright © 1972, by John Wiley & Sons, Inc.

All rights reserved. Published simultaneously in Canada.

No part of this book may be reproduced by any means,
nor transmitted, nor translated into a machine language without the written permission of the publisher.

Library of Congress Catalog Card Number: 70-180274

ISBN 0-471-94661-3

Printed in the United States of America.

10 9 8 7 6 5 4 3 2

QA39
.2
.W555
1972

To our loyal friends
Gai
Suzie
Meg
Maggie

212997

PREFACE

This revision of *Mathematics: The Alphabet of Science* is based on many letters, comments, reviews, and evaluations received by the authors, and upon the classroom experiences of many instructors and students since the first edition in 1967. The original text has been rewritten and supplemented with these comments, criticisms, and suggestions as a guide.

The book is intended for liberal arts students and other students who wish to know what mathematics is about, but who have no desire to be mathematicians. The presentation is designed so that even persons who have little or no high school mathematics, or those whose mathematics study is so far in the past that it is completely forgotten, will have no trouble following the explanations.

The topics selected for discussion are simple yet profound. Many have applications in fields other than mathematics. Explanations have been made as detailed as is reasonably possible. The manner in which the subjects are treated will cause no problems even to the most mathematically unsophistocated student. For a complete understanding of the topics presented, the reader should verify the calculations and ponder the arguments as he encounters them. But even if the reader reads this book as he reads a novel, he will find enough to acquaint him with some exciting topics of mathematics both ancient and modern.

Numerous illustrative examples are presented throughout the book to clarify the use of mathematical ideas. The number of exercises in all chapters has been substantially increased and the answers to all odd-numbered problems are included in the text.

Chapter 1 presents the basic notions of mathematical logic and gives the reader a respect for correct reasoning and a clear understanding of the type of deductive reasoning used in mathematics. In this edition a new section on negation of compound statements has been added and deMorgan's Laws are presented.

Chapters 2, 3, and 4 present topics from Number Theory. To justify the inclusion of these chapters, if such a justification is necessary, we quote an address of G. H. Hardy given before the American Mathematical Society.

> The elementary theory of numbers should be one of the very best subjects for early mathematical instruction. It demands very little previous knowledge; its subject matter is tangible and familiar; the processes of reasoning which it employs are simple, general and few; and it is unique among the mathematical sciences in its appeal to natural, human curiousity. A month's intelligent instruction in the Theory of Numbers ought to be twice as useful and at least ten times more entertaining than the same amount of calculus for engineers.

Chapter 5, Sets, Relations, and Functions, is completely new in this edition. Here we review sets, set operations, Cartesian products, and the Cartesian plane. Relations and functions are defined; function notation is presented; and linear and quadratic functions are discussed in detail and graphed.

Chapters 6 and 10 study abstract mathematical systems and stress the postulational method. Chapter 6, Groups and Fields, has been completely rewritten and supplemented. The modulo-seven system discussed in Chapter 6 is presented here as an example of a finite field (the F_7 field). Solution sets of linear equations over F_7 are found and graphed. Polynomials over F_7 are also discussed and graphed.

Chapter 10, Finite Geometry, gives the reader an opportunity to study a geometry with only a finite number of points as contrasted with the infinite geometry studied in high school. The numerous letters received by the authors praising this chapter have resulted in the presentation being virtually unchanged from the first edition.

Chapter 7, The Pythagorean Theorem, is substantually the same. The only change being the addition of a section on the distance formula from analytic geometry. This chapter, like Chapter 10, seems to have many fans among the users of the first edition.

Chapter 8 has, to a large extent, been rewritten. The fundamental counting principle and a discussion of permutations and combinations have been added. The discussion of these topics facilitates a deeper understanding of the probability theory presented and makes possible a slightly more sophistocated presentation of probability measure.

Chapter 9, Matrices, has been lengthened to include the solution of matrix equations, solutions of systems of linear equations using matrices, and determinants of square matrices.

Chapter 11, Analytic Geometry: The Straight Line and the Circle, is entirely new. This chapter includes the midpoint formula, the analytic geometry of the line and the circle, and analytic proofs of theorems from elementary geometry.

Chapters 12, 13, and 14, present some of the fundamental principles and the mathematical concepts related to the digital computer. The materials presented are those basics necessary in the effective use of a computer. These so-called "thinking" machines have become an integral part of life in this scientific age, and the authors feel that every educated person should understand their workings. The old adage "first things first" was primary in the selection of the materials included in these chapters.

One of the first prerequisites of a working knowledge of the sewing machine is how to thread it; similarly, to understand the workings of a computer one must have some idea of how the machines themselves operate and how they are programmed to solve problems. Many of the students using this text will someday have positions in which such knowledge will be invaluable as well as necessary. With these objectives in mind Chapters 12, 13, and 14 were written.

Chapter 12 covers systems of numeration used by computers and persons working with computers. All present-day, large-scale digital computers use the binary system of numeration to represent data. The binary system and other related systems (those based on powers of two) used in the computer field are covered. The decimal system is presented before the other positional numeration systems as an aid in the understanding of numeration systems with bases other than ten.

Chapter 13 attempts to explain some of the basics of the digital computer without becoming technical. Some topics covered are basic machine language (bit patterns) related to assembly language, complement arithmetic, and representation of real numbers versus integers. These topics show the need for studying the systems of numeration presented in Chapter 12.

Chapter 14 is a brief treatment of a higher level compiler language, FORTRAN. Compiler languages are used in most programs written today

PREFACE

for computers. They provide a relevant set of mnemonics to simulate the problem to be solved. Although these symbols must be translated into machine language by the compiling program, the user need not know details about either of these. FORTRAN was chosen to represent the world of compiler languages since its structure is very much like the structure of mathematical equations.

The authors would like to thank all those students and instructors who used the first edition of Mathematics: The Alphabet of Science and particularly those who wrote and made suggestions as to its improvement. We would like to thank Professor Walter J. Gleason, Bridgewater State College, Bridgewater, Massachusetts; Dr. Kelvin Casebeer, Southwestern State College, Weatherford, Oklahoma; Professor C. Ralph Verno, West Chester State College, West Chester, Pennsylvania; and Professor Henry Harmeling, Jr., North Shore Community College, Beverly, Massachusetts for their critical reviews of the manuscript and for their helpful suggestions for improving the book.

San Diego, California   MARGARET F. WILLERDING
February 1972   RUTH A. HAYWARD

CONTENTS

CONTENTS

CONTENTS

MATHEMATICS
The Alphabet
of Science

George Boole (1815–1864) discovered that the symbolism of algebra can be used not only for making statements about numbers, but also about mathematical logic. In his book called *The Mathematical Analysis of Logic*, Boole developed the idea of formal logic and introduced an algebra into the field of logic. Today, Boolean algebra is important not only in the field of logic, but also in the geometry of sets, the theory of probability, and other fields of mathematics.

LOGIC

1

1.
HISTORY

The importance of studying the procedures of correct reasoning—that is, **logic**—is obvious to all educated persons. The ancient Greeks developed logic as an area of study, and one of the leaders in this field, Aristotle, categorized arguments according to definite forms. These lines of reasoning are still identified as those of Aristotelian logic.

The growth of mathematics in the nineteenth century caused mathematicians to examine critically the reasoning procedures which had been used in developing mathematics. They discovered that many arguments used by earlier mathematicians were either incomplete or incorrect. Efforts to correct these mistakes and to avoid similar ones renewed interest in logic and resulted in a great extension and development of the study of logic. This new development is called **mathematical** or **symbolic logic** to distinguish it from the old Aristotelian logic.

Traditionally, logic is divided into two parts, **inductive logic** and **deductive logic.** A conclusion reached by repeated observations, as in a series of experiments, illustrates induction; a conclusion based on so-called "if–then" reasoning, as in mathematical proofs, represents deduction. Both Aristotelian logic and symbolic logic are parts of deductive logic. Since the deductive method is an important part of modern mathematical thinking, we shall discuss only deductive logic and refer to it simply as **logic.**

One of the principal concerns of logic is the analysis of the process of establishing the truth of general statements. In mathematics this process is called **proof.** Our objective is to introduce the most fundamental ideas of logic and an understanding of the nature of mathematical proof.

2.
DIAGRAMS IN DEDUCTIVE REASONING

The use of diagrams often is helpful in deciding whether our reasoning in an argument is correct. Suppose the following facts are assumed:

> All factory workers are union members.
> Kramer works in a factory.

What conclusion can be drawn? Figure 1.1 shows that the set of all factory workers (*F*) is included within the set of union members (*U*) as our first assumption states. The rectangle in Figure 1.1 represents the set of all people (*P*); the larger circle represents the set of all union members (*U*); and the smaller circle represents the set of all factory workers (*F*). If a person is a factory worker he must be represented by a point in the smaller circle. Moreover, any point outside the smaller circle cannot represent a factory worker. Since the smaller circle falls completely within the larger one, any point in the smaller circle also must be in the larger one. Thus we can conclude that Kramer belongs to a union. We write this **argument** as follows:

> All factory workers are union members.
> Kramer works in a factory.
> _____
> ∴ Kramer belongs to a union.

The symbol ∴ indicates the conclusion and is read "therefore." Notice that a line separates the assumptions, called **premises,** from the **conclusion** in the argument.

If the premises in the above argument are altered slightly, no valid conclusions can be drawn. Suppose we assume the following:

> All factory workers are union members.
> Kramer does not work in a factory.

Using Figure 1.1, we see that no valid conclusion can be drawn. A person who is not a factory worker may or may not be a union member.

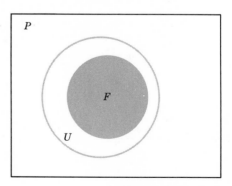

FIGURE 1.1

3

The following examples illustrate the use of diagrams to check the validity of an argument.

Example 1: Is the argument below a valid one? That is, does the conclusion follow from the premises?

> All mathematicians are intelligent.
> Some women are not intelligent.
> ──────────────────────────────
> ∴ Some women are not mathematicians.

Solution: First we draw a diagram representing the premises in the argument. We use a rectangle to represent the set of all people (*P*). The first premise states that the set of all mathematicians, represented by the circle *M*, must be included within the set of all intelligent people, represented by circle *I*, as shown in Figure 1.2.

The second premise states only that *some* women are not intelligent. This means that there are points in circle *W*, representing the set of all women, that are not points of circle *I*. In mathematical logic the word "some" means "at least one." Hence the second premise in our argument tells us that at least one point of circle *W* is not a point of circle *I*. We see that there are many possible placements for circle *W* in our diagram three of which are shown in Figure 1.3. In each of these diagrams at least one point of *W* is not a point of *I*. Examining these three diagrams, we see that in each case the conclusion, some women are not mathematicians (meaning that at least one woman is not a mathematician) is valid.

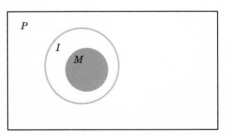

FIGURE 1.2

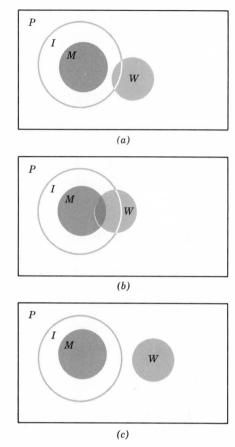

(a)

(b)

(c)

FIGURE 1.3

Example 2: Is the following argument valid? That is, does the conclusion follow from the premises?

> No ooks are acks.
> All kooks are ooks.
> All pocks are acks.
> _____
> ∴ No kooks are pocks.

Solution: The diagram in Figure 1.4 shows that the conclusion follows from the premises.

This argument shows that one may reason correctly even though the premises have no meaning.

5

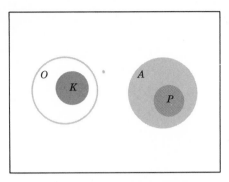

FIGURE 1.4

Diagrams are helpful in deciding the validity of a conclusion, but they must be used with caution. For example, using Figure 1.3a or Figure 1.3c, we might be tempted to conclude that no mathematicians are women, which is not necessarily true as shown in Figure 1.3b.

An argument in which the conclusion follows from the premises is said to be **valid.** The conclusion may be true or it may be false. The validity has nothing to do with the truth of the premises or the conclusion.

3.
UNDEFINED TERMS

The first requirement for an understanding of any subject, be it golf or mathematics, is to know the meaning of the terms that are used. As children we learned the meaning of a word, such as "cat," when someone pointed to an animal and called it "cat." Later we acquired the habit of looking up the definitions of unfamiliar words in a dictionary. A little experience in using a dictionary convinces us that some words must be **undefined.** Unless we understand some words, a dictionary will lead us in a circle. For example, suppose we look up the adjective "stubborn." We may find that, as we look up each definition given, we are ultimately led back to "stubborn" (Figure 1.5).

To avoid circular definitions we must accept a small number of words that will be undefined. The choice of these words depends on the subject. With these basic undefined words and nontechnical words from the

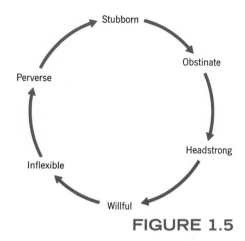

FIGURE 1.5

English language, we now define other words in terms of our undefined words and nontechnical English words. For example, if we accept "point," "line," and "betweenness" as undefined words, we may now define the line segment PQ, where P and Q are two points on a line, as the set consisting of the points P and Q and all the points on the line between P and Q.

4.
STATEMENTS

With our undefined words, nontechnical words, and the words we define, we form sentences called **statements** or **propositions.** These statements must have a truth value. That is, they must be either true or false, but not both. The following sentences are statements. The truth value of each is given in parentheses.

A cow has four legs.	(true)
Elephants can fly.	(false)
Hawaii is the fiftieth state of the U.S.A.	(true)
Water is wet.	(true)
July has forty days.	(false)
Two is an odd number.	(false)

The following sentences are not statements by our definition because as they stand they have no truth value. That is, we cannot tell as they are written whether they are true or false.

> That number is divisible by 6.
> He is an honor student in mathematics.
> She is fifteen pounds overweight.

In a deductive system such as mathematics we accept a set of statements, called **axioms** or **postulates,** as true. Once these initial propositions are stated, we combine them into more complicated statements. We attempt to prove these complicated statements true by a process called **logical reasoning.** Those propositions that turn out to be true are called **theorems.**

Exercise 1

1. Which of the following are statements? Give the truth value of each statement.
   (a) January is the first month of the year.
   (b) She is sixteen years old.
   (c) Texas is the largest state in the United States.
   (d) All cats have lavender eyes.
   (e) Ice is cold.
   (f) Buttermilk is a dairy product.
   (g) He is a juvenile delinquent.
   (h) The sum of 6 and 9 is 12.
   (i) That number is a prime number.
2. Using diagrams, draw a valid conclusion from each set of premises.
   (a) No even number is divisible by 3.
       Six is an even number.
   (b) All boys who drive sports cars are reckless drivers.
       John drives a sports car.
   (c) All rational numbers are real numbers.
       All real numbers can be graphed on the real number line.
3. Using diagrams, draw a valid conclusion from each set of premises.
   (a) Some girls are talkative.
       Talkative people are not popular.
   (b) All irrational numbers are real numbers.
       $\sqrt{2}$ is an irrational number.
   (c) All girls are beautiful.
       All career women are girls.
4. Using diagrams, draw a valid conclusion from each set of premises.

    (a) All Californians drink wine.
        Doris lives in California.
    (b) All peeps are creeps.
        No parfaits are creeps.
    (c) All rings are groups.
        All modular systems are rings.

5. Using diagrams, draw a valid conclusion from each set of premises.
    (a) All girls have two ears.
        Meg is a girl.
    (b) All handsome men are college students.
        No freshmen are college students.
    (c) No excellent talkers are bores.
        Some excellent talkers are college professors.

6. Test the validity of the following argument by use of a diagram.

All mathematicians are bores.
All bores are stupid.
_____

∴ All mathematicians are stupid.

7. Test the validity of the following argument by use of a diagram.

All shy creatures are finks.
Some shy creatures are rats.
Some students are shy creatures.
_____

∴ Some students are rat finks.

8. Draw a diagram for each of the following arguments. Use the diagram to tell whether the conclusion of each argument follows from the premises.

    (a) All church-goers are good people.
        Mrs. Lehman goes to church.
        _____

        ∴ Mrs. Lehman is a good person.

    (b) All San Franciscans are residents of California.
        All Hippies are residents of California.
        _____

        ∴ All Hippies are San Franciscans.

9. Draw a diagram for each of the following arguments. Use the diagram to tell whether the conclusion of each argument follows from the premises.

    (a) All women are intelligent animals.
        My cat, Madame Nu, is an intelligent animal.
        _____

        ∴ My cat is a woman.

    (b) All beer drinkers have red noses.
        Tom has a red nose.
        _____

        ∴ Tom is a beer drinker.

10. Draw a diagram for each of the following arguments. Use the diagram to tell whether the conclusion of each argument follows from the premises.

> (a) All senators are over twenty-one years of age.
> Fred is over twenty-one years of age.
> ───────────────────
> ∴ Fred is a senator.

> (b) All even numbers are integers.
> All odd numbers are integers.
> ───────────────────
> ∴ All odd numbers are even numbers.

5.
COMPOUND STATEMENTS

Logical reasoning is the process of combining given statements into other statements, and then doing this over and over and over again. There are many ways to combine statements, all of which are derived from four fundamental operations: (1) conjunction, (2) disjunction, (3) implication, and (4) negation.

In a **conjunction** we combine two given statements by placing an "and" between them. Thus the conjunction of the two statements

> (a) Ross is a physics major.
> (b) Wilson is a mathematics major.

is "Ross is a physics major and Wilson is a mathematics major."

A conjunction is defined as true if both of the statements, called **components,** used to form it are true. Thus, if either component is false, the conjunction is defined as false. Of course, if both components are false, the conjunction likewise is false.

It is customary to denote statements by lower-case letters of the alphabet such as p, q, r, and s. We denote the conjunction, whose components are p and q by the symbol

$$p \wedge q$$

read "p and q." We summarize the truth values of $p \wedge q$ below:

if p is true and q is true, then $p \wedge q$ is true
if p is false and q is true, then $p \wedge q$ is false
if p is true and q is false, then $p \wedge q$ is false
if p is false and q is false, then $p \wedge q$ is false

The above information can be given more compactly in the form of a **truth table.** Table 1.1 is a truth table giving the truth values of the conjunction. In the table, T means that the corresponding statement is true; F, that it is false. This table is merely an expansion of our description of conjunction. It says that if both p and q are true, then $p \wedge q$ is true; if one or the other or both p and q are false, then $p \wedge q$ is false.

To form a truth table all possible combinations of true and false are entered in the p and q columns, and the truth or falsity corresponding to each combination is entered in the column denoting this compound statement.

In a **disjunction** we combine two statements, called the **components** of the disjunction by placing "or" between them. The disjunction using the preceding statements (a) and (b) as components is "Ross is a physics major or Wilson is a mathematics major." The disjunction is defined to be true if either one or the other or both of its components are true and false otherwise. Notice that the connective "or" when used in a disjunction is used in the **inclusive** sense; that is "or" means one component or the other or both.

In everyday language we often use "or" in the **exclusive** sense; that is, "or" means one component or the other but not both. For example, when a student says "I shall get an A or a B in this course" he means that he will get a grade of A or a grade of B but not both grades in the course. Hereafter in this book, the word "or" will always be used in the inclusive sense unless stated otherwise.

TABLE 1.1

p	q	$p \wedge q$
T	T	T
T	F	F
F	T	F
F	F	F

TABLE 1.2

p	q	$p \vee q$
T	T	T
T	F	T
F	T	T
F	F	F

We denote the disjunction with components p and q by the symbol

$$p \vee q$$

read "p or q." Table 1.2 shows the truth values of the disjunction $p \vee q$.

6.
NEGATION

Any statement may be negated by preceding it by "it is not true that" or "it is false that." We call the denial of any statement the **negation** of that statement. If p is any statement, its negation is denoted by the symbol $\sim p$, read "not p" or "p is false."

If p denotes the statement "Lee is argumentative", then the negation of p, denoted by $\sim p$, is "It is not true that Lee is argumentative" or "It is false that Lee is argumentative."

When we negate statements by using words or phrases other than "it is not true that" or "it is false that", we must be extremely careful to obtain a statement that has *exactly* the same meaning as the statement obtained by prefixing "it is not true that" or "it is false that." For example, if q denotes the statement "Spiders have six legs", the negation of q, $\sim q$,

TABLE 1.3

p	$\sim p$
T	F
F	T

denotes the statement "It is not true that spiders have six legs." This negation asserts *only* that the statement "Spiders have six legs" is false. It does not assert that spiders have two legs or eight legs or one thousand legs. All we can conclude from the negation of the statement is that spiders definitely do not have six legs.

Clearly, p and $\sim p$ have opposite truth values. This fact is shown by the entries in Table 1.3.

7.
IMPLICATION

A very important compound statement in mathematics is the **implication.** It takes the form of an "if-then" sentence. The following statements are implications.

If I study hard, then I shall pass this course.
If it rains, then the picnic will be canceled.
If I get paid, then I shall go to the movies tonight.

From any two statements we can form two implications. From the statements "It is raining" and "There is a rainbow," we can form the implications:

If it is raining, then there is a rainbow.
If there is a rainbow, then it is raining.

In the implication "If it is raining, then there is a rainbow," "it is raining" is called the **antecedent,** and "there is a rainbow" is called the **conclusion.** In the implication "If there is a rainbow, then it is raining," "there is a rainbow" is the antecedent and "it is raining" is the conclusion.

In ordinary speech it is customary for the antecedent and the conclusion to be related, as:

If I go skiing, then I may break my leg.

It is generally meaningless to combine two apparently unrelated statements, as:

If $2 \times 2 = 4$, then the sky is blue.

13

In mathematical logic we shall not hesitate to combine unrelated statements in an implication.

If p and q are symbols denoting the antecedent and conclusion, respectively, we use the symbol $p \rightarrow q$ to denote the **implication** and read it "if p, then q." The symbol $p \rightarrow q$ may also be read

> p implies q
>
> q provided p
>
> q if p
>
> p is a sufficient condition for q
>
> p only if q
>
> q is a necessary condition for p

Table 1.4 defines the truth value of the implication $p \rightarrow q$.

Notice that the implication $p \rightarrow q$ is defined to be true in all cases except when the conclusion is false and the antecedent is true, in which case it is false.

A specific example of a conditional statement may help in accepting the truth values of the implication given in Table 1.4.

Suppose we are told by the postmaster: "If the package was wrapped properly, then it arrived safely at its destination." Under what conditions would we consider this statement untrue? Let p represent the antecedent and q the conclusion. Then

> p: The package was wrapped properly.
>
> q: The package arrived safely at its destination.

The first line in Table 1.4 is the case where p and q are both true. In terms of our given implication, this means that the package was wrapped properly and it did arrive safely at its destination. We agree that in this case the given implication is true.

TABLE 1.4

p	q	$p \rightarrow q$
T	T	T
T	F	F
F	T	T
F	F	T

In the second line of Table 1.4, p is true and q is false. This means that the package was wrapped properly but it did not arrive safely at its destination. In this case, the given implication is false.

The third line corresponds to the statement that the package was not wrapped properly but it arrived safely at its destination. This doesn't prove the given implication untrue, and we do not feel that the statement is false. It merely states what happens if the package were wrapped properly. The statement makes no prediction as to what would happen if the package were not wrapped properly. Therefore in this case we consider the given implication as true.

Would the statement be true if the package were not properly wrapped and that it did not arrive safely at its destination? Obviously not; it merely told us what would happen if the package were wrapped properly. There was no condition to be satisfied if the package were not properly wrapped. Thus, if p and q are both false, then the implication $p \rightarrow q$ is not incorrect or false. It must therefore be true.

We see, then, that the given implication is false if and only if the package is wrapped properly and it does not arrive safely at its destination. In all other cases it is true.

Example 1: Name the antecedent and the conclusion in the implication

Frank gets belligerent if he drinks too much.

Solution: In everyday speech we often invert the antecedent and the conclusion of an implication as we have done here. In "if–then" form the given implication reads "If Frank drinks too much, then he gets belligerent."
The antecedent is: Frank drinks too much.
The conclusion is: Frank gets belligerent.

Example 2: Give the truth values of the following implications.
(a) If $3 \times 4 = 7$, then $4 + 4 = 6$.
(b) If a cup of tea has less calories than a piece of cheese cake, then all candy bars have less calories than a cup of black coffee.
(c) If $2 \times 3 = 6$, then $5 \times 3 = 20$.
(d) June is the sixth month of the year, if February has exactly 30 days.

15

Solution: (a) True because both the antecedent and the conclusion are false.

(b) False because the conclusion is false and the antecedent is true.

(c) False because the conclusion is false and the antecedent is true.

(d) True because the conclusion is true.

Example 3: What is the antecedent of the implication "John dates Debby only if Gay is busy and Sandy snubs him."?

Solution: In "if–then" form this implication reads: "If John dates Debby then Gay is busy and Sandy snubs him." The antecedent is "John dates Debby."

8.
EQUIVALENCE

Two statements may be connected by the phrase "if and only if." When we connect two statements by the phrase "if and only if" the resulting statement is called the **biconditional** or the **equivalence.** Some examples of biconditionals are

A rectangle is a square if and only if all four of its sides are the same length.
A number is even if and only if it is divisible by 2.
You will understand the situation if and only if you hear both sides of the story.

TABLE 1.5

p	q	$p \leftrightarrow q$
T	T	T
T	F	F
F	T	F
F	F	T

If p and q are two statements used to form the biconditional, we denote this compound statement by the symbol

$$p \leftrightarrow q$$

read "p if and only if q."

The symbol $p \leftrightarrow q$ may also be read

p if q and q if p
if p then q and if q then p
q is a necessary and sufficient condition for p
p if q and p only if q

We see then that the equivalence p if and only if q is an abbreviated way of saying the two statements "if p then q and if q then p."

The biconditional $p \leftrightarrow q$ is defined to be true if p and q have the same truth values and false otherwise. The truth values of the biconditional are given in Table 1.5.

Exercise 2

1. Classify each of the following as a conjunction, a disjunction, an implication, or an equivalence.
   (a) Hawaii is the youngest state in the union and Rhode Island is the smallest.
   (b) If yesterday was Sunday, then today is certainly Monday.
   (c) All squares are rectangles and all rectangles are parallelograms.
   (d) He has survived the accident if and only if he was wearing a seat belt.
   (e) Either his lack of enthusiasm or his negative attitude will keep him from passing this course.
   (f) He is expected to be elected class president if he doesn't antagonize the Greeks in the class.
   (g) Douglas will get the job as life guard if and only if he can prove that he is qualified.
   (h) If I tried Tom Sawyer's tactics for getting my fence painted, all of my friends would see through me immediately.
   (i) Tom and Ron will open a poodle grooming shop if Ron can learn how to groom and Tom agrees to put up the money.
   (j) Ted and Lee are hurt if Jo has a party and doesn't invite them.
2. Write the following in symbolic notation.
   (a) The integer n is an even number. (Statement p)
   (b) The integer n is a composite number. (Statement q)

(c) The integer n is even and composite.

(d) If the integer n is even then it is composite.

(e) The integer n is not composite but it is not even.

3. Given: p: Mexico is south of the U.S.A.

   q: Canada is a democracy.

Write the English statements for the following.

(a) $p \rightarrow q$ (c) $\sim p$ (e) $(\sim p) \rightarrow (\sim q)$

(b) $p \leftrightarrow q$ (d) $p \rightarrow (\sim q)$ (f) $(\sim q) \rightarrow p$

4. Give the truth value of each of the following statements.

(a) All birds have wings and dogs have four legs.

(b) All cows have purple eyes and rats have tails.

(c) Two is an even number or 5 is divisible by 3.

(d) A square is a polygon and all polygons have four sides.

5. Form the negation of the following statements.

(a) It is foggy today.

(b) The waves are six feet high.

(c) A circle has a radius.

(d) An even number is divisible by 2.

(e) People work for a living.

(f) Yorkshire terriers are intelligent dogs.

6. Let p denote "It is raining," and q denote "The sky is gray." Translate the following into symbolic notation.

(a) If it is raining, then the sky is gray.

(b) If the sky is gray, then it is raining.

(c) It is raining if and only if the sky is gray.

(d) If the sky is gray, then it is not raining.

(e) It is not raining if the sky is not gray.

7. If p is true, give the truth values of the following.

(a) $(\sim p) \wedge q$

(b) $p \vee q$

(c) $q \rightarrow p$

(d) $(p \wedge q) \rightarrow (p \vee q)$

8. Which of the following implications are true?

(a) If $2 + 2 = 4$, then $7 \times 9 = 48$.

(b) If a triangle is a square, then a rectangle is a pentagon.

(c) If Lincoln was five feet tall, then John F. Kennedy was president of the United States.

(d) If cats purr, then dogs bark.

(e) Hawaii is not a state, only if Alabama is in the north.

9. Which of the following equivalences are true?

(a) Six is 2×3 if and only if $3 \times 3 = 9$.

(b) New York is a city if and only if Chicago is a state.

(c) The President of the United States is less than twenty-one years of age if and only if bears are citizens of the United States.

(d) A bluejay has red feathers if and only if all dogs read magazines.

(e) A mouse is a rodent if and only if a dog is a canine.

10. Write the following statements in symbolic form.
   (a) Leo likes Elaine. (Statement p.)
   (b) Elaine likes Leo. (Statement q.)
   (c) Elaine and Leo like each other.
   (d) Elaine likes Leo but Leo dislikes Elaine.
   (e) Elaine likes Leo if Leo likes her.
   (f) Elaine and Leo dislike each other.

11. Let p be "Property values are high" and let q be "Taxes are rising." Give an English statement for each of the following.
   (a) $p \wedge q$ (c) $p \rightarrow q$ (e) $\sim(p \vee q)$
   (b) $p \wedge (\sim q)$ (d) $\sim(p \wedge q)$ (f) $\sim[(\sim p) \wedge q]$

12. Write the following statements in symbolic form.
   (a) This quadrilateral is a rectangle. (Statement p)
   (b) This quadrilateral is a square. (Statement q)
   (c) This quadrilateral is a rectangle only if it is a square.
   (d) This quadrilateral is a square if and only if it is a rectangle.
   (e) This quadrilateral is a rectangle or a square.

13. Translate each of the following into English sentences. The letter p represents the statement "Nellie is lucky"; the letter q represents "Joan is intelligent"; and the letter r represents "Betty is stuffy."
   (a) $p \wedge (\sim q)$ (c) $[p \vee (\sim q)] \wedge r$
   (b) $(\sim p) \vee (q \wedge r)$ (d) $\sim[(p \vee q) \wedge r]$

14. Write the following in symbolic notation. Use the following symbolic notation.
   p: Deane is made of money.
   q: Money is made through hard work.
   (a) Deane is not made of money, and money is made through hard work.
   (b) Deane is made of money if money is not made through hard work and Deane is made of money only if money is not made through hard work.
   (c) A necessary condition for Deane to be made of money is that money is made through hard work.
   (d) If money is made through hard work, then Deane is not made of money.

9.
TAUTOLOGY

If a compound statement is true for all possible truth values of its components, it is called a **tautology.** For example, the statement

$$\sim[p \wedge (\sim p)]$$

TABLE 1.6

p	$\sim p$	$p \wedge (\sim p)$	$\sim[p \wedge (\sim p)]$
T	F	F	T
F	T	F	T

TABLE 1.7

p	$\sim p$	$p \vee (\sim p)$
T	F	T
F	T	T

is a tautology. Table 1.6 demonstrates that this statement is a tautology. In constructing the truth table to show that $\sim[p \wedge (\sim p)]$ is a tautology, we use the headings p, $\sim p$, $p \wedge (\sim p)$, and $\sim[p \wedge (\sim p)]$. We write all the possible truth values, T and F, under p and carry each line across to the right inserting the correct truth values. We find $\sim[p \wedge (\sim p)]$ is true for all possible truth values of p, and hence is a tautology. This tautology is called the **Law of Contradiction.**

Another tautology used in logical reasoning is the **Law of the Excluded Middle.** This law states that for every statement p, either p is true or $\sim p$ is true; that is, $p \vee (\sim p)$. Table 1.7 shows that $p \vee (\sim p)$ is a tautology.

Another important tautology is

$$[(p \rightarrow q) \wedge (q \rightarrow r)] \rightarrow (p \rightarrow r)$$

This is called the **Law of Syllogisms.** Table 1.8 verifies that the law of syllogisms is a tautology.

TABLE 1.8

p	q	r	$p \rightarrow q$	$q \rightarrow r$	$(p \rightarrow q) \wedge (q \rightarrow r)$	$p \rightarrow r$	$[(p \rightarrow q) \wedge (q \rightarrow r)] \rightarrow (p \rightarrow r)$
T	T	T	T	T	T	T	T
T	T	F	T	F	F	F	T
T	F	T	F	T	F	T	T
T	F	F	F	T	F	F	T
F	T	T	T	T	T	T	T
F	T	F	T	F	F	T	T
F	F	T	T	T	T	T	T
F	F	F	T	T	T	T	T

To construct Table 1.8 we begin with three columns because three statements are used to form the law of syllogisms. The eight lines show the possible combinations of truth values for p, q, and r.

p	q	r
T	T	T
T	T	F
T	F	T
T	F	F
F	T	T
F	T	F
F	F	T
F	F	F

On the basis of the truth values of p, q, and r we find the truth values of $p \rightarrow q$, $q \rightarrow r$, $(p \rightarrow q) \wedge (q \rightarrow r)$, and $p \rightarrow r$ as shown in Table 1.8. The law of syllogisms is also called the **chain rule.** Successive applications of the rule permit a chain of implications of any desired length. For example, if $p \rightarrow q$ and $q \rightarrow r$ and $r \rightarrow s$ and $s \rightarrow t$, then $p \rightarrow t$.

Using truth tables we can derive a series of tautologies known as the **Rules of Inference.** Some of the more important rules of inference were derived above.

10.
DERIVED IMPLICATIONS

From the implication $p \rightarrow q$ we can form several derived implications, which may or may not be true when the given implication is true. The most important ones are:

Converse	$q \rightarrow p$
Inverse	$(\sim p) \rightarrow (\sim q)$
Contrapositive	$(\sim q) \rightarrow (\sim p)$

Comparing truth tables for $p \rightarrow q$ and $q \rightarrow p$ (Table 1.9), we see that the converse may be false when the given implication is true. For example, the implication:

TABLE 1.9

p	q	$p \rightarrow q$	$q \rightarrow p$
T	T	T	T
T	F	F	T
F	T	T	F
F	F	T	T

If a quadrilateral is a rectangle, then it is a parallelogram.

is true, but its converse:

If a quadrilateral is a parallelogram, then it is a rectangle.

is false.

Sometimes the converse of a true implication may be true. For example, both the following implication and its converse are true.

Implication: If it is 8 o'clock P.S.T. in San Diego, then it is 10 o'clock C.S.T. in St. Louis.

Converse: If it is 10 o'clock C.S.T. in St. Louis, then it is 8 o'clock P.S.T. in San Diego.

Let us construct a truth table (Table 1.10) to investigate whether the inverse is true when the given implication is true and false when it is false. Since the last two columns of Table 1.10 differ, we can conclude that the inverse of a given implication is not always true when the given implication is true.

Now we consider Table 1.11 showing the truth values of the contra-positive of a given implication. Since the last two columns of the table are identical, we conclude that an implication and its contrapositive are simultaneously true or false; that is, they are **logically equivalent.** Likewise,

TABLE 1.10

p	q	$\sim p$	$\sim q$	$p \rightarrow q$	$(\sim p) \rightarrow (\sim q)$
T	T	F	F	T	T
T	F	F	T	F	T
F	T	T	F	T	F
F	F	T	T	T	T

TABLE 1.11

p	q	$\sim p$	$\sim q$	$p \rightarrow q$	$(\sim q) \rightarrow (\sim p)$
T	T	F	F	T	T
T	F	F	T	F	F
F	T	T	F	T	T
F	F	T	T	T	T

the inverse and the converse are logically equivalent. Any two statements that are simultaneously true or false are called **logically equivalent** or **equivalent** statements.

Exercise 3

1. Construct a truth table for $[\sim(p \vee q)] \vee [(q \vee p)]$.

2. Construct a truth table for $p \rightarrow (q \vee r)$.

3. Construct a truth table for $p \wedge (\sim p)$.

4. Construct a truth table for $[p \vee (\sim q)] \wedge r$.

5. Construct a truth table for $(p \rightarrow q) \vee [p \rightarrow (\sim p)]$.

6. Construct a truth table for $(p \rightarrow q) \leftrightarrow [(\sim p) \vee q]$.

7. Construct a truth table to show that $(p \rightarrow q) \rightarrow \sim[p \wedge (\sim q)]$ is a tautology.

8. Construct a truth table to show that $(p \vee q) \leftrightarrow \sim[(\sim p) \wedge (\sim q)]$ is a tautology.

9. Construct a truth table to show that $(p \wedge q) \rightarrow p$ is a tautology.

10. Construct a truth table to show that $(p \wedge q) \rightarrow (p \vee q)$ is a tautology.

11. Construct a truth table to show that $[p \wedge (\sim p)] \rightarrow q$ is a tautology.

12. Form the (1) converse, (2) inverse, and (3) contrapositive of the following implications.

(a) If n is an integer, then n is divisible by 1.

(b) If some college students are communists, then all college students are communists.

(c) If there is a depression, then prices go down.

(d) New Year's Day falls on Monday if Christmas falls on Monday.

(e) All girls are beautiful if a man is slightly tipsy.

13. Using a truth table show that $(\sim p) \vee (\sim q)$ is logically equivalent to $\sim(p \wedge q)$.

14. Construct a truth table to show that $p \wedge (\sim q)$ and $\sim(p \rightarrow q)$ are logically equivalent.

15. Construct a truth table to show that $p \rightarrow (\sim q)$ and $(\sim p) \vee (\sim q)$ are logically equivalent.

16. Construct a truth table to show that $(\sim p) \wedge (\sim q)$ and $\sim(p \vee q)$ are logically equivalent.

17. Construct a truth table to show that $p \rightarrow q$ and $(\sim p) \vee q$ are logically equivalent.

18. Is the inverse of the converse of an implication logically equivalent to the converse of the inverse of that implication?

19. If p and q are true and r is false, what is the truth value of:

$$[p \vee (\sim q)] \wedge (\sim r)$$

20. If p is true what are the truth values of the following?

(a) $p \vee q$ (c) $\sim p$
(b) $(\sim p) \wedge q$ (d) $\sim p \rightarrow p$

21. Prove that the conjunction of a statement with itself is logically equivalent to the statement.

22. Prove that the disjunction of a statement with itself is logically equivalent to the statement.

23. Prove that $\sim(\sim p)$ is logically equivalent to p.

11.
NEGATIONS OF COMPOUND STATEMENTS

It is important that we be able to negate statements. We know that the negation of p is $\sim p$. Now let us consider the negation of the conjunction $p \wedge q$. We know that $p \wedge q$ and its negation $\sim(p \wedge q)$ have opposite truth values. Since $p \wedge q$ is true only when both p and q are true, it will be false if either p or q is false. Therefore $\sim(p \wedge q)$ should be equivalent to $(\sim p) \vee (\sim q)$. Table 1.12 verifies that $(\sim p) \vee (\sim q)$ is equivalent to the negation of $p \wedge q$, $\sim(p \wedge q)$.

Now let us consider the negation of the disjunction $p \vee q$. We know

TABLE 1.12
The Negation of $p \wedge q$

p	q	$\sim p$	$\sim q$	$p \wedge q$	$\sim(p \wedge q)$	$(\sim p) \vee (\sim q)$
T	T	F	F	T	F	F
T	F	F	T	F	T	T
F	T	T	F	F	T	T
F	F	T	T	F	T	T

TABLE 1.13
The Negation of $p \lor q$

p	q	$\sim p$	$\sim q$	$p \lor q$	$\sim(p \lor q)$	$(\sim p) \land (\sim q)$
T	T	F	F	T	F	F
T	F	F	T	T	F	F
F	T	T	F	T	F	F
F	F	T	T	F	T	T

that $p \lor q$ is false only when both p and q are false. Hence $\sim(p \lor q)$ should be equivalent to $(\sim p) \land (\sim q)$. Table 1.13 verifies that $\sim(p \lor q)$ and $(\sim p) \land (\sim q)$ are indeed equivalent statements.

The two pairs of equivalent statements

$$\sim(p \land q) \quad \text{and} \quad (\sim p) \lor (\sim q)$$
$$\sim(p \lor q) \quad \text{and} \quad (\sim p) \land (\sim q)$$

are known as **DeMorgan's Laws.**

Now let us consider the implication $p \to q$. This statement is false when p is true and q is false. Hence its negation should be equivalent to $p \land (\sim q)$. Table 1.14 shows that $\sim(p \to q)$ and $p \land (\sim q)$ are equivalent statements.

The biconditional $p \leftrightarrow q$ is false when p and q have opposite truth values. That is, $p \leftrightarrow q$ is false when p is false and q is true or when p is true and q is false. We see then that the negation of $p \leftrightarrow q$ is equivalent to $[(\sim p) \land q] \lor [(\sim q) \land p]$. That $\sim(p \leftrightarrow q)$ and $[(\sim p) \land q] \lor [(\sim q) \land p]$ are equivalent is shown in Table 1.15.

TABLE 1.14
Negation of $p \to q$

p	q	$\sim q$	$p \to q$	$\sim(p \to q)$	$p \land (\sim q)$
T	T	F	T	F	F
T	F	T	F	T	T
F	T	F	T	F	F
F	F	T	T	F	F

TABLE 1.15
Negation of $p \leftrightarrow q$

p	q	$\sim p$	$\sim q$	$(\sim p) \wedge q$	$(\sim q) \wedge p$	$p \leftrightarrow q$	$\sim(p \leftrightarrow q)$	$[(\sim p) \wedge q] \vee [(\sim q) \wedge p]$
T	T	F	F	F	F	T	F	F
T	F	F	T	F	T	F	T	T
F	T	T	F	T	F	F	T	T
F	F	T	T	F	F	T	F	F

We summarize the above discussion below.

STATEMENT	NEGATION
p	$\sim p$
$p \wedge q$	$(\sim p) \vee (\sim q)$
$p \vee q$	$(\sim p) \wedge (\sim q)$
$p \rightarrow q$	$p \wedge (\sim q)$
$p \leftrightarrow q$	$[(\sim p) \wedge q] \vee [(\sim q) \wedge p]$

Example 1: Give the negation of the conjunction:

Debby likes baseball and Sandy likes football.

Solution: The negation is:

Debby does not like baseball or Sandy does not like football.

Example 2: Give the negation of the equivalence:

Al will get a promotion if and only if his father owns the company.

Solution: The negation is:

Al will get a promotion and his father does not own the company or he will not get a promotion and his father owns the company.

This sounds better written as follows.

Al will get a promotion without his father owning the company or Al will not get a promotion with his father owning the company.

11. NEGATIONS OF COMPOUND STATEMENTS

Example 3: Give the negation of the implication:

If Joe works hard and his funds hold out, then he will pass his courses and graduate.

Solution: Let us use the following symbolic notation:

p: Joe works hard
q: Joe's funds hold out
r: Joe will pass his courses
s: Joe will graduate

The given statement written in symbolic notation is

$$(p \wedge q) \rightarrow (r \wedge s)$$

The negation of this implication is

$$(p \wedge q) \wedge [\sim(r \wedge s)]$$

But $[\sim(r \wedge s)]$ is equivalent to $(\sim r) \vee (\sim s)$. The negation of the given implication is

$$(p \wedge q) \wedge [(\sim r) \vee (\sim s)]$$

Translating this into English we have:

Joe works hard and his funds hold out and he will not pass his courses or he will not graduate.

This sounds better written as:

Joe works hard and his funds hold out but he will not pass his courses or graduate.

Exercise 4

Negate each of the following statements.

1. John plays in the orchestra and in the marching band.
2. Either the price tag on this dress is wrong or I have been overcharged.
3. If Nellie picks out the date for the picnic, the sun will surely shine.
4. If Sue takes English and passes the course, then she will graduate in June.
5. My raise will come through if and only if the company receives a large defense contract.
6. If this rain doesn't stop, then the party will be called off and all this food will go to waste.

7. If he wins this law suit, then he will be an excellent candidate for attorney general.

8. If you are a seasoned traveler you have learned to carry as little baggage as needed and as much money as possible.

9. If you have the winning number and can produce your ticket, then you can collect the prize money.

10. Accept things as they are and quit complaining if you don't want to get fired.

12.
LAWS OF SUBSTITUTION AND DETACHMENT

Two more laws of logic that are used in deductive reasoning are the law of detachment and the law of substitution.

The **Law of Substitution** states that we may substitute at any point in the deductive process one statement for an equivalent statement. Thus if we know that $p \rightarrow q$ is true, we may at any point substitute the contrapositive, $(\sim q) \rightarrow (\sim p)$, because they are logically equivalent statements.

The **Law of Detachment** states that if an implication $p \rightarrow q$ is true and if the antecedent, p, is true, then the conclusion, q, is true. The proof of the law of detachment follows immediately from Table 1.4. Since we know $p \rightarrow q$ is true we are in lines 1, 3, or 4 of the table. Since p is also true we are in lines 1 or 2 of the table. Line 1 is the only one in which p and $p \rightarrow q$ are true as given; hence q is true.

13.
VALID ARGUMENTS

One of the most important tasks of a logician is the checking of the validity of arguments. By an argument we mean the assertion that a certain statement, called the **conclusion,** follows from other statements, called **premises.** An argument is **valid** if and only if the conjunction of the premises implies the conclusion. In other words, if all the premises are true, the conclusion is also true.

It is important to realize that the truth of the conclusion is irrelevant as a test for the validity of the argument. A true conclusion does not necessarily assure a valid argument, nor does a valid argument necessarily assure a true conclusion. The following examples show this. They also illustrate the form in which we write arguments.

Example 1: Test the validity of the following argument.

> If all triangles are polygons, then the sides of
>   all triangles are line segments.
> The sides of all triangles are line segments
> _____
> ∴ All triangles are polygons.

Solution: Let p represent the statement "All triangles are polygons" and q represent the statement "The sides of all triangles are line segments." Symbolically we can write this argument:

$$p \rightarrow q$$
$$\underline{q \qquad}$$
$$\therefore p$$

This argument is invalid because of truth of $p \rightarrow q$ and q does not assure the truth of p. Examining lines 1 and 3 in Table 1.4 (Section 7) we see that when $p \rightarrow q$ and q are true, p can be either true or false, hence the argument is invalid. In this argument the conclusion is a true statement even though the argument is invalid.

Example 2: Test the validity of the following argument.

> All owls have four legs.
> All four-legged animals are birds.
> _____
> ∴ All owls are birds.

Solution: Using a diagram (Figure 1.6) we see that the conclusion follows from the premises. The rectangle represents the set of all animals. The first premise assures us that the set of all owls, represented by the circle O, lies entirely in circle F, representing the set of all four-legged animals. The second premise tells us that circle F lies entirely within circle B,

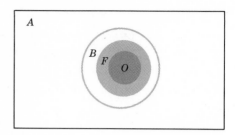

FIGURE 1.6

representing the set of all birds. From the diagram we see that the conclusion follows from the premises.

All of the statements in this argument can be translated into implications. Let p represent the statement "An animal is an owl"; q represent the statement "An animal is four-legged"; and r represent the statement "A bird is an animal." The first premise "All owls have four legs" means exactly the same as the statement "If an animal is an owl, then it is four-legged." The second premise "All four-legged animals are birds" means exactly the same thing as the statement "If an animal is four-legged, then it is a bird." The argument may be written

$$p \rightarrow q$$
$$q \rightarrow r$$
$$\overline{\therefore p \rightarrow r}$$

We see that the conclusion follows from the premises because of the law of syllogisms. Observing lines 1, 5, 7, and 8 in Table 1.8 (Section 9) we see that when $p \rightarrow q$ and $q \rightarrow r$ are true, $p \rightarrow r$ is also true. In this valid argument, the conclusion is true but both of the premises are false.

Example 3: Test the validity of the following argument.

> If I was born in April, then I was born under the sign of Taurus.
> I was not born under the sign of Taurus.
> _____
> ∴ I was not born in April.

Solution: Let p represent the statement "I was born in April," and

TABLE 1.16

p	q	$\sim p$	$\sim q$	$p \rightarrow q$
T	T	F	F	T
T	F	F	T	F
F	T	T	F	T
F	F	T	T	T

q represent the statement "I was born under the sign of Taurus." Then the argument can be written

$$p \rightarrow q$$
$$\sim q$$
$$\therefore \sim p$$

The validity of this argument is shown in the last line of Table 1.16. In this argument the conclusion is false (persons born on April 21 or later are born under the sign of Taurus, but if they were born before April 21 they are born under the sign of Aries), but the argument is valid since the conclusion follows from the premises. Notice that the first premise in the argument is false.

The construction of a truth table is one way to check the validity of an argument. Consider the following argument:

$$p \leftrightarrow q$$
$$p$$
$$\therefore q$$

The truth table for the argument is shown in Table 1.17.

Both premises are true only in the first row of the table. In this row the conclusion is also true, so the argument is valid.

TABLE 1.17

p	q	$p \leftrightarrow q$
T	T	T
T	F	F
F	T	F
F	F	T

Test the validity of the following arguments (Exercises 1–15).

1. No even number is exactly divisible by 7.
   Sixteen is an even number.

   ∴ Sixteen is not exactly divisible by 7.

2. If today is Wednesday, three days from today will be Saturday.
   Today is not Wednesday.

   ∴ Three days from today will be Saturday.

3. All vehicles have four wheels.
   A motor scooter is a vehicle.

   ∴ A motor scooter has four wheels.

4. If John passes the final examination, he will pass the course.
   John passes the final examination.

   ∴ John will pass the course.

5. If Sally lives in Los Angeles, then she lives in California.
   Sally lives in California.

   ∴ Sally lives in Los Angeles.

6. If prices are rising, inflation is coming.
   Inflation is coming.

   ∴ Prices are rising.

7. 1964 was a leap year.
   All leap years are election years.

   ∴ 1964 was an election year.

8. All intelligent men are Republicans.
   All Republicans are wealthy.

   ∴ All intelligent men are wealthy.

9. All mathematics books are interesting.
   All comic books are interesting.

   ∴ All mathematics books are comic books.

10. All men are handsome.
    All husbands are men.

    ∴ All husbands are handsome.

11. If I am satisfied only with the best of everything,

I am apt to get only the best of everything.
I am satisfied only with the best of everything.

∴. I am apt to get only the best of everything.

12. If a child has something to make with his hands, he will be perfectly happy.
This child is perfectly happy.

∴. He has something to make with his hands.

13. Either Buck took the money, or Frenchy is a liar.
Frenchy never lies.

∴. Buck didn't take the money.

14. I dislike cats or I own a cat.
I don't dislike cats.

∴. I don't own a cat.

15. Will will join the army if he is drafted.
Will is not drafted.

∴. Will will not join the army.

16. Using the law of detachment draw a valid conclusion from the following premises.
   (a) If it is 9 o'clock M.S.T. in Phoenix, it is 8 o'clock P.S.T. in San Francisco. It is 9 o'clock M.S.T. in Phoenix.
   (b) If 1972 is a leap year, it is an election year. 1972 is a leap year.

17. Using the law of detachment draw a valid conclusion from the following premises.
   (a) If X is the murderer, then he was at the scene of the crime. X is the murderer.
   (b) If $ABCD$ is a square, then it is a rectangle. $ABCD$ is a square.

18. Using the law of detachment draw a valid conclusion from the following premises.
   (a) x is less than y if w is greater than z. w is greater than z.
   (b) If all boys are intelligent, then Frank has two heads. All boys are intelligent.

19. Fill in the blanks with statements to make the following arguments valid.

   (a) If n is an integer, it is either even or odd.
. . . .

∴. n is either even or odd.

   (b) If $x > y$ and $y > z$, then $x > z$.
$x > y$ and $y > z$.

. . . .

20. Fill in the blanks with statements to make the following arguments valid.

(a) If *n* is a natural number, *n* is a whole number.

. . . .
_____

∴ *n* is a whole number.

(b) . . . .

I eat too much.
_____

∴ I shall get fat.

21. Fill in the blanks with statements to make the following arguments valid.

(a) . . . .

I have a 3.5 grade-point average.
_____

∴ I shall graduate with honors.

(b) If all mumbos are jumbos, then all booboos are kookoos.

. . . .
_____

∴ All booboos are kookoos.

22. Construct a truth table to show that the following arguments are invalid.

(a) $p \rightarrow q$

$\sim p$
_____

∴ $\sim q$

(b) $q \rightarrow (\sim p)$

$\sim p$
_____

∴ $\sim q$

14.
DIRECT PROOF

In English the word "proof" has many meanings. In photography, "proof" means a trial printing of a negative; in printing, "proof" means a trial impression; on a liquor bottle, "proof" means the relative strength of the beverage in the bottle; in law, "proof" means evidence operating to determine the judgment of a tribunal. In mathematics, "proof" means evidence that establishes the truth of a statement. Mathematical proof is proof by deduction. It consists of showing that the statement to be proved, called a **theorem,** is a logical consequence of the given premises.

The premises that are given in a theorem are called the **hypotheses.** The statement that we wish to establish is called the **conclusion.** A theorem takes the form of an implication $H \to C$, where H is the hypothesis and C is the desired conclusion.

Some examples of theorems from geometry are given below.
(1) If three sides of one triangle are respectively equal in length to three sides of another triangle, then the two triangles are congruent.
(2) If a diagonal of a parallelogram bisects the opposite angles, then the parallelogram is a rhombus.
(3) If a triangle is a right triangle, then the square of the length of the hypotenuse equals the sum of the squares of the lengths of the legs.
(4) The diagonals of a parallelogram bisect each other. (In implication form this theorem says: If a geometric figure is a parallelogram, then its diagonals bisect each other.)

In mathematics we use two kinds of proof, direct proof and indirect proof. A **direct proof** consists of starting with the hypotheses, establishing a chain of true implications, and ending with the desired conclusion. Each step in the proof must be justified by a premise, a definition, a previously proved theorem, or a law of logic.

We learn a great deal about the techniques used in proving theorems by reading proofs that others have made. Study the following examples of the use of direct proof in proving theorems.

Example 1: Prove: If the ground is dry, then it is not raining. If the ground is not dry, then the road is slippery. It is raining. Therefore, the road is slippery.

Solution: Let

> p represent the statement "It is raining."
> q represent the statement "The ground is dry."
> r represent the statement "The road is slippery."

The hypotheses are:

$$p$$
$$q \to (\sim p)$$
$$(\sim q) \to r$$

The conclusion is r.

Proof: Since $q \to (\sim p)$ is true, its contrapositive, $p \to (\sim q)$, is true.

We now have the following true implications

$$p \rightarrow (\sim q) \qquad \text{Contrapositive of a true implication}$$
$$(\sim q) \rightarrow r \qquad \text{Hypothesis}$$

We now know that $p \rightarrow r$ is a true implication by the law of syllogisms. Since $p \rightarrow r$ is true and p is true (by hypothesis), r is true by the law of detachment.

Example 2: Prove: Either Gay or Suzie is president. Suzie is not president. Therefore Gay is president.

Solution: Let

p represent the statement "Gay is president."
q represent the statement "Suzie is president."

The hypotheses are:

$$p \vee q$$
$$\sim q$$

The conclusion is p.

Proof: Since $\sim q$ is true, q is false. Since $p \vee q$ is true and q is false, p must be true by the definition of a disjunction.

Example 3: Prove: If I study hard, then I shall graduate. If I don't pass my exams, then I shall not graduate. Therefore, if I don't pass my exams, I don't study hard.

Solution: Let

p represent the statement "I study hard."
q represent the statement "I shall graduate."
r represent the statement "I pass my exams."

The hypotheses are:

$$p \rightarrow q$$
$$(\sim r) \rightarrow (\sim q)$$

The conclusion is $(\sim r) \rightarrow (\sim p)$.

Proof: Since $p \rightarrow q$ is true, its contrapositive $(\sim q) \rightarrow (\sim p)$ is true. We now have a chain of true implications

$$(\sim r) \rightarrow (\sim q) \qquad \text{Hypothesis}$$
$$(\sim q) \rightarrow (\sim p) \qquad \text{Contrapositive of a true implication}$$

Therefore, by the law of syllogisms, $(\sim r) \to (\sim p)$ is true.

Example 4: Prove: If this is a good book, it is worth reading. Either the reading is easy, or the book is not worth reading. The reading is not easy. Therefore this book is not worth reading.

Solution: Let

> p represent the statement "This is a good book."
> q represent the statement "The book is worth reading."
> r represent the statement "The reading is easy."

The hypotheses are:

$$p \to q$$
$$r \lor (\sim q)$$
$$\sim r$$

The conclusion is $\sim q$.

Proof: Since $\sim r$ is true, r is false. Since r is false and $r \lor (\sim q)$ is true, $\sim q$ is true by the definition of a disjunction.

15.
INDIRECT PROOF

The **indirect** method of proof is often called **proof by contradiction.** This method of proof relies on the fact that if $(\sim p)$ is false, p is true. Hence to prove p is true, we attempt to show that $(\sim p)$ is false. The best way to do this is to show that $(\sim p)$ is not consistent with the given premises. In other words, we add $(\sim p)$ to the list of premises and show that with this premise added we have a contradiction. When this contradiction is reached, we know that our assumption is not true; that is, $(\sim p)$ is false and p is true.

Study the following indirect proofs.

Example 1: Prove:

$$p$$
$$q \to (\sim p)$$
$$(\sim q) \to s$$
$$\overline{\therefore s}$$

Solution: Assume $(\sim s)$ is true Since $(\sim q) \rightarrow s$ is true, the contrapositive, $(\sim s) \rightarrow q$ is true. We now have a chain of syllogisms:

$$(\sim s) \rightarrow q \qquad \text{Contrapositive of a true implication}$$
$$q \rightarrow (\sim p) \qquad \text{Hypothesis}$$

Hence $(\sim s) \rightarrow (\sim p)$ by the law of syllogisms.

Now $(\sim s) \rightarrow (\sim p)$ is true and $(\sim s)$ is true by our assumption. Hence, by the law of detachment, $(\sim p)$ is true. But this leads to a contradiction because by the premises p is true. Hence our assumption is false, and s is true.

Example 2: Prove:

$$p \vee q$$
$$\underline{\sim q}$$
$$\therefore p$$

Solution: Assume $(\sim p)$ is true. Since, by assumption, $(\sim p)$ is true, p is false. Since p is false and $p \vee q$ is true, q must be true by the definition of a disjunction. But this leads to a contradiction since $(\sim q)$ is given true. Hence our assumption is false and p is true.

Example 3: Prove: Fisher is unemployed. He is on welfare. If Fisher is not unemployed, he has money in the bank. If Fisher is on welfare, he does not have money in the bank. Therefore Fisher does not have money in the bank.

Solution: Let p: Fisher is unemployed.

q: Fisher in on welfare.

s: Fisher has money in the bank.

Then the premises are

$$p$$
$$q$$
$$(\sim p) \rightarrow s$$
$$q \rightarrow (\sim s)$$

and the conclusion is $(\sim s)$.

Let us assume that $(\sim s)$ is false. Then s is true. But $q \rightarrow (\sim s)$ is true and q is true, therefore $(\sim s)$ is true by the law of detachment. We now have a contradiction, $(\sim s)$, both true and false. Therefore our assumption is false and $(\sim s)$ is true.

Exercise 6

Prove the following theorems.

1. p
    q
    $(p \wedge q) \rightarrow (r \vee q)$
    $(p \vee q) \rightarrow (r \wedge q)$
    _____
    $\therefore r$

2. $(p \wedge q) \rightarrow (r \wedge s)$
    $\sim s$
    q
    _____
    $\therefore \sim p$

3. p
    q
    $(p \vee q) \rightarrow r$
    _____
    $\therefore r$

4. $p \rightarrow q$
    $r \rightarrow (\sim q)$
    _____
    $\therefore p \rightarrow (\sim r)$

5. If this is a good course, then it is worth taking. Either math is easy or this course is not worth taking. Math is not easy. Therefore this is not a good course.

6. If I don't save my money, I shall not go to Europe. I shall go to Europe. Therefore I save my money.

7. If Carl is elected class president, then Bill is elected vice-president. If Bill is elected vice-president, then Betty is not elected secretary. Therefore if Betty is elected secretary, then Carl is not elected president.

8. Whenever it is not snowing the temperature is high. When the temperature is high, there is no ice on the streets. There is ice on the streets. Therefore it is snowing.

9. Vincent water-skis only if it is summer. Whenever Blanche is in town Vincent water-skis. It is not summer now. Therefore Blanche is not in town.

10. If Martin is on the dean's list, his father will buy him a car. If Martin's father buys him a car he will visit Mexico. If Martin does not visit Monterrey, he will not visit Mexico. Martin is on the dean's list. Therefore Martin will visit Monterrey.

11. John is a thief. Newton is a shoplifter. If John is not a thief, then Carl is guilty of car theft. If Newton is a shoplifter, then Carl is not guilty of car theft. Therefore Carl is not guilty of car theft.

12. If students have average intelligence, then they can understand the concepts of mathematics. If a student does not have average intelligence he is not a college capable student. These students can not understand the concepts of mathematics. Therefore these students are not college capable.

13. If this class is a bore, then the instructor is not interesting. If this class isn't a bore, then the subject is worthwhile. The instructor is interesting. Therefore the subject is worthwhile.

14. Men stare at Kathy. Kathy is attractive. If men do not stare at Kathy, they stare at Frances. If Kathy is attractive, the men do not stare at Frances. Therefore men do not stare at Frances.

Pierre De Fermat (1601–1665), the greatest writer of number theory, was a modest counselor of the parliament of Toulouse who devoted his leisure time to mathematics. Most famous of all Fermat's work is his so-called Last Theorem which states that $x^n + y^n = z^n$ has no solution in integers x, y, and z, all different from zero, when $n \geq 3$. Fermat customarily recorded his thoughts in marginal notes, and about his famous theorem he wrote, "I have a truly remarkable demonstration which this margin is too narrow to contain." Whether Fermat possessed a proof of his last theorem probably never will be known. To date no one has proved the theorem for all values of $n \geq 3$.

DIVISIBILITY

2

1.
THE INTEGERS

Mathematical statements that we assume to be true without further justification are called **axioms** or **postulates.** In this chapter we shall be considering various properties and theorems pertaining to the set I of **integers**

$$I = \{\ldots, -2, -1, 0, 1, 2, \ldots\}$$

and their operations. We shall assume that the operations of addition, subtraction, multiplication, and division of these numbers are understood.

We now state the axioms of the system of integers. In the axioms below, a, b, and c are symbols for integers.

Axioms of the Integers

I-1: **(Closure Property of Addition)** *For all a and b, their sum, denoted by $a + b$, is an integer.*

I-2: **(Closure Property of Multiplication)** *For all a and b, their product, denoted by ab, is an integer.*

I-3: **(Commutative Property of Addition)** *For all a and b,*

$$a + b = b + a$$

I-4: **(Commutative Property of Multiplication)** *For all a and b,*

$$ab = ba$$

I-5: **(Associative Property of Addition)** *For all a, b, and c,*

$$(a + b) + c = a + (b + c)$$

I-6: **(Associative Property of Multiplication)** *For all a, b, and c,*

$$(ab)c = a(bc)$$

I-7: **(Additive Identity Axiom)** *There exists in I an element, 0, called* *the* **identity element of addition** *or the* **additive identity,** *with the property that for all a,*

$$a + 0 = 0 + a = a$$

I-8: **(Multiplicative Identity Axiom)** *There exists in I an element, 1, called the* **identity element of multiplication** *or the* **multiplicative identity** *with the property that for all a,*

$$a \cdot 1 = 1 \cdot a = a$$

I-9: **(Additive Inverse Axiom)** *Corresponding to every integer a, there is a unique integer called the* **opposite** *or* **additive inverse** *of a, denoted by* $-a$, *such that*

$$a + (-a) = (-a) + a = 0$$

I-10: **(Distributive Property)** *For all a, b, and c,*

$$a(b + c) = ab + ac$$

I-11: **(Multiplication Property of Zero)** *For all a,*

$$a \cdot 0 = 0 \cdot a = 0$$

I-12: **(Cancellation Property of Addition)** *For all a, b, and c, if* $a + c = a + b$ *or* $c + a = b + a$, *then* $c = b$.

I-13: **(Cancellation Property of Multiplication)** *For all a, b, and c,* $a \neq 0$, *if* $ab = ac$ *or* $ba = ca$, *then* $b = c$.

I-14: **(Closure Property of Subtraction)** *For all a and b, their difference, denoted by* $a - b$, *is an integer.*

Let us consider three subsets of the set of integers:

$$I_p = \{1, 2, 3, 4, \ldots\}$$
$$I_0 = \{0\}$$
$$I_n = \{-1, -2, -3, \ldots\}$$

The set I_p is called the set of **positive integers** or the **natural numbers.** Set I_p together with the set I_0 is called the set of **nonnegative integers** or the **whole numbers.** Set I_n is called the set of **negative integers.** The integer 0 is neither positive nor negative.

2.
MULTIPLES
AND DIVISORS

Any integer a is said to be **divisible** by an integer b, $b \neq 0$, if there is a third integer c such that

$$a = bc$$

If both a and b are positive, c is necessarily positive. If a is divisible by b, we say that b is a **divisor** or **factor** of a. We also say that a is a **multiple** of b. We express the fact that a is divisible by b by the symbol

$$b \mid a$$

For example,

$$2 \mid 6 \text{ since } (2)(3) = 6$$
$$5 \mid 10 \text{ since } (5)(2) = 10$$
$$7 \mid 21u \text{ since } (7)(3u) = 21u$$

We use the symbol

$$b \nmid a$$

to express the negation† of $b \mid a$. Thus

$$2 \nmid 7 \qquad \text{since there is no integer } c \text{ such that } 2c = 7$$
$$3 \nmid 11 \qquad \text{since there is no integer } c \text{ such that } 3c = 11$$
$$4 \nmid 21 \qquad \text{since there is no integer } c \text{ such that } 4c = 21$$

If a is an integer, the multiples of a are all numbers of the form na where n is an integer. For example, the multiples of 2 are the **even integers**

$$\ldots, -6, -4, -2, 0, 2, 4, 6, \ldots$$

When we defined divisibility of a by b, we stated that b was not zero. Because of its importance, we shall discuss zero and division.

Division is the inverse operation of multiplication. The division of a by b denoted $\frac{a}{b}$ or $a \div b$ asks the question: What number multiplied

---

† See Chapter 1, Section 6.

by b gives a? That is, if a and b are integers, then $a \div b$ is the integer c if $a = bc$.

Observe that

$$0 \div 5 = n \rightarrow n \cdot 5 = 0\dagger$$
$$0 \div (-2) = n \rightarrow n \cdot (-2) = 0$$
$$0 \div (-3) = n \rightarrow n \cdot (-3) = 0$$

Since these statements are true if and only if $n = 0$, we conclude that 0 divided by any nonzero integer is 0.

We always say that *division by zero is impossible*. To see why we specify that we can only divide by nonzero integers, let us look at the following:

$$6 \div 0 = n \rightarrow n \cdot 0 = 6$$
$$(-3) \div 0 = n \rightarrow n \cdot 0 = -3$$

Since zero times any number is zero, there is no integer n that makes the above statements true. Hence, *division by zero is impossible*. When we define $a \div b = c$ as $bc = a$, we must state $b \neq 0$.

There is one other case to be considered: that is $0 \div 0$. In this case

$$0 \div 0 = n \rightarrow n \cdot 0 = 0$$

Since zero times any integer is zero, n may be replaced by any integer in $n \cdot 0 = 0$ to make the statement true. Because of this, we call $0 \div 0$ an **indeterminate symbol.**

We may summarize this discussion by saying that $a \div 0$ is impossible for all integers. That is, *division by zero is impossible*.

We now prove some important theorems about the divisibility of integers. In the following theorems, although not stated, it is assumed that the divisors are not zero.

THEOREM 2.1: *If $a|k$ and $a|h$, then $a|h + k$.*

      Proof: Since $a|k$ and $a|h$, we may write, by the definition of divisibility,

$$k = am$$
$$h = an$$

where m and n are integers. Adding these two equations member by member we have

$\dagger$ See Chapter 1, Section 7.

$$k + h = am + an$$
$$= a(m + n)$$

by the distributive property. Since m and n are integers, their sum $m + n$ is an integer, let us call it N. Then

$$k + h = aN$$

Hence $a \mid k + h$ by the definition of divisibility.

THEOREM 2.2: *If $a \mid k$ and $a \mid h$, then $a \mid k - h$.*

Proof: Since $a \mid k$ and $a \mid h$, we may write, by the definition of divisibility,

$$k = am$$
$$h = an$$

where m and n are integers. Subtracting these two equations member by member we have

$$k - h = am - an$$
$$= a(m - n)$$

Since m and n are integers, their difference is an integer, call it M. Then

$$k - h = aM$$

Hence

$$a \mid k - h$$

by the definition of divisibility.

THEOREM 2.3: *If $a \mid k$ and $a \mid h$, then $a \mid kh$.*

Proof: Since $a \mid k$ and $a \mid h$, we may write, by the definition of divisibility,

$$k = am$$
$$h = an$$

where m and n are integers. Multiplying these two equations member by member we have

$$kh = (am)(an)$$
$$= a(man)$$

by the associative property of multiplication

Since a, m, and n are integers, their product is an integer, call it L. Then

$$kh = aL$$

Hence

$$a \mid kh$$

by the definition of divisibility.

As an illustration of the above theorems, notice that since $3 \mid 27$ and $3 \mid 36$,

$$3 \mid 63 \text{ because } 63 = 27 + 36$$
$$3 \mid 9 \text{ because } 36 - 27 = 9$$
$$3 \mid -9 \text{ because } 27 - 36 = -9$$
$$3 \mid 972 \text{ because } 27 \cdot 36 = 972$$

We observe that:

1. *Every integer n is divisible by $+1$ and -1, since $n = (+1)(n) = (-1)(-n)$.*
2. *Every integer $n \neq 0$ is divisible by $+n$ and $-n$, since $n = (+n)(1) = (-n)(-1)$.*
3. *Zero is a multiple of every integer n since $0 = n \cdot 0$.*
4. *For every integer $c \neq 0$, if $c = ab$, a and b both integers, $\pm a$ and $\pm b$ are factors of c since $c = (+a)(+b) = (-a)(-b)$.*

We now prove:

THEOREM 2.4: *If $d \mid c$ and $c \mid a$, then $d \mid a$.*

Proof: Since $d \mid c$ and $c \mid a$, we may write, by the definition of divisibility,

$$c = de$$
$$a = cb$$

where b and e are integers. Using the law of substitution† we have

$$a = cb = (de)b = d(eb)$$

by the associative property of multiplication. Since b

---

† See Chapter 1, Section 12.

and e are integers, their product eb is an integer, call it N. Then

$$a = dN$$

and hence

$$d \,|\, a$$

by the definition of divisibility.

Theorem 2.4 assures us that if m and n are integers and $m \,|\, n$, then any divisor of m also divides n. For example, since $32 \,|\, 256$, 2, 4, 8 and 16 also divide 256, since they are all divisors of 32.

We can find all the positive divisors of an integer $n > 0$ by writing it as the product of two positive factors in as many ways as possible. In writing a positive integer n as the product of two positive factors a and b, the factors a and b cannot both be greater than $\sqrt{n}$. If both a and b were greater than $\sqrt{n}$, we would have their product, ab, greater than $\sqrt{n} \cdot \sqrt{n} = n$, which is impossible since $ab = n$. Hence either a or b must be less than or equal to $\sqrt{n}$. We can assume, therefore, in looking for pairs of factors a and b of n, that we will have $a \leq \sqrt{n}$ and $b \geq \sqrt{n}$. This limits the possible numbers we have to try in determining the divisors of n.

Example 1:   Find all the positive divisors of 72.

Solution: Since $\sqrt{72} < 9$, in writing 72 as a product ab, we know that $a \leq 9$ and $b \geq 9$. Then

$$
\begin{aligned}
72 &= 1 \cdot 72 \\
&= 2 \cdot 36 \\
&= 3 \cdot 24 \\
&= 4 \cdot 18 \\
&= 6 \cdot 12 \\
&= 8 \cdot 9
\end{aligned}
$$

All the positive divisors of 72 are 1, 2, 3, 4, 6, 8, 9, 12, 18, 24, 36, and 72.

Example 2:   Find all the integers that are factors of 99.

Solution: First we find all the positive factors of 99. Since $\sqrt{99} < 10$, in writing 99 as the product ab, $a > 0$ and $b > 0$, we must have $a \leq 10$ and $b \geq 10$. Then

$$99 = 1 \cdot 99$$
$$= 3 \cdot 33$$
$$= 9 \cdot 11$$

All the positive factors of 99 are 1, 3, 9, 11, 33, and 99. If a positive integer a divides b, then $-a$ also divides b, hence -1, -3, -9, -11, -33, and -99 also divide 99. Hence all the integral factors of 99 are ±1, ±3, ±9, ±11, ±33, and ±99.

Exercise 1

1. Write the first integer given in each pair of integers below as a multiple of the second.

(a) 48; 2
(b) 56; 7
(c) 81; -3
(d) 120; -5

(e) 324; 4
(f) 725; -25
(g) 688; 16
(h) 1536; -32

2. Name two integers that are factors of every integer.

3. Find all the positive divisors of the following.

(a) 12
(b) 56

(c) 96
(d) 144

(e) 180
(f) 220

(g) 270
(h) 160

4. Name an integer that is a multiple of every integer.

5. Name all the integral factors of the following.

(a) 2
(b) 3

(c) 11
(d) 23

(e) 41
(f) 53

(g) 83
(h) 101

6. An integer n is divisible by 24. Name seven other positive divisors of n.

7. Write 144 as the product of two positive integers in as many ways as possible.

8. What whole number greater than 1 is a factor of every even integer?

9. Which of the following are true statements?

(a) Every positive integer is divisible by 1.
(b) Zero is a multiple of every positive integer.
(c) For integers a, b, and c, if a is divisible by b and b is divisible by c, then a is divisible by c.
(d) For integers k and m, if k divides m, then $k = mp$, p an integer.
(e) For integers a, b, c, and k, if k divides a, b, and c, then k divides their product abc.
(f) If an integer k is divisible by 9, then it is divisible by 1, 3, and 6.
(g) Every integer except 0 is divisible by itself.
(h) Every integer $k \neq 0$ has at least four integral factors, ±1 and $\pm k$.

10. Every even integer may be represented in the form $2n$ where n is an integer. Every odd integer may be represented in the form $2n + 1$ where n is an integer.

DIVISIBILITY

For example, $18 = 2 \cdot 9$; $27 = 2 \cdot 13 + 1$. Write each of the following as $2n$ or $2n + 1$; in each case give the value of n.

(a) 17	(c) −53	(e) 6472	(g) −4113
(b) −86	(d) 98	(f) −346	(h) 7613

11. Every integer may be written in the form $3n$, $3n + 1$ or $3n + 2$ where n is an integer. For example

$$16 = 3 \cdot 5 + 1$$
$$18 = 3 \cdot 6$$
$$-29 = 3(-10) + 1$$

Write the following as $3n$, $3n + 1$, or $3n + 2$.

(a) 6	(c) 14	(e) 20	(g) −58
(b) 32	(d) 27	(f) 35	(h) 39

12. Every integer may be written as $5n$, $5n + 1$, $5n + 2$, $5n + 3$, or $5n + 4$ where n is an integer. For example

$$26 = 5 \cdot 5 + 1$$
$$-12 = 5 \cdot (-3) + 3$$

Write the following as $5n$, $5n + 1$, $5n + 2$, $5n + 3$ or $5n + 4$.

(a) 29	(c) −34	(e) 523	(g) −444
(b) 127	(d) −267	(f) 788	(h) −874

13. Justify each statement below by one of the axioms of the set of integers and their operations.

(a) $3 + (-4) = (-4) + 3$
(b) $-4 + 0 = -4$
(c) $5[3 + (-2)] = 5 \cdot 3 + 5(-2)$
(d) $7 \cdot (3 \cdot 8) = (7 \cdot 3) \cdot 8$
(e) $-16 \cdot 1 = -16$
(f) $(3 + 4) + [5 + (-2)] = [5 + (-2)] + (3 + 4)$
(g) $17 \cdot 0 = 0$

14. Is the set of positive integers closed under the operation of addition, that is, is the sum of two positive integers always a positive integer?

15. Name the additive inverse of each of the following integers.

(a) 7	(e) x, x an integer
(b) −3	(f) $-y$, y an integer
(c) 12	(g) $x + y$, x and y integers
(d) −302	(h) $-(x + y)$, x and y integers.

16. For a, b, and c integers, is it always true that $a(b - c) = ab - ac$?

3.
DIVISION
AND REMAINDERS

Let $b \neq 0$ be any nonnegative integer. Every other integer a will either be a multiple of b or will fall between two consecutive multiples of b. That is, there is a unique integer q such that

$$a = bq$$

or

$$bq < a < b(q + 1)$$

Geometrically, this may be visualized as follows. Starting at 0, the number line may be partitioned into intervals b units in length. Any point with an integer for a coordinate either lies within one of these intervals or is an endpoint of an interval (Figure 2.1).

For example, let $a = 18$ and $b = 5$. Since 18 is not a multiple of 5, it must fall between two consecutive multiples of 5. In fact,

$$5 \cdot 3 < 18 < 5 \cdot (3 + 1) = 5 \cdot 4$$

Similarly, if $a = -26$ and $b = 6$, then

$$6(-5) < -26 < 6(-5 + 1).$$

For any two integers a and b, $b > 0$, we can find unique integers q and r such that

$$a = bq + r$$

and r is one of the nonnegative integers $0, 1, 2, \ldots, b - 1$. That is $r \geq 0$ and $r < b$. We usually write this conjunction $0 \leq r < b$, which is read "r is greater than or equal to zero and less than b." In the equation $a = bq + r$, a is called the **dividend,** b is called the **divisor,** q is called the

FIGURE 2.1

quotient, and *r* is called the **remainder.** The process of finding *q* and *r*, given *a* and *b*, is called **division.** This division property is called the **division algorithm.** We shall accept the division algorithm without proof.

DIVISION ALGORITHM. **For every pair *a* and *b* of integers, *b* > 0, there exist unique integers *q* and *r* such that $a = bq + r, \ 0 \leq r < b.$**

Let us now consider only the set of nonnegative integers

$$\{0, 1, 2, 3, 4, \ldots\}$$

We recall that the numbers in this set are called the **whole numbers.** When any whole number is divided by a whole number $m \neq 0$, the possible remainders are $0, 1, 2, 3, \ldots, (m-1)$. We see then that the set of whole numbers fall into *m* classes when they are divided by a whole number $m \neq 0$. For example, if we divide each of the whole numbers by 5, the possible remainders are 0, 1, 2, 3, or 4. All of the whole numbers that have remainders 0 when divided by 5 we put into the same class; all of the whole numbers that have remainders 1 we put into a second class; all of the whole numbers that have remainders 2 we put into a third class; all of the whole numbers that have remainders 3 we put into a fourth class; and all of the whole numbers that have remainders 4 we put into a fifth class. Table 2.1 shows the five classes when the whole numbers are divided by 5 together with some members of each class.

Notice that all of the numbers that have remainders 0 when divided by 5 are of the form $5k$, *k* a whole number; those with remainder 1 are of the form $5k + 1$; those with remainder 2 are of the form $5k + 2$; those with remainder 3 are of the form $5k + 3$; and those with remainder 4 are of the form $5k + 4$.

We see that every whole number may be represented by one of the

TABLE 2.1

Remainders	Whole numbers in the class
0	$0, 5, 10, 15, 20, 25, \ldots$
1	$1, 6, 11, 16, 21, 26, \ldots$
2	$2, 7, 12, 17, 22, 27, \ldots$
3	$3, 8, 13, 18, 23, 28, \ldots$
4	$4, 9, 14, 19, 24, 29, \ldots$

forms $5k$, $5k + 1$, $5k + 2$, $5k + 3$, or $5k + 4$ where k is a whole number.

In general every whole number may be represented by one of the form

$$mk, \ mk + 1, \ mk + 2, \ mk + 3, \ \ldots, \ mk + (m - 1)$$

m and k whole numbers and $m \neq 0$. We find which form represents a particular whole number by dividing the whole number by m to find the quotient and the remainder. If the quotient is k and the remainder is r, the number is of the form $mk + r$.

The division algorithm leads us to many important properties of whole numbers. All whole numbers are either **even** or **odd**. Even numbers are of the form $2k$ and odd numbers are of the form $2k + 1$. When even numbers are squared we have

$$(2k)^2 = 4k^2$$

which is a multiple of 4. When odd numbers are squared we have

$$(2k + 1)^2 = 4k^2 + 4k + 1$$
$$= 4(k^2 + k) + 1$$

by the distributive property. Since k is an integer, $k^2 + k$ is an integer, call it M. Then we have

$$(2k + 1)^2 = 4M + 1$$

We see from this that the square of a whole number is either a multiple of 4 (if it is even), or of the form $4M + 1$, that is, has a remainder of 1 when it is divided by 4 (if it is odd).

We can also show that every square is divisible by 3 or is of the form $3k + 1$. All whole numbers may be represented by one of the forms $3k$, $3k + 1$, or $3k + 2$. Let us square each of these in turn:

(1) $$(3k)^2 = 9k^2 = 3(3k^2)$$

Since k is an integer, $3k^2$ is an integer. Let $3k^2 = M$. Then

$$(3k)^2 = 3M$$

(2) $$(3k + 1)^2 = 9k^2 + 6k + 1$$
$$= 3(3k^2 + 2k) + 1$$

Since k is an integer, $3k^2 + 2k$ is an integer. Let $3k^2 + 2k = M$. We have

$$(3k + 1)^2 = 3M + 1$$

(3)
$$(3k + 2)^2 = 9k^2 + 12k + 4$$
$$= 9k^2 + 12k + 3 + 1$$
$$= 3(3k^2 + 4k + 1) + 1$$

Since k is an integer, $3k^2 + 4k + 1$ is an integer. Let $3k^2 + 4k + 1 = M$. We obtain

$$(3k + 2)^2 = 3M + 1$$

From this we see that the square of any whole number is a multiple of 3 or has a remainder 1 when divided by 3 (that is, it is of the form $3k + 1$).

We can also show that the square of an odd number is of the form $8M + 1$. We know from the division algorithm that each whole number is represented by one of the forms $4k$, $4k + 1$, $4k + 2$, or $4k + 3$. Of these, only those numbers of the forms $4k + 1$ and $4k + 3$ are odd. Squaring these we see that

$$(4k + 1)^2 = 16k^2 + 8k + 1$$
$$= 8(2k^2 + k) + 1$$

Since k is an integer, so is $2k^2 + k$. Let $2k^2 + k = M$. We obtain

$$(4k + 1)^2 = 8M + 1$$

Similarly

$$(4k + 3)^2 = 16k^2 + 24k + 9$$
$$= 16k^2 + 24k + 8 + 1$$
$$= 8(2k^2 + 3k + 1) + 1$$
$$= 8M + 1$$

We summarize the above results in the theorems stated below.

THEOREM 2.5: *The square of a whole number is either a multiple of 4 or has a remainder of 1 when divided by 4.*

THEOREM 2.6: *The square of any whole number is either a multiple of 3 or has a remainder of 1 when divided by 3.*

THEOREM 2.7: *The square of any odd number is of the form $8M + 1$.*

We shall now prove that the sum of two even numbers is an even number.

THEOREM 2.8: *The sum of two even numbers is an even number.*
        Proof: Every even number may be written in the form $2k$

where k is a whole number. Let $2k$ and $2h$ represent two even numbers. Then

$$2k + 2h = 2(k + h) \quad \text{Distributive Property}$$
$$= 2M \quad\quad\quad k + h \text{ is an integer, call it } M$$

But any number of the form $2M$ is an even number. Hence the sum of two even numbers is an even number.

We can also prove the following.

THEOREM 2.9: *The sum of an even number and an odd number is an odd number.*

Proof: Any even number may be written in the form $2k$. Any odd number may be written in the form $2m + 1$. Then

$$
\begin{aligned}
2k + (2m + 1) & \quad \text{Associative Property} \\
= (2k + 2m) + 1 & \quad \text{of Addition} \\
= 2(k + m) + 1 & \quad \text{Distributive Property} \\
= 2L + 1 & \quad k + m \text{ is an integer} \\
& \quad \text{call it } L
\end{aligned}
$$

But any number of the form $2L + 1$ is odd. Hence the sum of an even number and an odd number is an odd number.

Exercise 2

1. Given a and b as follows, find q and r such that $a = bq + r$, $0 \leq r < b$.
   (a) $a = 26$, $b = 8$
   (b) $a = 39$, $b = 7$
   (c) $a = 126$, $b = 15$
   (d) $a = 256$, $b = 27$
   (e) $a = 369$, $b = 21$
   (f) $a = 1274$, $b = 97$
   (g) $a = 8$, $b = 12$
2. If every whole number is divided by 7, the possible remainders are 0, 1, 2, 3, 4, 5, and 6. Hence all whole numbers may be represented by one of seven forms. What are they?
3. Prove: The product of two even numbers is an even number.
4. Prove: The product of two odd numbers is an odd number.
5. Prove: The sum of two odd numbers is an even number.

6. Prove: The fourth power of a number that is not divisible by 5 is of the form $5k + 1$.

7. Prove: If 3 divides a and b (that is, a and b are multiples of 3) then 3 divides $a + b$.

8. Every whole number may be represented by one of the forms:

$$
\begin{array}{ll}
12k & 12k + 6 \\
12k + 1 & 12k + 7 \\
12k + 2 & 12k + 8 \\
12k + 3 & 12k + 9 \\
12k + 4 & 12k + 10 \\
12k + 5 & 12k + 11
\end{array}
$$

(a) Which of the numbers named above are not divisible by 2?
(b) Which of the numbers named above are not divisible by 3?
(c) Which of the numbers named above are not divisible by 2 or 3?
(d) Using the result you found in (c), prove that the square of a number not divisible by 2 or 3 is of the form $12k + 1$.

9. All whole numbers may be represented by one of the forms

$$6k \quad 6k + 1 \quad 6k + 2 \quad 6k + 3 \quad 6k + 4 \quad 6k + 5$$

(a) Which of the numbers named above are odd?
(b) Which of the numbers named above are even?
(c) Which of the numbers named above are divisible by 3?

10. Prove: Every whole number of the form $12k + 6$ is divisible by 2, 3, and 6.

4.
MORE THEOREMS ON DIVISIBILITY

We now prove the following theorems concerning divisibility. In all cases a, b, and c represent integers.

THEOREM 2.10: *If $a \neq 0$, then $a\,|\,0$ and $a\,|\,a$.*

Proof: By the multiplication property of zero

$$0 = a \cdot 0$$

Hence by the definition of divisibility $a\,|\,0$. By the multiplicative identity axiom $a = 1 \cdot a$, and, by the definition of divisibility, $a\,|\,a$.

4. MORE THEOREMS ON DIVISIBILITY

THEOREM 2.11: $1 \mid a$ *for all a.*

Proof: Since $a \cdot 1 = a$ (why?) the definition of divisibility assures us that $1 \mid a$.

THEOREM 2.12: *If $a \mid b$, $a \neq 0$, then $a \mid bc$ for any c.*

Proof: Since $a \mid b$ we know that for some integer k

$$ak = b$$

Multiplying each member of this equation by c we have

$$(ak)c = bc$$
$$a(kc) = bc \qquad \text{Associative property of multiplication}$$

and $a \mid bc$ by the definition of divisibility.

THEOREM 2.13: *If $a \mid b$ and $a \mid c$, $a \neq 0$, for any integers x and y, $a \mid bx + cy$.*

Proof: Since $a \mid b$ and $a \mid c$ we know that there exist integers k and n such that

$$ak = b \qquad \text{and} \qquad an = c$$

Multiplying each member of the first equation by x and each member of the second equation by y and adding the resulting equations member by member we obtain

$$bx + cy = (ak)x + (an)y$$
$$= a(kx) + a(ny) \qquad \text{Associative property of multiplication}$$
$$= a(kx + ny) \qquad \text{Distributive property}$$

and $a \mid bx + cy$ by the definition of divisibility.

Exercise 3

Prove the following theorems.

1. For integers a and b, $a \neq 0$, if $a + b = c$ and $a \mid b$, then $a \mid c$.
2. For integers a and b, $a \neq 0$, if $a + b = c$ and $a \mid c$, then $a \mid b$.
3. For d a positive integer if $d \mid 1$, then $d = 1$.

4. For any integer n, $n(n + 1)(2n + 1)$ is divisible by 6.
5. For integers a and d, $a \neq 0$, if $d \mid a$, then $-d \mid a$.
6. One of three consecutive integers is divisible by 3.
7. The sum of three consecutive integers is divisible by 3.
8. For whole numbers x, y, and z, both x and y cannot be odd if $x^2 + y^2 = z^2$.
9. The sum of two even integers is an even integer.
10. The sum of an even integer and an odd integer is an odd integer.
11. For integers a and b, $a \neq 0$ and $b \neq 0$, if $a \mid b$ and $b \mid a$ then $a = b$ or $a = -b$.
12. The sum of an even number of integers is even.
13. The product of n integers is even unless they are all odd.
14. The product of three integers of the form $4k + 1$ is also of the form $4k + 1$.
15. Every odd integer is either of the form $4k + 1$ or of the form $4k + 3$.
16. If p and q are integers and $p^2 = 2q^2$, then p and q are both even.
17. If two odd integers are both of the form $4k + 3$, then their difference is divisible by 4.
18. For integers a, b, c, q, r, s, and t, $c \neq 0$, if $a = cq + r$ and $b = ct + s$, and $c \mid a - b$, then $c \mid r - s$.
19. For integers a, b, c, q, r, s, and t, $c \neq 0$ if $a = cq + r$, $b = ct + s$ and $c \mid a + b$, then $c \mid r + s$.
20. If n is an odd integer, then $n(n^2 - 1)$ is divisible by 24.

5.
RULES OF DIVISIBILITY

The decimal numeral of every whole number, n, with $k + 1$ digits may be written in **expanded form** as

$$n = a_0 10^k + a_1 10^{k-1} + a_2 10^{k-2} + \cdots + a_{k-2} 10^2 + a_{k-1} 10 + a_k$$

where each of the coefficients $a_0, a_1, a_2, \ldots, a_k$ is one of the digits 0, 1, 2, 3, 4, 5, 6, 7, 8, or 9, and $a_0 \neq 0$. For example

$$3{,}467 = 3 \cdot 10^3 + 4 \cdot 10^2 + 6 \cdot 10 + 7$$
$$123{,}895 = 1 \cdot 10^5 + 2 \cdot 10^4 + 3 \cdot 10^3 + 8 \cdot 10^2 + 9 \cdot 10 + 5$$
$$8{,}000{,}000 = 8 \cdot 10^6 + 0 \cdot 10^5 + 0 \cdot 10^4 + 0 \cdot 10^3 + 0 \cdot 10^2 + 0 \cdot 10 + 0$$

With this in mind, rules for divisibility by 2, 3, 4, 5, 7, 9, and 10 can be derived.

THEOREM 2.14: *A number n is divisible by 2 if and only if the units digit of its numeral is 0, 2, 4, 6, or 8.*

The phrase "if and only if" used in Theorem 2.14 combines two statements into one. In Theorem 2.14, the two statements are: (1) If a number is divisible by 2 then the units digit of its numeral is 0, 2, 4, 6, or 8; and (2) If the units digit of the numeral of a number is 0, 2, 4, 6, or 8, then the number is divisible by 2. In looking at these two statements, we see that the second is the converse of the first. In proving Theorem 2.14 we must prove statement (1) and its converse, statement (2).

Proof of Theorem 2.14: We first prove statement (1). Since n is divisible by 2,

$$a_0 10^k + a_1 10^{k-1} + \cdots + a_{k-1} 10 + a_k$$

is divisible by 2, that is, it is an even number. Now

$$
\begin{aligned}
a_0 10^k &+ a_1 10^{k-1} + \cdots + a_{k-1} 10 + a_k \\
&= 10(a_0 10^{k-1} + a_1 10^{k-2} + \cdots + a_{k-1}) + a_k \\
&= [2 \cdot 5(a_0 10^{k-1} + a_1 10^{k-2} + \cdots + a_{k-1})] + a_k
\end{aligned}
$$

is an even number. Since the sum $[(2 \cdot 5)(a_0 10^{k-1} + a_1 10^{k-2} + \cdots + a_{k-1})] + a_k$ is even and $(2 \cdot 5)(a_0 10^{k-1} + a_1 10^{k-2} + \cdots + a_{k-1})$ is even (why?), a_k must be even.† The possible values of a_k are 0, 1, 2, 3, 4, 5, 6, 7, 8, or 9. Of these only 0, 2, 4, 6, and 8 are even. Hence for a_k to be even it must be 0, 2, 4, 6, or 8. We now prove the second part of Theorem 2.14. That is, we prove the converse of statement (1) which was proved above. We are given that a_k is 0, 2, 4, 6, or 8. We wish to prove that n is divisible by 2. Since each of $a_0 10^k$, $a_1 10^{k-1}$, ..., $a_{k-1} 10$, and a_k are divisible by 2, their sum

$$n = a_0 10^k + a_1 10^{k-1} + \cdots + a_{k-1} 10 + a_k$$

is divisible by 2 by Theorem 2.1. Hence n is divisible by 2.

THEOREM 2.15: *A number is divisible by 5 if and only if the units digit of its numeral is 0 or 5.*

---

† If it were odd the sum would be odd since an even number plus an odd number is odd.

Proof: We prove Theorem 2.15 using an argument similar to that used to prove Theorem 2.14. Since n is divisible by 5 and

$$a_0 10^k + a_1 10^{k-1} + \cdots + a_{k-1} 10$$
$$= 10(a_0 10^{k-1} + a_1 10^{k-2} + \cdots + a_{k-1})$$
$$= 5 \cdot 2(a_0 10^{k-1} + a_1 10^{k-2} + \cdots a_{k-1})$$

is divisible by 5, their difference

$$n - [5 \cdot 2(a_0 10^{k-1} + a_1 10^{k-2} + \cdots + a_{k-1})] = a_k$$

is divisible by 5 by Theorem 2.2. The possible values of a_k are 0, 1, 2, 3, 4, 5, 6, 7, 8, or 9. Of these only 0 and 5 are divisible by 5. Hence if a number is divisible by 5, its units digit is 0 or 5.

We now prove the converse. If $a_k = 0$ or $a_k = 5$, we have

$$n = a_0 10^k + a_1 10^{k-1} + \cdots + a_{k-1} 10 + 0$$

or

$$n = a_0 10^k + a_1 10^{k-1} + \cdots + a_{k-1} 10 + 5$$

But

$$a_0 10^k + a_1 10^{k-1} + \cdots + a_{k-1} 10 + 0$$
$$= 10(a_0 10^{k-1} + a_1 10^{k-2} + \cdots + a_{k-1} + 0)$$
$$= 5 \cdot 2(a_0 10^{k-1} + a_1 10^{k-2} + \cdots + a_{k-1} + 0)$$

which is divisible by 5, hence n is divisible by 5. Also

$$a_0 10^k + a_1 10^{k-1} + \cdots + a_{k-1} 10 + 5$$
$$= 10(a_0 10^{k-1} + a_1 10^{k-2} + \cdots + a_{k-1}) + 5$$
$$= 5 \cdot 2(a_0 10^{k-1} + a_1 10^{k-2} + \cdots + a_{k-1}) + 5$$
$$= 5[2(a_0 10^{k-1} + a_1 10^{k-2} + \cdots + a_{k-1}) + 1]$$

which is divisible by 5, hence n is divisible by 5.

THEOREM 2.16: *A number is divisible by 10 if and only if the units digit of its numeral is 0.*

Proof: Since $10 = 2 \cdot 5$, a number that is divisible by 10 must be divisible by both 2 and 5. By Theorem 2.14 if a number is divisible by 2 the units digit of its numeral is 0, 2, 4, 6, or 8. By Theorem 2.15 if a number is divisible by 5 the units digit of its numeral is 0 or 5. Since 0 is the only units digit that satisfies the condition that a number is divisible by both 2 and 5, that is by 10, we see that if a number is divisible by 10 its units digit must be 0.

We now prove the converse. If the units digit of a number is 0 we have

$$n = a_0 10^k + a_1 10^{k-1} + \cdots + a_{k-1} 10 + 0$$
$$= a_0 10^k + a_1 10^{k-1} + \cdots + a_{k-1} 10$$
$$= 10(a_0 10^{k-1} + a_1 10^{k-2} + \cdots + a_{k-1})$$

and we see that 10 divides n.

THEOREM 2.17: *A number is divisible by 3 if and only if the sum of the digits of its numeral is divisible by 3.*

Proof: Proving Theorem 2.17 requires the rearrangement of the expanded form of the numeral of the number. Let us first look at the powers of 10:

$$10^1 = 9 + 1$$
$$10^2 = 99 + 1$$
$$10^3 = 999 + 1$$
$$10^4 = 9999 + 1$$
$$\vdots$$
$$10^k = \underbrace{999 \ldots 9}_{k \, 9s} + 1$$

Now

$$n = a_0 10^k + a_1 10^{k-1} + \cdots + a_{k-1} 10 + a_k$$
$$= a_0(99 \ldots 9 + 1) + a_1(99 \ldots 9 + 1)$$
$$+ \cdots + a_{k-1}(9 + 1) + a_k$$
$$= [a_0 \cdot 99 \ldots 9 + a_1 \cdot 99 \ldots 9 + \cdots + a_{k-1} \cdot 9]$$
$$+ [a_0 \cdot 1 + a_1 \cdot 1 + \cdots + a_{k-1} \cdot 1 + a_k]$$

This rearrangement is possible using the distributive property and the associative and commutative properties of addition. Now, applying the multiplicative identity axiom we have

$$n = [a_0 \cdot 99 \ldots 9 + a_1 \cdot 99 \ldots 9 + \cdots + a_{k-1} \cdot 9]$$
$$+ [a_0 + a_1 + \cdots + a_{k-1} + a_k]$$

Since n is divisible by 3 and

$$a_0 \cdot 999 \ldots 9 + a_1 \cdot 99 \ldots 9 + \cdots + a_{k-1} \cdot 9$$
$$= 9[a_0 \cdot 111 \ldots 1 + a_1 \cdot 11 \ldots 1 + \cdots + a_{k-1} \cdot 1]$$
$$= 3 \cdot 3(a_0 \cdot 11 \ldots 1 + a_1 \cdot 11 \ldots 1 + \cdots + a_{k-1} \cdot 1]$$

is divisible by 3, their difference

$$n - 3 \cdot 3(a_0 \cdot 11 \ldots 1 + a_1 \cdot 11 \ldots 1$$
$$+ \cdots + a_{k-1} \cdot 1]$$
$$= a_0 + a_1 + a_2 + \cdots + a_{k-1} + a_k$$

is divisible by 3 by Theorem 2.2. But their difference

$$a_0 + a_1 + \cdots + a_{k-1} + a_k$$

is the sum of the digits of the numeral of n. Hence if a number is divisible by 3, the sum of the digits of the numeral of the number must be divisible by 3. Conversely, if $a_0 + a_1 + \cdots + a_k$ is divisible by 3, then

$$n = [a_0 \cdot 99 \ldots 9 + a_1 \cdot 99 \ldots 9 + \cdots + a_{k-1} \cdot 9]$$
$$+ [a_0 + a_1 + \cdots + a_k]$$

is divisible by 3 since $a_0 + a_1 + \cdots + a_k$ is divisible by 3 and

$$a_0 \cdot 99 \ldots 9 + a_1 \cdot 99 \ldots 9 + \cdots + a_{k-1} \cdot 9$$
$$= 9(a_0 \cdot 11 \ldots 1 + a_1 \cdot 11 \ldots 1 + \cdots + a_{k-1} \cdot 1)$$

is divisible by 3 as shown above. Hence their sum, n, is divisible by 3 by Theorem 2.1.

THEOREM 2.18: *A number is divisible by 9 if and only if the sum of the digits in its numeral is divisible by 9.*

This theorem is proved in exactly the same way as we prove Theorem 2.19. Observe that

$$a_0 \cdot 99 \ldots 9 + a_1 \cdot 99 \ldots 9 + \cdots + a_{k-1} \cdot 9$$

is divisible by 9 as well as by 3. The proof of this theorem is left to the reader.

THEOREM 2.19: *A number is divisible by 4 if and only if the number named by the last two digits in its numeral is divisible by 4.*

Proof: First we prove that if a number is divisible by 4, then the number named by the last two digits of its numeral is divisible by 4. We are given that 4 divides n. Then

$$n = a_0 10^k + a_1 10^{k-1} + \cdots + a_{k-1}10 + a_k$$
$$= 10^2(a_0 10^{k-2} + a_1 10^{k-3} + \cdots + a_{k-2})$$
$$+ (a_{k-1}10 + a_k)$$

Since $4 \mid n$ and $4 \mid 10^2$ $(a_0 10^{k-2} + a_1 10^{k-3} + \cdots + a_{k-2})$, because $4 \mid 10^2$, their difference

$$n - 10^2(a_0 10^{k-2} + a_1 10^{k-3} + \cdots + a_{k-2})$$
$$= a_{k-1}10 + a_k$$

is divisible by 4 by Theorem 2.2. But $a_{k-1}10 + a_k$ is the number named by the last two digits of the numeral of n.

We now prove that if $4 \mid (a_{k-1}10 + a_k)$, then $4 \mid n$.

n
$$= a_0 10^k + a_1 10^{k-1} + \cdots + a_{k-2}10^2 + a_{k-1}10 + a_k$$
$$= 10^2(a_0 10^{k-2} + a_1 10^{k-3} + \cdots + a_{k-2})$$
$$+ (a_{k-1}10 + a_k)$$

Since $4 \mid 10^2$ it divides $10^2(a_0 10^{k-2} + a_1 10^{k-3} + \cdots + a_{k-2})$. Now 4 divides $10^2(a_0 10^{k-2} + a_1 10^{k-3} + \cdots + a_{k-2})$ and $a_{k-1}10 + a_k$, hence 4 divides their sum n by Theorem 2.1.

Example 1: Is 326,454 divisible by 2? by 3? by 6?

Solution: Since the units digit of 326,454 is 4, 326,454 is divisible by 2 by Theorem 2.14.
Since

$$3 + 2 + 6 + 4 + 5 + 4 = 24$$

and $3\,|\,24$, 326,454 is divisible by 3 by Theorem 2.17.
To be divisible by 6 a number must be divisible both by
2 and by 3 since $6 = 2 \cdot 3$. Since 326,454 is divisible by 2
and by 3 it is divisible by 6.

Example 2: Is 439,710 divisible by 3? by 5? by 15?
Solution: Since

$$4 + 3 + 9 + 7 + 1 + 0 = 24$$

and $3\,|\,24$, 439,710 is divisible by 3. Since the units digit of
439,710 is 0, the number is divisible by 5. To be divisible
by 15, a number must be divisible both by 3 and by 5. Since
439,710 is divisible by 3 and by 5 it is divisible by 15.

Exercise 4

1. Which of the following are divisible by 2?
   (a) 168,764
   (b) 97,842
   (c) 43,105
   (d) 876,000
   (e) 7,659,430
   (f) 578,999
   (g) 876,042
   (h) 178,401
2. Which of the numbers named in problem 1 are divisible by 3?
3. Which of the numbers named in problem 1 are divisible by 5?
4. Which of the numbers named in problem 1 are divisible by 4?
5. Which of the numbers named in problem 1 are divisible by 9?
6. If a number is divisible by 7, then the difference obtained by subtracting
two times the units digit from the number named by the remaining digits in its
numeral is divisible by 7. To illustrate let $n = 132{,}342$. We delete the units digit,
132,34$\cancel{2}$. We multiply the deleted units digit by 2, $2 \cdot 2 = 4$ and subtract this
product from 13,234: $13234 - 4 = 13230$. Continuing in this fashion:

$$
\begin{array}{rr}
 & 132{,}34\cancel{2} \\
2 \times 2 = 4 & 4 \\
\hline
 & 132\ 3\cancel{0} \\
2 \times 0 = 0 & 0 \\
\hline
 & 132\ \cancel{3} \\
2 \times 3 = 6 & 6 \\
\hline
 & 12\cancel{6} \\
2 \times 6 = 12 & 12 \\
\hline
 & 0
\end{array}
$$

$7\,|\,0$, hence $7\,|\,132{,}342$

Which of the following are divisible by 7?

(a) 143,724   (d) 808,669

(b) 367,909   (e) 999,107

(c) 876,421   (f) 777,777

7. Which of the following are divisible by 6?

(a) 143,724   (d) 837,102

(b) 67,301   (e) 366,999

(c) 82,524   (f) 421,212

8. Give a rule for divisibility by 30.

9. Give a rule for divisibility by 45.

10. A number is divisible by 12 if and only if it is divisible by 3 and 4. Which of the numbers named in problem 7 are divisible by 12?

11. A leap year is a year whose date is divisible by 4, except century years in which case the date must be divisible by 400. Which of the following are leap years?

(a) 1492   (d) 1969

(b) 1812   (e) 2432

(c) 1917   (f) 2858

12. What are the possible units digits for the numeral of an odd number?

13. If a number is divisible by 4 it must be even. Why? If a number is even is it necessarily divisible by 4?

14. Suppose a number is divisible by 9. If you reverse the digits in its numeral is the resulting number divisible by 9? Why?

15. Replace each blank with a single digit so that the resulting number is divisible by 3. (There may be more than one digit in each case.)

(a) 43,26_?_   (d) 813,4_?_5

(b) 30,2_?_5   (e) 7,321,27_?_

(c) 4_?_3,478   (f) 10,_?_34,243

16. Replace each blank with a single digit so that the resulting number is divisible by 6. (There may be more than one digit in each case.)

(a) 346_?_   (c) 123,_?_50

(b) 23,1_?_4   (d) 53_?_,616

17. Can a number be divisible by 4 if the units digit of its numeral is 1, 3, 5, 7, or 9? Why?

18. A number n is divisible by 13 if

$$\frac{n - a_k}{10} - 9a_k$$

is divisible by 13 (a_k is the units digit of the numeral of the number n). Using this rule, which of the following are divisible by 13?

(a) 169   (c) 1104   (e) 1612

(b) 741   (d) 1213   (f) 3300

19. A number is divisible by 11 if the difference between the sum of the odd ordered digits (starting with the units digit, add every other digit) and the sum of the even ordered digits (starting with the tens digit, add every other digit) is divisible by 11. This difference may be a positive integer, a negative integer, or zero. To illustrate let $n = 32,654$. Then

$$4 + 6 + 3 = 13 \qquad 5 + 2 = 7 \qquad 13 - 7 = 6$$

$11 \nmid 6$, hence $11 \nmid 32,654$. Using this rule, test the following for divisibility by 11.

(a) 346,987   (c) 89,157,486

(b) 463,127   (d) 3,421,154

20. Attempt to discover a rule for divisibility by 8. (Use a method similar to that to derive the rule for divisibility by 4.)

Leonhard Euler (1707–1783) is probably the greatest man of science that Switzerland ever produced. It has been said that "Euler calculated without apparent effort, as men breathe or as eagles sustain themselves in the wind." Euler was certainly one of the most prolific mathematicians in history. It has been estimated that Euler's work would fill one hundred books. Euler was taught mathematics by his father who intended that his son study theology and succeed him as pastor in the village church. Euler obeyed his father and studied theology. Nevertheless, he was sufficiently talented in mathematics to attract the attention of Johann, Daniel, and Nikolaus Bernoulli who became his close friends. Euler could work anywhere under any conditions. He often would compose his memoirs while one of his thirteen children sat on his lap and the others played at his feet. It has been said that he would dash off a mathematical paper in the half hour or so between the first and second calls for dinner.

PRIME
NUMBERS

3

PRIME AND COMPOSITE NUMBERS

In the previous chapter we discussed some rules for divisibility. The concept of divisibility introduced indicates the possibility of breaking down or "decomposing" some numbers in terms of others. This leads us to the discussion of prime and composite numbers.

Let us consider all the positive integers greater than 1. These integers fall into two classes, prime numbers and composite numbers. An integer $p > 1$ is called a **prime number,** or simply a **prime,** when its only divisors† are 1 and p. The first prime number is 2 because its only divisors are 1 and 2. The first few primes are

$$2, 3, 5, 7, 11, 13, 17, 19, 23, 29, 31, 37$$

The only *even* prime is 2 because every other even number is divisible by 2, and hence has at least one divisor other than 1 and itself.

An integer $m > 1$ which has divisors greater than 1 and less than m is called a **composite number.** The first composite number is 4. It has divisors 1, 2, and 4. The first few composite numbers are

$$4, 6, 8, 9, 10, 12, 14, 15, 16$$

A characteristic property of a composite number consists of the possibility of representing it as a product of two factors a and b,

$$m = ab$$

each of which is greater than 1. For example,

$$4 = 2 \cdot 2 \qquad 27 = 3 \cdot 9$$
$$6 = 2 \cdot 3 \qquad 56 = 7 \cdot 8$$
$$8 = 2 \cdot 4 \qquad 110 = 10 \cdot 11$$

Such a representation is impossible for a prime.

We can show that *every composite number is divisible by a prime.* Of all the divisors of a given composite number m, let us select the smallest,

---

† In this chapter when we say "divisor" we will mean "positive divisor."

p, which is still greater than 1. Now p must be a prime, otherwise it, too, would have a divisor q greater than 1 and less than p, and q would be a divisor of m. This contradicts the assumption that p, of all the divisors greater than 1 of m, was the smallest.

Every composite number m is divisible by a prime $p \leq \sqrt{m}$. Since m is a composite number it can be represented in the form

$$m = ab$$

where $a > 1$ and $b > 1$. We suppose $a \leq b$, and then $a \leq \sqrt{m}$. Now if $a > 1$ has a prime factor $p \leq a \leq \sqrt{m}$, p will be a divisor of m.

We now have a practical test to ascertain whether a given number is a prime. It suffices to divide it by the primes less than or equal to its square root. If one divisor succeeds without a remainder, then the number is composite; otherwise it is a prime.

Example 1: Is 661 a prime or a composite number?
Solution: Since $\sqrt{661}$ is between 25 and 26 ($25^2 = 625$, $26^2 = 676$), we need test only those primes not exceeding 25 and determine whether any of them divide 661. These primes are 2, 3, 5, 7, 11, 13, 17, 19 and 23. Dividing 661 by each of these in turn we find that none is a divisor of 661. Hence 661 is a prime number.

Example 2: Is 387 a prime or a composite number?
Solution: Since $19 < \sqrt{387} < 20$ ($19^2 = 361, 20^2 = 400$), we need test only those primes not exceeding 19. These primes are 2, 3, 5, 7, 11, 13, 17, and 19. Since 387 is odd, it is not divisible by 2. Since $3 + 8 + 7 = 18$ and 18 is divisible by 3, 387 is divisible by 3 and hence is not a prime.

2.
THE SIEVE OF ERATOSTHENES

We have just given a method for determining whether or not a number is a prime. This method works well if the number to be tested is not large. When the number to be tested is large, however, the trials become too numerous and burdensome. Innocent as it may seem, the problem of determining whether a given integer is a prime has no general solution.

A simple approach to the problem is called the **Sieve of Eratosthenes,** after the Greek mathematician Eratosthenes (266–194 B.C.). This method consists of writing down all the integers from 2 to the number n, which is to be tested, and sieving out the composite numbers. Two is the smallest prime, and the multiples of 2

$$2 \cdot 2, 2 \cdot 3, 2 \cdot 4, \ldots, 2k, \ldots$$

occur in the list of positive integers at intervals of two following 2. Thus we scratch from the list every second number after 2, all of which are composite numbers. Now, 3, the next integer not scratched out, is a prime. Again multiples of 3 occur in the list of positive integers at intervals of three following 3, so we scratch out every third number after 3. We continue in this fashion. Since every composite number must have a prime factor not exceeding its square root, every composite number in the list must have a prime factor not exceeding $\sqrt{n}$. Thus, by the time we have deleted all multiples of all primes less than or equal to $\sqrt{n}$, we have sieved out all the composite numbers and all those that remain will be primes not exceeding n.

Table 3.1 shows the completed sieve for $n = 100$. Note that since

TABLE 3.1
Sieve of Eratosthenes for $n = 100$

	2	3	4	5	6	7	8	9	10
11	12	13	14	15	16	17	18	19	20
21	22	23	24	25	26	27	28	29	30
31	32	33	34	35	36	37	38	39	40
41	42	43	44	45	46	47	48	49	50
51	52	53	54	55	56	57	58	59	60
61	62	63	64	65	66	67	68	69	70
71	72	73	74	75	76	77	78	79	80
81	82	83	84	85	86	87	88	89	90
91	92	93	94	95	96	97	98	99	100

$10^2 = 100$, the process is completed by the time all the multiples of 7 (largest prime less than $\sqrt{100} = 10$) have been struck out. The primes in the table have been circled.

Variations on the sieve method provide the most effective means for computing tables of primes. The best tables of primes are those of D. N. Lehmer which extend beyond ten million.

3.
INFINITUDE OF PRIMES

It is quite natural to ask: Is the set of prime numbers an infinite set? This question was answered by the ancient Greeks. Euclid gave a very simple proof that shows there are infinitely many primes.

THEOREM 3.1: *There are infinitely many primes.*

Proof: Let us assume that there is a finite number of primes and let p be the greatest prime. Then, by our assumption, the primes

$$2, 3, 5, 7, \ldots, p$$

taken in their natural order compose the complete set.

Now let us form the number

$$N = (2 \cdot 3 \cdot 5 \cdot 7 \cdot \ \cdots \ \cdot p) + 1$$

Since N is a natural number greater than 1, it is either prime or composite. If N is a prime number, then our list of primes is not complete, since $N > p$. If N is a composite number it has a prime divisor q which is different from $2, 3, \ldots, p$. For if q were one of these primes both N and $2 \cdot 3 \cdot 5 \cdot \ \cdots \ \cdot p$ as well as their difference, 1, would be divisible by it, which is impossible. We see, in either case, there is at least one other prime number and our list is not complete. Thus no finite list of primes is complete and the number of primes is infinite.

Exercise 1

1. Using the method of the Sieve of Eratosthenes, find all the primes less than 200.

2. List the even prime numbers.

3. Can the numeral of a prime number greater than 2, end in 0, 2, 4, 6, or 8? Why?

4. Can the numeral of a prime number greater than 5 end in 0 or 5? Why?

5. What are the possible units digits of the numeral of a prime number greater than 5?

6. Form $N = (2 \cdot 3 \cdot 5 \cdot 7 \cdot \cdots \cdot p) + 1$, when p equals each of the following.

(a) 7   (c) 13   (e) 23

(b) 11   (d) 19   (f) 31

7. Give a prime that divides the following.

(a) 893   (c) 9999   (e) 861

(b) 365   (d) 1460   (f) 17,503

8. Name three primes of the following forms.

(a) $6k + 1$ (b) $6k + 5$

9. Name three primes of the following forms.

(a) $4k + 1$ (b) $4k + 3$

10. Some primes can be written in the form $1 + n^2$, n a positive integer. For example, $5 = 1 + 2^2$, $17 = 1 + 4^2$. Find three more primes of the form $1 + n^2$.

11. Some primes can be written in the form $n^2 - 1$, n a positive integer. For example, $3 = 2^2 - 1$. Can you find other primes of this form? Can n be an odd number?

12. Some primes are one more than a power of 2. For example, $5 = 2^2 + 1$. Find two other primes that are one more than a power of 2.

13. Some primes are one less than a power of 2. For example, $3 = 2^2 - 1$. Find three other primes that are one less than a power of 2.

14. Which of the following statements are true?

(a) The number 1 is a prime number.

(b) All prime numbers are odd numbers.

(c) If n is a positive integer, then $2n$ is a composite number.

(d) There are infinitely many primes.

(e) The only positive divisors of a prime number p are 1 and p.

15. Define a prime number.

16. Define a composite number.

17. Which of the following are prime numbers?

(a) 97   (c) 645   (e) 769

(b) 345   (d) 701   (f) 7843

18. Every prime number of the form $4k + 1$ may be represented as the sum of two squares. For example, $5 = 4 \cdot 1 + 1$ and $5 = 2^2 + 1^2$. Write the following primes as the sum of two squares.

(a) 13   (c) 73   (e) 89
(b) 41   (d) 61   (f) 101

19. Determine whether the following statements are true or false. If a statement is false, give a counterexample.

(a) If p is a prime and $p|x$ and $p|x^2 + y^2$, then $p|y$

(b) If p is a prime and $p|a^2$, then $p|a$.

(c) If p is a prime and $p|a^2 + b^2$ and $p|b^2 + c^2$, then $p|a^2 + c^2$.

(d) If p is a prime of the form $3k + 1$, it is also of the form $6k + 1$.

(e) If n is a composite number, it has a prime divisor $p \leq \sqrt{n}$.

20. If $2^n - 1$ is a prime number, then n is a prime number. Demonstrate the truth of this statement for the following primes.

(a) 3   (c) 7
(b) 31   (d) 127

4.
GREATEST COMMON DIVISOR

Let a and b be two positive integers. If a number c divides both a and b, it is called a **common divisor** of a and b. Among the common divisors of a and b there is a greatest one that is divisible by *all* the other common divisors of a and b and is called the **greatest common divisor (g.c.d.)** of a and b. It is usually denoted by the symbol (a, b). If $(a, b) = 1$, we say that a and b are **relatively prime.**

For example, let $a = 8$ and $b = 12$. The divisors of 8 are 1, 2, 4, and 8. The divisors of 12 are 1, 2, 3, 4, 6, and 12. The common divisors of 8 and 12 are 1, 2, and 4. Since 4 is the greatest of the common divisors, it is the greatest common divisor of 8 and 12, thus $(8, 12) = 4$.

In observing the common divisors of 8 and 12, notice that the greatest common divisor is divisible by *all* the other common divisors of the two numbers.

5.
EUCLID'S ALGORITHM

We now discuss an orderly, systematic process called **Euclid's Algorithm** for finding the greatest common divisor of two positive integers.

Euclid's Algorithm is based on the division algorithm: if a and b are integers, $b > 0$, unique integers q and r can be found such that $a = bq + r$, $0 \leq r < b$.

We will demonstrate Euclid's Algorithm by means of an example. Suppose we wish to find the g.c.d. of 368 and 88. We let $a = 368$ and $b = 88$ and use the division algorithm to find q and r:

$$
\begin{array}{r}
4 \\
88\overline{)368} \\
352 \\
\hline
16
\end{array}
$$

Then

$$368 = 88 \cdot 4 + 16 \quad \text{or} \quad 368 - 88 \cdot 4 = 16.$$

Any number that divides 88 certainly divides $88 \cdot 4$. Any number that divides 368 and $88 \cdot 4$ divides their difference, 16, by Theorem 2.2. This means that we can reduce the problem of finding the g.c.d. of 368 and 88 to finding the g.c.d. of 88 and 16. Now

$$
\begin{array}{r}
5 \\
16\overline{)88} \\
80 \\
\hline
8
\end{array}
$$

So

$$88 = 16 \cdot 5 + 8 \quad \text{or} \quad 88 - 16 \cdot 5 = 8$$

Using the same reasoning as above we see that any number that divides 88 and $16 \cdot 5$ divides their difference, 8. Now the problem is reduced to finding the g.c.d. of 16 and 8. Again, using the division algorithm we find $16 = 8 \cdot 2$.

$$
\begin{array}{r}
2 \\
8\overline{)16} \\
16 \\
\hline
0
\end{array}
$$

That is, 8 divides 16 and there can be no greater divisor of 16 and 8. Therefore, $(16, 8) = 8$. But

$$(8, 16) = (16, 88) = (368, 88)$$

as detailed in the preceding steps. Hence

$$(368, 88) = 8$$

The above work usually is shortened as follows:

$$\begin{array}{r}
4 \\
88\overline{)368} \\
352 \\
\hline
16
\end{array}
\begin{array}{r}
5 \\
\overline{)88} \\
80 \\
\hline
8
\end{array}
\begin{array}{r}
2 \\
\overline{)16} \\
16 \\
\hline
0
\end{array}$$

The g.c.d. of the two given numbers is the last *nonzero* remainder in the division process (8 in this case).

Example 1: Use Euclid's algorithm to find $(564, 27)$.
Solution:

$$\begin{array}{r}
20 \\
27\overline{)564} \\
540 \\
\hline
24
\end{array}
\begin{array}{r}
1 \\
\overline{)27} \\
24 \\
\hline
3
\end{array}
\begin{array}{r}
8 \\
\overline{)24} \\
24 \\
\hline
0
\end{array}$$

Hence $(564, 27) = 3$.

If $(a, b) = d$, *we can always find integers x and y such that $ax + by = d$.* If a and b are small, we can find x and y by inspection or by trial and error. For example, if $a = 5$ and $b = 3$, then $(a, b) = (3, 5) = 1$. We can easily find integers x and y such that $5x + 3y = 1$. For example,

$$5(2) + 3(-3) = 1$$
$$5(5) + 3(-8) = 1$$
$$5(-7) + 3(12) = 1$$

The reader should find other values for x and y.

When a and b are large, it is not always obvious that x and y can be found. In order to find values for x and y we use Euclid's Algorithm. The computation for finding $(368,88) = 8$ by Euclid's Algorithm gives

1. $368 = 88(4) + 16$
2. $88 = 16(5) + 8$
3. $16 = 8(2) + 0$

The italics identify a and b in the division algorithm. We will now reverse these steps to find x and y such that $368x + 88y = 8$. We begin by expressing 8 in terms of 16 and 88 using step 2:

$$8 = 88 - 16(5)$$

We do not simplify the right member of the equality (we would just get 8), but instead leave it intact.

Using step 1 we can write

$$16 = 368 - 88(4)$$

We now substitute this expression for 16 in the expression for 8:

$$8 = 88 - [368 - 88(4)](5)$$
$$8 = 88 - 5[368 - 88(4)]$$

Keeping 88 and 368 (which are b and a respectively) intact, we have

$$8 = 88 - 5(368) + 20(88)$$
$$= (-5)(368) + 88(20 + 1)$$
$$= (-5)(368) + (21)(88)$$

We now have integers x and y that satisfy

$$368x + 88y = 8$$

We can clearly check our answer:

$$(368)(-5) = -1840$$
$$(88)(21) = 1848$$

and $1848 - 1840 = 8$.

Example 2: Determine $(288, 51)$ and find integers x and y such that

$$288x + 51y = (288, 51)$$

Solution: Using Euclid's Algorithm we have

1. $288 = 51(5) + 33$
2. $51 = 33(1) + 18$
3. $33 = 18(1) + 15$
4. $18 = 15(1) + 3$
5. $15 = 3(5) + 0$

Thus $(288, 51) = 3$.

Reversing the process to find x and y we have:
Solving step 4 for 3 we have

(a) $$3 = 18 - 15(1)$$

Solving step 3 for 15 we have

(b) $$15 = 33 - 18(1)$$

Substituting this value for 15 in (a) we have

(c)
$$3 = 18 - (1)[33 - 18(1)]$$
$$3 = (-1)(33) + (2)(18)$$

Solving step 2 for 18 we have

$$18 = 51 - 33(1)$$

Substituting this value for 18 in (c) we have

(d)
$$3 = (-1)(33) + 2[51 - 33(1)]$$
$$3 = (2)(51) + (33)(-3)$$

Solving step 1 for 33 we have

$$33 = 288 - (51)(5)$$

Substituting this value for 33 in (d) we have

$$3 = (2)(51) - 3[288 - 51(5)]$$
$$= 288(-3) + 51(17)$$

We now have integers x and y that satisfy

$$288x + 51y = 3$$

We can easily check our answers:

$$288(-3) = -864$$
$$51(17) = 867$$
$$867 + (-864) = 3$$

We are now able to prove a very important theorem of number theory.

THEOREM 3.2: *If $(a, m) = 1$ and $a \mid mk$, $a \mid k$.*

Proof: Since $(a, m) = 1$, we can find integers x and y such that

$$ax + my = 1$$

Multiplying each member of the above equation by k we have

$k(ax) + k(my) = k$

$a(kx) + (km)y = k$ Associative and commutative properties of multiplication

Since $a|a$ and $a|km$, $a|akx + kmy = k$.

Exercise 2

Use Euclid's Algorithm to find the g.c.d. of the following pairs of numbers (Exercises 1–10).

1. (56, 24)   **6.** (1472, 1124)

2. (221, 143)   **7.** (76084, 63030)

3. (272, 98)   **8.** (18416, 17296)

4. (139, 49)   **9.** (20132, 1472)

5. (629, 357)   **10.** (839648, 2848)

Determine integers x and y to make the following true statements (Exercises 11–16).

11. $629x + 357y = (357, 629)$

12. $272x + 98y = (272, 98)$

13. $139x + 49y = (139, 49)$

14. $126x + 98y = (126, 98)$

15. $1472x + 1124y = (1472, 1124)$

16. $4284x + 2888y = (4284, 2888)$

17. Which of the following pairs of numbers are relatively prime?

(a) (4, 6)   (e) (171, 183)

(b) (8, 9)   (f) (121, 227)

(c) (50, 63)   (g) (1728, 1512)

(d) (20, 35)   (h) (6912, 20101)

18. If p and q are prime number, what is (p, q)?

19. What is the g.c.d. of 0 and n, n a positive integer?

20. Find integers x and y such that $56x + 72y = (56, 72)$. Find integers p and q such that $56p + 72q = 40$.

6.
THE FUNDAMENTAL THEOREM OF ARITHMETIC

We now state the fundamental theorem of arithmetic.

THEOREM 3.3: FUNDAMENTAL THEOREM OF ARITHMETIC. *Every composite number can be expressed as the product of prime factors in one and only one way except for the order of the factors.*

Before we prove the fundamental theorem of arithmetic, we prove the following.

THEOREM 3.4: *If a prime p divides the product ab, then p divides a or p divides b.*

Proof: Suppose $p \mid b$, then the theorem is true. Now suppose p does not divide b. Then $(p, b) = 1$ because the only divisors of p are ± 1 and $\pm p$. Since p and b are relatively prime we can find integers x and y such that $bx + py = 1$. Multiplying each member of $bx + py = 1$ by a we have $abx + apy = a$. Since p divides ab and p divides p, p divides a by Theorem 2.1.

THEOREM 3.5: *If p divides the product $a_1 a_2 \ldots a_n$, then p must divide at least one of the factors $a_1, a_2, \ldots, a_n$.*

This theorem is a direct consequence of Theorem 3.4. The proof is left to the reader.

We now prove Theorem 3.3, The Fundamental Theorem of Arithmetic.

Proof: Every composite number N can be written as the product of prime factors. We have shown that there exists a prime, p_1, such that $N = p_1 n_1$. If n_1 is composite, we can find a further prime p_2 such that $n_1 = p_2 n_2$. This process can be continued with decreasing numbers n_1, $n_2, \ldots, n_k$ until n_k is a prime.

Now that the existence of a prime factorization of N has been established, we must prove that this factorization is unique.

Let us assume that there exists two different prime factorizations of N:

$$N = p_1 p_2 \cdots p_r = q_1 q_2 \cdots q_s$$

where $p_1, p_2, \ldots, q_1, q_2, \ldots, q_s$ are primes and $r < s$. Since p_1 divides N, it divides the product $q_1 q_2 \cdots q_s$. Since p_1 divides the product $q_1 q_2 \cdots q_s$, it divides at least one of the factors. Let p_1 divide q_1. Since q_1 is a prime, it has only two divisors, 1 and itself. Since p_1 is a prime, it is greater than 1. Hence

$$p_1 = q_1$$

Continuing in this fashion we find

$$p_1 = q_1$$
$$p_2 = q_2$$
$$\vdots$$
$$p_r = q_r$$

Dividing both members of

$$p_1 p_2 \cdots p_r = q_1 q_2 \cdots q_s$$

by $(p_1 p_2 \cdots p_r)$, we have

$$1 = q_{r+1} q_{r+2} \cdots q_s$$

which is impossible since all of the factors q_{r+1}, q_{r+2}, $q_{r+3}, \ldots q_s$, are whole numbers greater than 1. Hence $r = s$, and $q_1 = p_1$, $q_2 = p_2$, $\ldots$, and the prime factorization of N is unique.

A composite number may be factored into its prime factors by a method called the **consecutive primes method.** We shall illustrate this method by an example. Let us factor 144 into its prime factors; that is, let us find the **complete factorization** of 144. We begin with the smallest prime, 2,

6. THE FUNDAMENTAL THEOREM OF ARITHMETIC

to see whether or not it is a factor of 144. We see by inspection that 144 is divisible by 2:

$$144 = 2 \cdot 72$$

Since 2 is a factor of 72, we have

$$144 = 2 \cdot 2 \cdot 36$$

We see that 2 is a factor of 36, hence

$$144 = 2 \cdot 2 \cdot 2 \cdot 18$$

Again, 2 is a factor of 18, so we have

$$144 = 2 \cdot 2 \cdot 2 \cdot 2 \cdot 9$$

Since 2 is not a factor of 9, we try the next prime, 3, and we see that

$$144 = 2 \cdot 2 \cdot 2 \cdot 2 \cdot 3 \cdot 3 = 2^4 \cdot 3^2$$

Since all of the factors in the above product are primes, we have found the **complete factorization** of 144. The essential results of this method may be written in this shortened form:

$$
\begin{array}{r|l}
2 & 144 \\
\hline
2 & 72 \\
\hline
2 & 36 \\
\hline
2 & 18 \\
\hline
3 & 9 \\
\hline
 & 3
\end{array}
\qquad
\begin{aligned}
144 &= 2 \cdot 2 \cdot 2 \cdot 2 \cdot 3 \cdot 3 \\
&= 2^4 \cdot 3^2
\end{aligned}
$$

Example 1: Factor 388 completely.

Solution:

$$
\begin{array}{r|l}
2 & 388 \\
\hline
2 & 194 \\
\hline
 & 97
\end{array}
$$

The complete factorization of 388 is $2^2 \cdot 97$.

Example 2: Find the complete factorization of 7968.

Solution:

$$
\begin{array}{r}
2\overline{)\,7968} \\
2\overline{)\,3984} \\
2\overline{)\,1992} \\
2\overline{)\,996} \\
2\overline{)\,498} \\
3\overline{)\,249} \\
83
\end{array}
$$

The complete factorization of 7888 is $2^5 \cdot 3 \cdot 83$.

Exercise 3

1. Find the prime factors of each of the following.
   (a) 156   (d) 288
   (b) 365   (e) 450
   (c) 404   (f) 1688
2. Find the complete factorization of each of the following.
   (a) 93   (d) 188
   (b) 72   (e) 707
   (c) 118   (f) 3455
3. State the Fundamental Theorem of Arithmetic.
4. What is the greatest prime that divides the following?
   (a) 4325   (c) 8080
   (b) 589   (d) 252
5. If a number is a prime, what are its divisors?
6. If N is composite number whose complete factorization is

$$ N = p_1^{\alpha_1} \cdot p_2^{\alpha_2} \cdot p_3^{\alpha_3} \cdots p_k^{\alpha_k} $$

where $p_1, p_2, \ldots, p_k$ are distinct primes and $\alpha_1, \alpha_2, \ldots, \alpha_k$ are positive integers, it can be proved that the number of positive divisors of N is

$$ (\alpha_1 + 1)(\alpha_2 + 1)(\alpha_3 + 1) \ldots (\alpha_k + 1) $$

For example, $12 = 2^2 \cdot 3 = 2^2 \cdot 3^1$. According to the above theorem 12 has $(2 + 1)(1 + 1) = 3 \cdot 2 = 6$ positive divisors. We see that this is true since the positive divisors of 12 are 1, 2, 3, 4, 6, and 12. Using this theorem find the number of positive divisors of 144.

7. Using the theorem of problem 6, find the number of positive divisors of 320. Check by listing the divisors.

8. Using the theorem of problem 6 find the number of positive divisors of 5,217,520.

9. A positive integer is called a perfect number when it equals the sum of all its positive divisors other than itself. Six is a perfect number since $1 + 2 + 3 = 6$. Show that 28 is a perfect number.

10. Show that if n is a perfect number (see problem 9) with positive divisors $d_1, d_2, \ldots, d_{k-1}, d_k = n$, then $d_1 + d_2 + \cdots + d_{k-1} + d_k = 2n$.

7.

SOME UNSOLVED PROBLEMS ABOUT PRIME NUMBERS

There are many questions that have been asked about prime numbers; some of these questions have been answered but many of them still remain unanswered.

Let us look at the first thirty odd primes:

3	37	79
5	41	83
7	43	89
11	47	97
13	53	101
17	59	103
19	61	107
23	67	109
29	71	113
31	73	127

An examination of this list reveals the presence of several pairs of consecutive primes that differ by 2. Such primes are called **twin primes.** Some of the questions that arise when we think of twin primes are: How many pairs of twin primes are there? Is there an infinity of these primes or is there a largest pair?

The answers to these questions have never been found. Emperical evidence points to the conclusion that there is an infinity of twin primes. It has been shown that there are fifteen pairs of twin primes between 999,999,990,000 and 1,000,000,000,000, and that there are twenty pairs between 1,000,000,000,000, and 1,000,000,010,000. No one to date, however, has succeeded in solving the problem of whether or not there is an infinity of twin primes.

Another unanswered question about primes was asked by Goldbach, a Russian mathematician. In about 1742 Goldbach made the following conjecture: **every even number greater than 2 can be written as the sum of two primes.**

Being unable to prove his conjecture, Goldbach sought the assistance of the Swiss mathematician Euler (1707–1783). Euler was convinced of the truth of this conjecture, but was unable to prove it.

We can easily verify Goldbach's conjecture for small numbers. For example,

$$4 = 2 + 2$$
$$6 = 3 + 3$$
$$8 = 3 + 5$$
$$10 = 3 + 7$$

When numbers are fairly large, there usually will be several ways of representing them as the sum of two primes. For example,

$$48 = 5 + 43$$
$$= 7 + 41$$
$$= 11 + 37$$
$$= 17 + 31$$
$$= 19 + 29$$

Goldbach's conjecture has been verified for numbers up to 100,000. In 1931, an unknown Russian mathematician, Schnirelmann, succeeded in proving that every even number can be represented as the sum of not more than 300,000 primes. In 1937, the Russian mathematician Vinogradoff proved that every even number "beyond a certain point" is the sum of four primes. Even though progress has been made toward the solution, Goldbach's conjecture is still unproved.

Some other unsolved problems about primes are stated below.

1. There are infinitely many primes of the form $p = 4m + 3$ such that $q = 2p + 1$ is also a prime. Thus

$$p = 11 = 4 \cdot 2 + 3 \qquad \text{and} \qquad q = 2 \cdot 11 + 1 = 23$$

2. There are infinitely many primes of the form $n^2 - 2$. For example,

7. SOME UNSOLVED PROBLEMS ABOUT PRIME NUMBERS

$$2^2 - 2 = 2 \qquad 3^2 - 2 = 7$$

3. There are infinitely many positive integers n for which $n^2 - 2$ is twice a prime. For example,

$$4^2 - 2 = 2 \cdot 7 \qquad 6^2 - 2 = 2 \cdot 17$$

4. There are infinitely many primes p such that $q = 2p + 1$ is also a prime. Thus

$$p = 11 \qquad \text{and} \qquad q = 2 \cdot 11 + 1 = 23$$
$$p = 2 \qquad \text{and} \qquad q = 2 \cdot 2 + 1 = 5$$

Exercise 4

1. Find six pairs of twin primes between 127 and 233.
2. Write the following as the sum of two primes.
   - (a) 20
   - (b) 30
   - (c) 32
   - (d) 50
   - (e) 110
   - (f) 150
   - (g) 226
   - (h) 200
3. Find six primes of the form $n^2 - 2$.
4. Find three numbers n such that $n^2 - 2$ is twice a prime.
5. Find three primes p such that $q = 2p + 1$ is also a prime.
6. It has been conjectured that every even integer can be represented as the difference of two primes in infinitely many ways. For example: $12 = 19 - 7$ or $29 - 17$ or $23 - 11$. Write 28 as the difference of two primes in three ways.
7. It has been conjectured that every even integer is the difference of two *consecutive* primes in an infinitude of ways. For example: $6 = 29 - 23$; $37 - 31$; $53 - 47$; $99929 - 99923$. Write 10 as the difference of two consecutive primes in three ways.
8. Between n^2 and $n^2 - n$ there exists at least one prime. Find at least one prime between n^2 and $n^2 - n$ when n has the following values:
   - (a) 2
   - (b) 4
   - (c) 5
   - (d) 10
   - (e) 15
   - (f) 27
9. There are at least four primes between the squares of consecutive primes greater than 3. Find the primes between 5^2 and 7^2.
10. If a prime p is of the form $2^n - 1$, then n is also a prime. Find three primes of the form $2^n - 1$ and exhibit in each case that n is a prime.

Karl Friedrich Gauss (1777–1855) was born the son of a day laborer in Germany. It was Gauss who said "Mathematics is the queen of the sciences and the theory of numbers is the queen of mathematics." While still in his teens Gauss constructed a regular polygon of seventeen sides, thus settling a 2000-year-old question. During his university career he conceived the idea of least squares. He introduced the theory of congruences in his *Disquisitiones Arithmeticae*, his most important work on number theory.

CONGRUENCES

4

1.
FINITE
NUMBER SYSTEMS

A **mathematical system**† is any nonempty set of elements together with one or more operations defined on the elements of the set. Most people are familiar with (1) the system of whole numbers and their operations; (2) the system of rational numbers and their operations; and (3) the system of integers and their operations. All these familiar systems are **infinite systems,** that is, the set of elements in the system is an infinite set. We are now going to study a mathematical system that is a **finite system,** that is, the number of elements in the system is finite.

The particular finite system that we shall study is called a **modular system.** This system results from the use of a set of numbers named on the face of a clock or a timer.

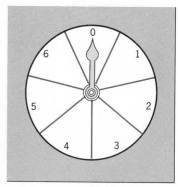

FIGURE 4.1

For simplicity, let us examine a modular system by using the seven-hour clock shown in Figure 4.1. This clock resembles a timer. The elements in this system are 0, 1, 2, 3, 4, 5, and 6. Counting in this system repeats these numbers over and over: 1, 2, 3, 4, 5, 6, 0, 1, 2, 3, 4, 5, 6, 0, 1, ...

We now define the operations of addition and multiplication in this modular system, called the **modulo-seven system.**

† We shall study more mathematical systems in Chapter 6.

TABLE 4.1
Modulo-seven addition table

+	0	1	2	3	4	5	6
0	0	1	2	3	4	5	6
1	1	2	3	4	5	6	0
2	2	3	4	5	6	0	1
3	3	4	5	6	0	1	2
4	4	5	6	0	1	2	3
5	5	6	0	1	2	3	4
6	6	0	1	2	3	4	5

To find the sum of two elements a and b, denoted by $a + b$, we proceed as follows. The hand of the clock starts at 0, it turns a spaces in a clockwise direction, and then it turns b more spaces in the same direction. Thus to find $2 + 3$, we start at 0, turn two spaces in a clockwise direction, and then turn three more spaces in a clockwise direction. Thus

$$2 + 3 = 5$$

Using this definition of addition in the modulo-seven system we see that

$$3 + 5 = 1$$
$$4 + 3 = 0$$
$$6 + 6 = 5$$
$$5 + 4 = 2$$

We can now find the sum of every pair of elements in the system. These sums are shown in Table 4.1, called an **addition table** for the modulo-seven system. We call addition in the modulo-seven system **addition modulo 7,** which is abbreviated **addition mod 7.** The system is called the **modulo-seven system** because there are seven elements in the system.

Addition modulo 7 has some of the same properties as addition of integers. Notice that in the addition table only the numbers 0, 1, 2, 3, 4, 5, and 6 appear. In other words, if we consider the set of seven numbers 0, 1, 2, 3, 4, 5, and 6 and combine them under the rule of addition mod 7, the sum of any two numbers is again an element of this set. We say that the modulo-seven system is closed under the operation of addition mod 7. This property is called the **closure property of addition.**

If 0 is added to any number in this system, the result is the given number; that is, if a is any element in the system,

$$a + 0 = 0 + a = a$$

The number 0 is the only number in the system which has this property and is called the **additive identity** of the system.

As we look at the addition table we see that each row and column of the table contains 0 exactly once. This corresponds to another property of addition mod 7. For example, in the fourth row, 0 occurs in the fifth column. This corresponds to the fact that

$$3 + x = 0$$

has the solution $x = 4$ and this is the only solution. Similarly, for any a in the system,

$$a + x = 0$$

has one and only one solution. We express this by saying for any element a in the system, there exists an element called the **additive inverse of a,** denoted by $-a$, such that

$$a + (-a) = (-a) + a = 0$$

The addition table is symmetric about the diagonal from the upper left corner to the lower right corner. That is, the entry in the ith row and the kth column is the same as the entry in the kth row and the ith column. This means that, for all elements a and b in the system,

$$a + b = b + a$$

We express this fact by saying that addition mod 7 is **commutative.**

There is another property that is not apparent from the table, but which is very important. Notice

(1) $(2 + 3) + 6 = 5 + 6 = 4$
$2 + (3 + 6) = 2 + 2 = 4$

$\therefore (2 + 3) + 6 = 2 + (3 + 6)$

(2) $(4 + 5) + 6 = 2 + 6 = 1$
$4 + (5 + 6) = 4 + 4 = 1$

$\therefore (4 + 5) + 6 = 4 + (5 + 6)$

In general, this is true: if a, b, and c are elements in the system, then

$$(a + b) + c = a + (b + c)$$

This is called the **associative property of addition.**

We find the product of two elements a and b, denoted by ab, in the modulo-seven system in the following manner. The hand of the clock starts at 0 and makes a turns of b spaces each in a clockwise direction. Thus to find the product $4 \cdot 3$, the hand of the clock starts at 0 and makes four turns of three spaces each in a clockwise direction. Thus

$$4 \cdot 3 = 3 + 3 + 3 + 3 = 5$$

Using this definition of multiplication in the modulo-seven system, called **multiplication mod 7,** we have

$$2 \cdot 3 = 6 \text{ (2 clockwise turns of 3 spaces each)}$$
$$4 \cdot 5 = 6 \text{ (4 clockwise turns of 5 spaces each)}$$
$$6 \cdot 6 = 1 \text{ (6 clockwise turns of 6 spaces each)}$$

A multiplication table for the modulo-seven system is shown in Table 4.2.

Studying the multiplication table for the modulo-seven system, we discover:

1. The set of elements is **closed** under multiplication.
2. Multiplication is **commutative;** that is, for all elements a and b, $ab = ba$.
3. There is an element, 1, called the **multiplicative identity,** such that for all elements a in the system, $a \cdot 1 = 1 \cdot a = a$.
4. Every element a, except 0, has a **multiplicative inverse** denoted by a^{-1} such that $a \cdot a^{-1} = a^{-1} \cdot a = 1$.

TABLE 4.2
Modulo-seven
multiplication table

×	0	1	2	3	4	5	6
0	0	0	0	0	0	0	0
1	0	1	2	3	4	5	6
2	0	2	4	6	1	3	5
3	0	3	6	2	5	1	4
4	0	4	1	5	2	6	3
5	0	5	3	1	6	4	2
6	0	6	5	4	3	2	1

5. Again, it is not obvious from the table but it is true that, for all a, b, and c in the system, $(ab)c = a(bc)$. This is the **associative property of multiplication.**

The two operations of addition and multiplication mod 7 are connected by the **distributive property,** which states that if a, b, and c are elements of the system, then

$$a(b + c) = ab + ac$$

For example,

$$2(3 + 5) = 2 \cdot 1 = 2$$
$$2 \cdot 3 + 2 \cdot 5 = 6 + 3 = 2$$
$$\therefore 2(3 + 5) = 2 \cdot 3 + 2 \cdot 5$$

$$5(4 + 6) = 5 \cdot 3 = 1$$
$$5 \cdot 4 + 5 \cdot 6 = 6 + 2 = 1$$
$$\therefore 5(4 + 6) = 5 \cdot 4 + 5 \cdot 6$$

Exercise 1

1. Use Table 4.1 to find the following.
   (a) $(3 + 4) + 5$
   (b) $(6 + 3) + 5$
   (c) $(5 + 3) + 2$
   (d) $(4 + 6) + (6 + 3)$
   (e) $(5 + 2) + (6 + 5)$
   (f) $(2 + 3) + (6 + 1)$
2. What is the additive inverse of each of the following elements of the modulo-seven system?
   (a) 2
   (b) 3
   (c) 6
   (d) 1
   (e) 5
   (f) 4
3. Using Table 4.2, find:
   (a) $(3 \cdot 2) \cdot 6$
   (b) $(4 \cdot 5) \cdot 2$
   (c) $(6 \cdot 6) \cdot 3$
   (d) $(6 \cdot 5) \cdot (5 \cdot 5)$
   (e) $(3 \cdot 3) \cdot (4 \cdot 5)$
   (f) $(6 \cdot 3) \cdot (6 \cdot 2)$
4. Using Tables 4.1 and 4.2, compute the following.
   (a) $(6 \cdot 4) + (3 \cdot 2)$
   (b) $6(5 + 4)$
   (c) $(3 \cdot 2) + (6 \cdot 4)$
   (d) $(6 \cdot 6) + (5 \cdot 1)$
5. Using Tables 4.1 and 4.2, show that the following are true.
   (a) $3(5 + 4) = (3 \cdot 5) + (3 \cdot 4)$
   (b) $6(4 + 6) = (6 \cdot 4) + (6 \cdot 6)$
   (c) $3(1 + 5) = (3 \cdot 1) + (3 \cdot 5)$
   (d) $6(5 + 2) = (6 \cdot 5) + (6 \cdot 2)$

6. What modular system is based on a clock with numerals 0, 1, 2, 3, and 4?
7. Construct an addition table for a modulo-five system.
8. Construct a multiplication table for a modulo-five system.
9. Answer the following with reference to the modulo-five system.
    (a) Show that $(4 + 2) + 3 = 4 + (2 + 3)$.
    (b) Show that $2(4 + 3) = (2 \cdot 4) + (2 \cdot 3)$.
    (c) What is the additive inverse of each element in this system?
    (d) What is the multiplicative inverse of every element $a \neq 0$ of the system?
    (e) Solve the equation $2 + x = 1$.
    (f) Solve the equation $3x = 2$.
10. Construct addition and multiplication tables for a modulo-eight system.
11. Answer the following with reference to the modulo-eight system.
    (a) What is the additive inverse of 2?
    (b) What is the multiplicative inverse of 2?
    (c) Does every element in this system have a multiplicative inverse? If not, which do not?
    (d) Solve the equation $4x = 3$.

2.
CLASSIFICATION
OF THE INTEGERS

In Section 1 of this chapter we studied finite number systems called modular systems. We shall now show a connection between the elements of these systems and ordinary integers. For example, for the elements 0, 1, 2, 3, 4, 5, and 6 of the modulo-seven system, we consider a circle with seven equally spaced divisions as shown in Figure 4.2. Each point of division is marked with one of the numbers 0, 1, 2, 3, 4, 5, and 6, inclusive.

Now let us mark off segments on a line equal in length to one-seventh of the circumference of this circle. We label the endpoints of these segments with integers as shown in Figure 4.3. Placing the circle on the line as shown in Figure 4.3, we roll it along the line as indicated in the figure. Because of our choice of unit on the line, the point representing 1 on the circle will strike the point labeled 1 on the line; the point representing 2 on the circle will strike the point labeled 2 on the line; and so on. After the first complete turn of the circle, the point representing 0 on the circle will strike the point labeled 7 on the line, and so on. If we roll the circle

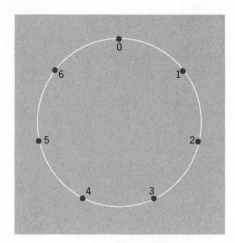

FIGURE 4.2

to the left, the point representing 6 on the circle will strike the point labeled −1 on the line and so on.

Thus we see that every point on the line will have a "label" on the circle. Let us see what points on the line correspond to the numbers named on the circle.

0 on the circle represents: . . . , −21, −14, −7, 0, 7, 14, . . .
1 on the circle represents: . . . , −20, −13, −6, 1, 8, 15, . . .
2 on the circle represents: . . . , −19, −12, −5, 2, 9, 16, . . .
3 on the circle represents: . . . , −18, −11, −4, 3, 10, 17, . . .
4 on the circle represents: . . . , −17, −10, −3, 4, 11, 18, . . .
5 on the circle represents: . . . , −16, −9, −2, 5, 12, 19, . . .
6 on the circle represents: . . . , −15, −8, −1, 6, 13, 20, . . .

We call this a **classification** of the integers modulo 7. In the 0-class are

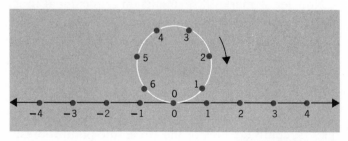

FIGURE 4.3

all the integers to which the point labeled 0 on the circle corresponds; in the 1-class are all the integers to which the point labeled 1 on the circle corresponds; and so forth.

We notice that

1. **Every integer is in one and only one class.**
2. **The difference between any two numbers in the same class is a multiple of 7.**
3. **Any integer k is in the**

> **0-class if k is of the form $7n$**
> **1-class if k is of the form $7n + 1$**
> **2-class if k is of the form $7n + 2$**
> **3-class if k is of the form $7n + 3$**
> **4-class if k is of the form $7n + 4$**
> **5-class if k is of the form $7n + 5$**
> **6-class if k is of the form $7n + 6$.**

In general a classification of integers mod m consists of m classes. Each class is designated by that whole number in the class that is less than m.

Example 1: To which class mod 7 does 386 belong?
 Solution: Since $386 = (7)(55) + 1$, 386 belongs to the 1-class.

Example 2: To which class mod 7 does -145 belong?
 Solution: Since $-145 = (7)(-21) + 2$, -145 belongs to the 2-class.

Exercise 2

1. Mark a circle into eight parts. Use it to classify integers mod 8. Give three integers in each of the eight classes: 0-class, 1-class, 2-class, 3-class, 4-class, 5-class, 6-class, and 7-class.
   2. If the integers are classified mod 8, to which class will each of the following belong?
   (a) 36 (c) 85 (e) -34
   (b) 76 (d) 107 (f) -100
   3. If two numbers belong to the same class for a particular modular system, they are called **congruent.** Which of the following are congruent to 4 for the indicated modular systems?

(a) 9 (mod 5) (e) 53 (mod 7)
(b) 39 (mod 9) (f) 78 (mod 8)
(c) 56 (mod 12) (g) 137 (mod 9)
(d) 126 (mod 8) (h) 336 (mod 16)

4. The classes when the integers are classified mod 12 are designated 0, 1, 2, 3, 4, 5, 6, 7, 8, 9, 10, and 11. To which class mod 12 do the following belong?

(a) 48 (d) 110
(b) 59 (e) −3
(c) 67 (f) − 106

5. If the integers are classified mod 16 how many classes will there be? To which class do the following belong mod 16?

(a) 27 (d) − 16
(b) 39 (e) −34
(c) 48 (f) −67

6. If an integer belongs to the 0-class mod 16, will it also belong to the 0-class mod 4? Why?

7. If an integer is of the form $7k + 3$, to what class does it belong mod 7?

8. If an integer is of the form $4k + 1$, to what class does it belong mod 4?

9. If an integer is of the form $12k + 8$, to what class does it belong mod 12?

10. Let a circle of m divisions roll on a line as indicated in Figure 4.3. How would you find what number marked on the circle strikes a large number?

3.
DEFINITION
OF CONGRUENCE

Whenever the question of divisibility of integers by a fixed positive integer m occurs, the concept and notation of **congruence** serves to simplify and clarify the reasoning.

To introduce the concept of congruence let us examine the remainders when integers are divided by 3. Using the division algorithm we find

$$0 = 0 \cdot 3 + 0 \qquad 6 = 2 \cdot 3 + 0 \qquad -1 = -1 \cdot 3 + 2$$
$$1 = 0 \cdot 3 + 1 \qquad 7 = 2 \cdot 3 + 1 \qquad -2 = -1 \cdot 3 + 1$$
$$2 = 0 \cdot 3 + 2 \qquad 8 = 2 \cdot 3 + 2 \qquad -3 = -1 \cdot 3 + 0$$
$$3 = 1 \cdot 3 + 0 \qquad 9 = 3 \cdot 3 + 0 \qquad -4 = -2 \cdot 3 + 2$$
$$4 = 1 \cdot 3 + 1 \qquad 10 = 3 \cdot 3 + 1 \qquad -5 = -2 \cdot 3 + 1$$
$$5 = 1 \cdot 3 + 2 \qquad 11 = 3 \cdot 3 + 2 \qquad -6 = -2 \cdot 3 + 0$$

$$\vdots \qquad\qquad \vdots$$

Observe that the remainder when any integer is divided by 3 is one of the integers 0, 1, or 2. We say two integers a and b are **congruent** modulo 3 when they have the same remainder when divided by 3. Thus 2, 5, 8, 11, . . . , -1, -4, . . . , are all congruent modulo 3. In general, we say that two integers a and b are **congruent modulo m,** where m is a fixed positive integer, if a and b have the same remainder when divided by m. We write

$$a \equiv b \,(\text{mod } m)$$

to express the fact that a and b are congruent modulo m. If a and b are integers and a is not congruent to b, modulo m, we write

$$a \not\equiv b \,(\text{mod } m)$$

Since $a \equiv b \,(\text{mod } m)$ means that a and b have the same remainder when divided by m, the following statements are equivalent to the statement "a is congruent to b modulo m."

$$a = b + mk \text{ for some integer } k$$
$$a - b \text{ is divisible by } m$$

To emphasize that $a \equiv b \,(\text{mod } m)$ is equivalent to $a = b + mk$ for some integer k, and that $a - b$ is divisible by m, notice

$5 \equiv 2 \,(\text{mod } 3)$	$5 = 2 + 3 \cdot 1$	$3 \mid (5 - 2) = 3$
$7 \equiv 1 \,(\text{mod } 6)$	$7 = 1 + 6 \cdot 1$	$6 \mid (7 - 1) = 6$
$28 \equiv 3 \,(\text{mod } 5)$	$28 = 3 + 5 \cdot 5$	$5 \mid (28 - 3) = 25$

Congruences occur frequently in daily life. For example, the statement that Easter comes on Sunday determines the day modulo 7; the hand on a clock indicates the hour modulo 12; the odometer on a car gives the total mileage traveled modulo 100,000.

Example 1: Find all the integers $0 \le x < 8$ that make $x + 3 \equiv 5$ (mod 8) a true statement.

Solution: Trying 0, 1, 2, 3, 4, 5, 6, and 7 in turn we find

$0 + 3 \equiv 3 \not\equiv 5 \,(\text{mod } 8)$	$4 + 3 \equiv 7 \not\equiv 5 \,(\text{mod } 8)$
$1 + 3 \equiv 4 \not\equiv 5 \,(\text{mod } 8)$	$5 + 3 \equiv 8 \not\equiv 5 \,(\text{mod } 8)$
$2 + 3 \equiv 5 \,(\text{mod } 8)$	$6 + 3 \equiv 9 \not\equiv 5 \,(\text{mod } 8)$
$3 + 3 \equiv 6 \not\equiv 5 \,(\text{mod } 8)$	$7 + 3 \equiv 10 \not\equiv 5 \,(\text{mod } 8)$

The only integer satisfying the given conditions is 2.

CONGRUENCES

Example 2: Find all integers $0 \leq x < 8$ that make $x^2 \equiv 0 \pmod 8$ a true statement.

Solution: Trying 0, 1, 2, 3, 4, 5, 6, and 7 in turn we have

$$0^2 \equiv 0 \pmod 8 \qquad 4^2 \equiv 16 \equiv 0 \pmod 8$$
$$1^2 \equiv 1 \not\equiv 0 \pmod 8 \qquad 5^2 \equiv 25 \not\equiv 0 \pmod 8$$
$$2^2 \equiv 4 \not\equiv 0 \pmod 8 \qquad 6^2 \equiv 36 \not\equiv 0 \pmod 8$$
$$3^2 \equiv 9 \not\equiv 0 \pmod 8 \qquad 7^2 \equiv 49 \not\equiv 0 \pmod 8$$

The integers satisfying the given conditions are 0 and 4.

Exercise 3

1. Which of the following are true statements?
   (a) $6 \equiv 12 \pmod 4$
   (b) $58 \equiv 8 \pmod{10}$
   (c) $127 \equiv 2 \pmod 5$
   (d) $387 \equiv 64 \pmod{17}$
   (e) $-86 \equiv -57 \pmod{17}$
   (f) $-37 \equiv 26 \pmod 2$

2. Find all the integers $4 \leq a \leq 8$ which make the following congruences true.
   (a) $a \equiv 0 \pmod 3$
   (b) $a \equiv 2 \pmod 3$
   (c) $a \equiv 1 \pmod 3$
   (d) $a \equiv 16 \pmod{11}$
   (e) $a^2 \equiv 1 \pmod 8$
   (f) $a^2 \equiv 0 \pmod 5$

3. Find all the integers $4 \leq a \leq 8$ that make the following congruences true.
   (a) $3a \equiv 0 \pmod 9$
   (b) $a + 6 \equiv 25 \pmod 7$
   (c) $3a + 5 \equiv 7 \pmod{11}$
   (d) $a^2 \equiv 0 \pmod 6$
   (e) $a^2 - 1 \equiv 4 \pmod 4$
   (f) $a^2 \equiv 0 \pmod 8$

4. For each of the following, give two values of m for which each congruence is true and two for which it is false.
   (a) $16 \equiv 8 \pmod m$
   (b) $19 \equiv 7 \pmod m$
   (c) $-10 \equiv 36 \pmod m$
   (d) $14 \equiv 28 \pmod m$
   (e) $-15 \equiv -9 \pmod m$
   (f) $-8 \equiv 32 \pmod m$

5. Find all the integers $0 \leq x < 8$ which make the following congruences true.
   (a) $x \equiv 3 \pmod 8$
   (b) $5x \equiv 2 \pmod 8$
   (c) $12x \equiv 16 \pmod 8$
   (d) $36x \equiv 44 \pmod 8$

6. Find all the integers $0 \leq x < 12$ which make the following congruences true.
   (a) $x \equiv 5 \pmod{12}$
   (b) $4x \equiv 8 \pmod{12}$
   (c) $x^2 \equiv 0 \pmod{12}$

7. Find all the integers $0 \leq x < 16$ which make the following congruences true.
   (a) $x \equiv 8 \pmod{16}$
   (b) $5x \equiv 9 \pmod{16}$
   (c) $x^2 \equiv 4 \pmod{16}$

8. Find all integers $0 \leq x < 11$ which make the following congruences true.
  (a) $5x \equiv 3 \pmod{11}$
  (b) $x^2 \equiv 1 \pmod{11}$
  (c) $5x \equiv 7 \pmod{11}$

9. Give three positive integers and three negative integers congruent to 3 mod 8.

10. Give five positive integers and five negative integers congruent to 8 mod 12.

4.
PROPERTIES
OF CONGRUENCES

The usefulness of the congruence notation lies in the fact that congruence modulo m has many of the same properties as the equality relation. The most important properties of the equality relation are:

1. $a = a$ for all a.
2. If $a = b$, then $b = a$.
3. If $a = b$ and $b = c$, then $a = c$.
4. If $a = b$ and $c = d$, then $a + c = b + d$.
5. If $a = b$ and $c = d$, then $a - c = b - d$.
6. If $a = b$ and $c = d$, then $ac = bd$.

These properties are also true when the equality relation, denoted by $=$, is replaced by the congruence relation, denoted by $\equiv$.

1'. $a \equiv a \pmod{m}$ for all a.
2'. If $a \equiv b \pmod{m}$, then $b \equiv a \pmod{m}$.
3'. If $a \equiv b \pmod{m}$ and $b \equiv c \pmod{m}$, then $a \equiv c \pmod{m}$.
4'. If $a \equiv b \pmod{m}$ and $c \equiv d \pmod{m}$, then $a + c \equiv b + d$ $\pmod{m}$.
5'. If $a \equiv b \pmod{m}$ and $c \equiv d \pmod{m}$, then $a - c \equiv b - d$ $\pmod{m}$.
6'. If $a \equiv b \pmod{m}$ and $c \equiv d \pmod{m}$, then $ac \equiv bd \pmod{m}$.

From 4', 5', and 6', we see that congruences with respect to the same modulus may be added, subtracted, and multiplied, member by member.

CONGRUENCES

We shall now prove some of the properties of the congruence relation. These properties are proved using the properties of the equality relation and the axioms of the system of integers.

THEOREM 4.1: *If $a \equiv b \,(mod\ m)$ and $c \equiv d \,(mod\ m)$, then $a + c \equiv b + d \,(mod\ m)$.*

    Proof: Since $a \equiv b \,(\mathrm{mod}\ m)$ and $c \equiv d \,(\mathrm{mod}\ m)$, by the definition of congruence we can write

$$a = b + mk$$
$$c = d + mh$$

where h and k are integers. Adding these two equations member by member we have

$a + c = (b + mk) + (d + mh)$	Property 4 of the equality relation
$\quad = (b + mk) + (mh + d)$	Commutative property of addition of integers
$\quad = [(b + mk) + mh] + d$	Associative property of addition of integers
$\quad = [b + (mk + mh)] + d$	Associative property of addition of integers
$\quad = d + [b + (mk + mh)]$	Commutative property of addition of integers
$\quad = (d + b) + (mk + mh)$	Associative property of addition of integers
$\quad = (b + d) + (mk + mh)$	Commutative property of addition of integers
$\quad = (b + d) + m(k + h)$	Distributive property of integers

Since h and k are integers, their sum, $k + h$, is an integer by the closure property of addition. Hence by the definition of congruence

$$a + c \equiv b + d \,(\mathrm{mod}\ m)$$

THEOREM 4.2: *If $a \equiv b \,(mod\ m)$ and $c \equiv d \,(mod\ m)$, then $a - c \equiv b - d \,(mod\ m)$*

    Proof: Since $a \equiv b \,(\mathrm{mod}\ m)$ and $c \equiv d \,(\mathrm{mod}\ m)$, using the definition of congruence we can write

$$a = b + mk$$
$$c = d + mh$$

where h and k are integers. Subtracting these equations member by member we have

$$a - c = (b + mk) - (d + mh)$$
$$= (b - d) + m(k - h)$$

using the definition of subtraction of integers, the associative and commutative properties of addition of integers, and the distributive property. Since k and h are integers, their difference is an integer because of the closure property of subtraction. Hence, by the definition of congruence,

$$a - c \equiv b - d \,(\text{mod } m)$$

THEOREM 4.3: *If $a \equiv b \,(mod\ m)$ and $c \equiv d \,(mod\ m)$, then $ac \equiv bd$ (mod m).*

Proof: Since $a \equiv b \,(\text{mod } m)$ and $c \equiv d \,(\text{mod } m)$,

$$a = b + mk$$
$$c = d + mh$$

where k and h are integers. Multiplying these two equations member by member we have

$$ac = (b + mk)(d + mh)$$

Using the associative and commutative properties of multiplication of integers and the distributive property, the right member of the above equation can be simplified and we have

$$ac = bd + bmh + mkd + m^2kh$$
$$= bd + m(bh + kd + mkh)$$

Since $bh + kd + mkh$ is an integer we have

$$ac \equiv bd \,(\text{mod } m)$$

We now prove some additional properties of congruences.

THEOREM 4.4: *If $a \equiv b \,(mod\ m)$ and n is a positive factor of m, then $a \equiv b \,(mod\ n)$.*

Proof: Since $a \equiv b \,(\text{mod } m)$ we know that

$$a = b + mk$$

Now n is a factor of m, so $m = nr$ where r is an integer. Substituting nr for m in the above equation we have

$$a = b + (nr)k$$
$$= b + n(rk)$$

and hence

$$a \equiv b \,(\text{mod } n)$$

THEOREM 4.5: *If* $ab \equiv ac$ *(mod m), and* $(a, m) = 1$, *then* $b \equiv c$ *(mod m).*

Proof: Since $ab \equiv ac \,(\text{mod } m)$ we have

$$ab = ac + mk$$

and

$$a(b - c) = mk$$

We see from the preceding statement that m divides $a(b - c)$. Since $(a, m) = 1$, m divides $b - c$, by Theorem 3.1. Hence

$$b \equiv c \,(\text{mod } m)$$

THEOREM 4.6: *If* $(a, m) = d$ *and* $ab \equiv ac$ *(mod m), then* $b \equiv c$ *(mod $\frac{m}{d}$).*

Proof: Since d is a factor of a, for an integer a_1

$$a = a_1 d$$

Since $ab \equiv ac \,(\text{mod } m)$

$$ab = ac + mk$$

and

$$a(b - c) = mk$$

Since a and m both have a factor d we can divide both members of the preceding equation by d.

$$a_1(b - c) = \frac{m}{d}k$$

Since $\frac{m}{d}$ and a_1 are relatively prime, $\frac{m}{d}$ divides $b - c$ by Theorem 3.1 and hence

$$b \equiv c \left(\text{mod } \frac{m}{d}\right)$$

Example 1: Given $36a \equiv 54 \pmod{27}$, show that $4a \equiv 6 \pmod{3}$ and $a \equiv 0 \pmod{3}$.

Solution: Since $(36, 27) = 9$, and $54 = 9 \cdot 6$, by Theorem 4.6 we have

$$\frac{36}{9} a \equiv \frac{54}{9} \left(\mathrm{mod}\ \frac{27}{9} \right)$$

$$4a \equiv 6 \pmod{3}$$

Since $4 \equiv 1 \pmod{3}$ and $6 \equiv 0 \pmod{3}$, we have

$$a \equiv 0 \pmod{3}$$

Example 2: Given $a \equiv 3 \pmod{9}$, show that $a^2 \equiv 0 \pmod{9}$

Solution: Since $a \equiv 3 \pmod{9}$, using Theorem 4.3, we have

$$a \cdot a \equiv 3 \cdot 3 \pmod{9}$$
$$a^2 \equiv 9 \pmod{9}$$

But $9 \equiv 0 \pmod{9}$, hence

$$a^2 \equiv 0 \pmod{9}$$

Example 3: Given $5a \equiv 31 \pmod{9}$, show that $5a \equiv 31 \pmod{3}$.

Solution: Since $5a \equiv 31 \pmod{9}$ and $3|9$, we have by Theorem 4.4

$$5a \equiv 31 \pmod{3}$$

Example 4: Given $6x \equiv 30 \pmod{7}$, show that $x \equiv 5 \pmod{7}$.

Solution: Since $(6, 7) = 1$, by Theorem 4.5, we have

$$\frac{6}{6} x \equiv \frac{30}{6} \pmod{7}$$

$$x \equiv 5 \pmod{7}$$

5.
RESIDUE CLASSES

It is customary when dealing with congruences to use only small, positive integers. If two numbers are congruent for a modulus m, each is called a **residue** of the other modulo m. For example, since

107

$$8 \equiv 2 \ (\text{mod } 6)$$

8 is called the residue of 2, modulo 6.

Every integer is congruent modulo m to one and only one of the numbers

$$0, 1, 2, 3, \ldots, (m-1)$$

When an integer a is divided by m, the remainder will be congruent to $a \ (\text{mod } m)$, and will be contained in the set

$$0, 1, 2, 3, \ldots, (m-1)$$

For example, if we divide 362 by 8, the remainder is 2. Hence

$$362 \equiv 2 \ (\text{mod } 8)$$

That every integer is congruent to only one of the numbers $0, 1, \ldots,$ $(m-1)$ follows from the fact that no two distinct numbers in the set

$$\{0, 1, 2, \ldots, (m-1)\}$$

are congruent modulo m since their difference is less than m and, being different from zero, cannot be divisible by m.

Since every integer is congruent to one and only one of the numbers $0, 1, 2, 3, \ldots, (m-1)$, the integers are distributed into m classes, called **residue classes,** modulo m. For example, if $m = 5$, the integers are distributed into five residue classes as follows:

$\vdots$	$\vdots$	$\vdots$	$\vdots$	$\vdots$
-10	-9	-8	-7	-6
-5	-4	-3	-2	-1
0	1	2	3	4
5	6	7	8	9
10	11	12	13	14
$\vdots$	$\vdots$	$\vdots$	$\vdots$	$\vdots$

If, from each of the m classes into which all integers are distributed modulo m, we pick one member, the numbers selected

$$r_1, r_2, r_3, \ldots, r_m$$

are representative of the m residue classes and constitute a **complete set of residues** modulo m. The set

$$0, 1, 2, 3, 4, \ldots, (m-1)$$

is called the set of **least positive residues** modulo m.

For example,

$$0, 1, 2, 3, 4, 5$$

is the set of least positive residues modulo 6, and the set

$$12, 19, 26, 9, 10, 35$$

is a complete set of residues modulo 6 since

$$
\begin{array}{ll}
12 \equiv 0 & 9 \equiv 3 \\
19 \equiv 1 & 10 \equiv 4 \ (\text{mod } 6) \\
26 \equiv 2 & 35 \equiv 5
\end{array}
$$

If a is an integer relatively prime to m, that is, $(a, m) = 1$, and $r_1, r_2,$ $\ldots, r_m$ is a complete set of residues mod m, then the numbers

$$ar_1, ar_2, ar_3, \ldots, ar_m$$

form another complete set of residues mod m. We can prove that this set is a complete set of residues by showing that no distinct numbers of the set are congruent modulo m. If

$$ar_i \equiv ar_k \ (\text{mod } m)$$

then, by Theorem 4.5,

$$r_i \equiv r_k \ (\text{mod } m)$$

which is impossible since r_i and r_k belong to different residue classes. In particular, if $(a, m) = 1$

$$0, a, 2a, 3a, \ldots, (m-1)a$$

represent a complete set of residues modulo m.

Because each integer is congruent modulo m to one of the integers

$$0, 1, 2, \ldots, (m-1)$$

it is unnecessary to work with integers greater than m in solving congruences. For example, the congruence

$$27x \equiv 126 \ (\text{mod } 15)$$

can be reduced to

$$12x \equiv 6 \;(\text{mod } 15)$$

because $27 \equiv 12 \;(\text{mod } 15)$ and $126 \equiv 6 \;(\text{mod } 15)$.

Example 1: Find the least positive residue modulo 7 of $36 \cdot 58 + 396$.
   Solution: Since $36 \equiv 1 \;(\text{mod } 7)$, $58 \equiv 2 \;(\text{mod } 7)$ and $396 \equiv 4 \;(\text{mod } 7)$,

$$
\begin{aligned}
36 \cdot 58 + 396 &\equiv 1 \cdot 2 + 4 \;(\text{mod } 7) \\
&\equiv 2 + 4 \;(\text{mod } 7) \\
&\equiv 6 \;(\text{mod } 7)
\end{aligned}
$$

Example 2: Find the least positive residue modulo 9 of 4^5.
   Solution:

$$
\begin{aligned}
4^2 &\equiv 4 \cdot 4 \equiv 16 \equiv 7 \;(\text{mod } 9) \\
4^3 &\equiv 4^2 \cdot 4 \equiv 7 \cdot 4 \equiv 28 \equiv 1 \;(\text{mod } 9) \\
4^5 &\equiv 4^3 \cdot 4^2 \equiv 1 \cdot 7 \equiv 7 \;(\text{mod } 9)
\end{aligned}
$$

Exercise 4

1. Prove: $a \equiv a \;(\text{mod } m)$ for all a.

2. Prove: If $a \equiv b \;(\text{mod } m)$, then $b \equiv a \;(\text{mod } m)$.

3. Prove: If $a \equiv b \;(\text{mod } m)$ and $b \equiv c \;(\text{mod } m)$, then $a \equiv c \;(\text{mod } m)$.

4. Which of the following are complete sets of residues for the given moduli (moduli is the plural of modulus)?

   (a) 3, 9, 11 (mod 3)
   (b) 8, 14, 25, 35 (mod 4)
   (c) 0, 41, 52, 64, 71 (mod 5)
   (d) $0, 1, 2^2, 2^3, 2^4, 2^5, 2^6, 2^7, 2^8$ (mod 9)
   (e) $0, 1, 3, 3^2, 3^3, \ldots, 3^{15}$ (mod 17)

5. Find the least positive residue congruent to each of the following.

   (a) $(14)(8^2)$ (mod 5)
   (b) $(8)(7^2)$ (mod 9)
   (c) $45 + 38$ (mod 11)
   (d) $16 - 37 + 19$ (mod 8)

6. Give the set of all integers congruent to 3 modulo 9.

7. Find all the integers x such that:

   (a) $1 \leq x \leq 50$ and $x \equiv 6 \;(\text{mod } 11)$
   (b) $-20 \leq x \leq 20$ and $x \equiv 12 \;(\text{mod } 17)$
   (c) $-50 \leq x \leq 50$ and $x \equiv -87 \;(\text{mod } 18)$

8. Replace each congruence by an equivalent one involving numbers from the set of least positive residues.

   (a) $56x \equiv 18 \;(\text{mod } 11)$
   (b) $37x \equiv -15 \;(\text{mod } 9)$
   (c) $139x \equiv -87 \;(\text{mod } 23)$
   (d) $278x \equiv 126 \;(\text{mod } 101)$

9. Find the least positive residue modulo 7 of $22 \cdot 51 + 698$.

10. Find a complete set of residues modulo 5 composed entirely of multiples of 9.

11. Show that m consecutive integers form a complete residue system modulo m.

12. Prove that $(a + b)^3 \equiv a^3 + b^3 \,(\text{mod } 3)$

13. Show that $2, 4, 6, 8, \ldots, 2m$ is a complete set of residues modulo m if m is odd.

14. Prove that $(a + b)^5 \equiv a^5 + b^5 \,(\text{mod } 5)$

15. Show that $b^2 - 4ac$ is congruent to 0 or 1 modulo 4 for all integers a, b, and c. (*Hint:* Recall that the square of an integer is either of the form $4k$ or $4k + 1$; Theorem 2.5)

6.
SOLUTION OF CONGRUENCES

We now turn to the problem of solving congruences involving a single variable x. In general we will have a congruence of the form

$$F(x) \equiv 0 \,(\text{mod } m)$$

where $F(x)$ (read: F at x) is an **integral polynomial** in x; that is, $F(x)$ is an expression built up from the integers and the variable x by addition, subtraction, and multiplication. For example,

$$x^3 - 3x^2 - 4 \equiv 0 \,(\text{mod } 9)$$
$$5x^2 + 2x + 3 \equiv 0 \,(\text{mod } 7)$$

A congruence such as

$$x^3 + 5x^2 \equiv 3 \,(\text{mod } 8)$$

may be handled in the form

$$x^3 + 5x^2 - 3 \equiv 0 \,(\text{mod } 8)$$

By a **solution** of a congruence, we mean an integer a such that when x is replaced by a, a true congruence results. Thus 2 is a solution of

111

$$x^2 + 2x + 1 \equiv 0 \,(\text{mod } 9)$$

since

$$2^2 + 2(2) + 1 \equiv 0 \,(\text{mod } 9)$$

but 3 is not a solution since

$$3^2 + 2(3) + 1 \equiv 16 \equiv 7 \not\equiv 0 \,(\text{mod } 9)$$

Because each integer is congruent to one of the integers

$$0, 1, 2, \ldots, (m - 1)$$

it is unnecessary to work with integers greater than m in solving congruences. For example,

$$9x^2 + 7x + 14 \equiv 0 \,(\text{mod } 6)$$

is equivalent to

$$3x^2 + x + 2 \equiv 0 \,(\text{mod } 6)$$

because $9 \equiv 3 \,(\text{mod } 6)$, $7 \equiv 1 \,(\text{mod } 6)$, and $14 \equiv 2 \,(\text{mod } 6)$.

We will not always use a reduction, however. For example, the solution of $x^2 \equiv 7 \,(\text{mod } 9)$ can be found by trial and error, that is, by substituting the numbers $0, 1, \ldots, 8$ for x and determining which of these numbers are solutions. But $x^2 \equiv 7 \,(\text{mod } 9)$ is equivalent to $x^2 \equiv 16 \,(\text{mod } 9)$ since $7 \equiv 16 \,(\text{mod } 9)$, and this congruence has two obvious solutions, $x \equiv 4$ and $x \equiv -4 \,(\text{mod } 9)$.

A second simplification of our problem is most important in the solution of congruences. That is: *if a is a solution of the congruence $F(x) \equiv 0$ (mod m), then all the numbers in the set*

$$\ldots, a - 2m, a - m, a, a + m, a + 2m, \ldots$$

are also solutions. Hence, in looking for solutions of congruences, we look only for solutions among the numbers

$$0, 1, \ldots, (m - 1)$$

because all other integers are congruent to one of these. These solutions can be found by a finite number of calculations. Although this is not the most efficient method of solution, it is one way to solve congruences.

Let us solve some congruences using this method.

Example 1: Solve $x^3 \equiv 3 \pmod 5$.

Solution: We test the numbers 0, 1, 2, 3, and 4 and list those that are solutions.

$$0^3 \not\equiv 3 \pmod 5$$
$$1^3 \not\equiv 3 \pmod 5$$
$$2^3 \equiv 8 \equiv 3 \pmod 5$$
$$3^3 \equiv 27 \equiv 2 \not\equiv 3 \pmod 5$$
$$4^3 \equiv 64 \equiv 4 \not\equiv 3 \pmod 5$$

Thus $x \equiv 2 \pmod 5$ is the only solution. If we are asked for *all* solutions x such that $x^3 \equiv 3 \pmod 5$, the answer is

$$x = 2 + 5t$$

where t is an arbitrary integer. This gives the set

$$\{\ldots, -13, -8, -3, 2, 7, \ldots\}$$

Example 2: Solve $x^2 + 2x - 1 \equiv 0 \pmod 3$.

Solution: We test the numbers 0, 1, and 2.

$$0^2 + (2)(0) - 1 \not\equiv 0 \pmod 3$$
$$1^2 + (2)(1) - 1 \not\equiv 0 \pmod 3$$
$$2^2 + (2)(2) - 1 \not\equiv 0 \pmod 3$$

Therefore this congruence has no solution.

Example 3: Solve $x^2 \equiv 1 \pmod 8$.

Solution: We test the numbers 0, 1, 2, 3, 4, 5, 6, and 7.

$$0^2 \not\equiv 1 \pmod 8$$
$$1^2 \equiv 1 \pmod 8$$
$$2^2 \equiv 4 \not\equiv 1 \pmod 8$$
$$3^2 \equiv 9 \equiv 1 \pmod 8$$
$$4^2 \equiv 16 \not\equiv 1 \pmod 8$$
$$5^2 \equiv 25 \equiv 1 \pmod 8$$
$$6^2 \equiv 36 \not\equiv 1 \pmod 8$$
$$7^2 \equiv 49 \equiv 1 \pmod 8$$

Therefore the solutions are

$$x \equiv 1 \pmod 8$$
$$x \equiv 3 \pmod 8$$
$$x \equiv 5 \pmod 8$$
$$x \equiv 7 \pmod 8$$

and all solutions are given by

$$x = 1 + 8t$$
$$x = 3 + 8t$$

$$x = 5 + 8t$$
$$x = 7 + 8t$$

where t is an arbitrary integer.

Exercise 5

1. By trial and error find all the solutions $0 \le x \le (m - 1)$ of the following.
   (a) $x^2 \equiv 1 \pmod 2$ (d) $x^2 \equiv 5 \pmod 7$
   (b) $x^2 \equiv 1 \pmod 3$ (e) $x^3 \equiv 0 \pmod 8$
   (c) $x^2 \equiv 2 \pmod 3$

2. By trial and error find all the solutions $0 \le x \le (m - 1)$ of the following.
   (a) $x^2 + x - 1 \equiv 0 \pmod 3$
   (b) $2x^2 + 1 \equiv 0 \pmod 5$
   (c) $x^2 - 5x + 6 \equiv 0 \pmod 9$
   (d) $x^3 \equiv 1 \pmod 3$
   (e) $x^3 \equiv 2 \pmod 3$

3. By trial and error find all the solutions $0 \le x \le (m - 1)$ of the following.
   (a) $x^3 \equiv 3 \pmod 7$
   (b) $x^2 - 3x + 1 \equiv 0 \pmod 5$
   (c) $2x^2 + 4x + 2 \equiv 0 \pmod 6$
   (d) $x^3 \equiv 0 \pmod 9$
   (e) $5x^2 \equiv 1 \pmod 7$

4. By trial and error find all the solutions $0 \le x \le (m - 1)$ of the following.
   (a) $2x^2 \equiv 5 \pmod 6$
   (b) $3x \equiv 7 \pmod{12}$
   (c) $5x - 2 \equiv 6 \pmod 9$
   (d) $7x \equiv 11 \pmod{13}$
   (e) $9x \equiv 15 \pmod 6$

5. Find all the solutions of $x^2 + 3x + 2 \equiv 0 \pmod 6$.
6. Find all the solutions of $x^2 + x - 2 \equiv 0 \pmod 5$.
7. Find all the solutions of $x^4 - 2 \equiv 0 \pmod{13}$.
8. Find all the solutions of $2x^3 + 3x^2 + x \equiv 0 \pmod 6$.
9. Find all the solutions of $2x^3 - 3x + 9 \equiv 0 \pmod{11}$.
10. Find all the solutions of $3x^3 - 6x^2 + 5x - 3 \equiv 0 \pmod 7$.

7.
SOLUTION OF
LINEAR CONGRUENCES

A **linear congruence** is one that involves only the first power of the variable and constants. Examples of linear congruences are

$$3x \equiv 7 \ (\text{mod } 9)$$
$$x - 1 \equiv 0 \ (\text{mod } 2)$$
$$4x \equiv 6 \ (\text{mod } 8)$$

Linear congruences can always be put in the form

$$ax \equiv b \ (\text{mod } m)$$

where $0 \le a < m$ and $0 \le b < m$. For example,

$$12x - 7 \equiv 6 \ (\text{mod } 8)$$

is equivalent to

$$12x \equiv 13 \ (\text{mod } 8)$$

and

$$4x \equiv 5 \ (\text{mod } 8)$$

since $12 \equiv 4 \ (\text{mod } 8)$ and $13 \equiv 5 \ (\text{mod } 8)$.

If $a \equiv 0 \ (\text{mod } m)$ and if $b \equiv 0 \ (\text{mod } m)$, every integer is a solution of $ax \equiv b \ (\text{mod } m)$; if $a \equiv 0 \ (\text{mod } m)$ and $b \not\equiv 0 \ (\text{mod } m)$, there is no solution to the congruence. This situation is of no interest and we will always assume that $a \not\equiv 0 \ (\text{mod } m)$.

In solving linear congruences we distinguish two cases: (1) $(a, m) = 1$ and (2) $(a, m) = d \neq 1$. If $(a, m) = 1$, then the numbers

$$0, a, 2a, 3a, \ldots, (m - 1)a$$

form a complete residue system modulo m. Consequently one and only one of them is in the same residue class as b, that is, one and only one of them is congruent to b modulo m. If this number is ax_0, then x_0 is the unique solution of $ax \equiv b \ (\text{mod } m)$, and all other roots are congruent to x_0 modulo m; that is, all solutions are given by

$$x = x_0 + mt$$

where t is an arbitrary integer.

If $(a, m) = d$, then the proposed congruence is impossible (that is, it has no solutions) if b is not divisible by d. This can easily be seen:

$$ax \equiv b \,(\mathrm{mod}\ m)$$

is equivalent to

$$ax = b + mk$$

or

$$ax - mk = b$$

Since $d\,|\,a$ and $d\,|\,m$, d must divide b for $ax - mk = b$ to be a true statement.

If $(a, m) = d$, and $d\,|\,b$, then the given congruence is equivalent to the congruence

$$\frac{a}{d} x \equiv \frac{b}{d} \left(\mathrm{mod}\ \frac{m}{d} \right)$$

in which $\frac{a}{d}$ and $\frac{m}{d}$ are relatively prime numbers by Theorem 4.6. Hence this congruence has a unique solution x_0, (modulo $\frac{m}{d}$). That is, all its roots as well as the roots of congruence

$$ax \equiv b \,(\mathrm{mod}\ m)$$

are

$$x = x_0 + \frac{m}{d} t$$

where t is an arbitrary integer. Taking $t = 0, 1, \ldots, (d-1)$, we get exactly d distinct modulo m roots,

$$x_0, \ x_0 + \frac{m}{d}, \ x_0 + 2\frac{m}{d}, \ldots, x_0 + (d-1)\frac{m}{d}$$

for two values of t congruent modulo d lead to two values of x which are congruent modulo m, while the exhibited d numbers belong to different residue classes modulo m. We conclude, thereby, that

$$ax \equiv b \,(\mathrm{mod}\ m)$$

is possible if and only if $(a, m) = d$ divides b, and in this case has exactly d roots.

7. SOLUTION OF LINEAR CONGRUENCES

Finding solutions of linear congruences, that is, finding solutions of congruences of the form

$$(1) \qquad\qquad ax \equiv b \,(\text{mod } m)$$

is equivalent to solving the equation

$$ax = b + mk$$

which is equivalent to

$$ax - mk = b$$

Since $(a, m) = d$ and $d \,|\, b$ (d must divide b for the congruence to have a solution), we can write

$$a = a_1 d, \qquad b = b_1 d, \qquad m = m_1 d$$

We know by Theorem 4.6 that congruence (1) is equivalent to

$$(2) \qquad\qquad a_1 x \equiv b_1 \,(\text{mod } m_1)$$

and $(a_1, m_1) = 1$. Congruence (2) has only one solution since $(a_1, m_1) = 1$.
Congruence (2) may be written

$$a_1 x = b_1 + m_1 k$$

which is equivalent to

$$(3) \qquad\qquad a_1 x - m_1 k = b_1$$

Since a_1 and m_1 are relatively prime, that is, $(a_1, m_1) = 1$, we know that we can find integers P and Q such that

$$(4) \qquad\qquad a_1 Q + m_1 P = 1\dagger$$

If we multiply each member of (4) by b, we have

$$a_1(Qb_1) + m_1(Pb_1) = b_1$$
$$a_1(Qb_1) - m_1(-Pb_1) = b_1$$

From this we see that, if we set $x = Qb_1$ and $k = (-Pb_1)$ in equation (3), we have a solution. Thus the solution to congruence (2) is

$$x \equiv Qb_1 \,(\text{mod } m_1)$$

---

† See Chapter 3, Section 5.

CONGRUENCES

All the solutions to the given congruence are given by

$$x \equiv Qb_1 + m_1t \qquad t = 0, 1, 2, 3, \ldots, (d-1)$$

Study the following examples.

Example 1: Solve $3x \equiv 5 \pmod 7$.

Solution: Since $(3, 7) = 1$, this congruence has only one solution. The given congruence is equivalent to the equation

$$3x - 7m = 5$$

Values for x and m can be found by inspection. For example,

$$3(4) + 7(-1) = 5$$

Hence

$$x \equiv 4 \pmod 7$$

All solutions are given by

$$x = 4 + 7t$$

where t is an arbitrary integer.

Example 2: Solve $18x \equiv 30 \pmod{48}$.

Solution: Since $(18, 48) = 6$, and $6 \mid 30$, this congruence has six solutions. By Theorem 4.6 the given congruence is equivalent to the congruence

$$3x \equiv 5 \pmod 8$$

Since $(3, 8) = 1$, this congruence has only one solution. The congruence $3x = 5 \pmod 8$ is equivalent to the equation

$$3x - 8m = 5$$

We know that we can find integers P and Q such that

$$3Q + 8P = 1$$

because if $(a, b) = 1$, integers x and y can be found such that $ax + by = 1$. Using Euclid's Algorithm we have

$$3 = 8(0) + 3$$
$$8 = 3(2) + 2$$
$$3 = 2(1) + 1$$
$$2 = 1(2)$$

Reversing this process we have

$$1 = 3 - 2(1)$$
$$1 = 3 - 1[8 - 3(2)] = 3(3) + 8(-1)$$

Hence $Q = 3$ and hence

$$x \equiv 3 \cdot 5 \equiv 15 \equiv 7 \pmod{8}$$

The solutions of the congruence $18x \equiv 30 \pmod{48}$ are

$$x \equiv 7 + 8t \qquad t = 0, 1, 2, \ldots, 5 \pmod{48}$$

or

$$x \equiv 7 \pmod{48}$$
$$x \equiv 15 \pmod{48}$$
$$x \equiv 23 \pmod{48}$$
$$x \equiv 31 \pmod{48}$$
$$x \equiv 39 \pmod{48}$$
$$x \equiv 47 \pmod{48}$$

All solutions are given by

$$x = 7 + 48t \qquad t \text{ an arbitrary integer}$$
$$x = 15 + 48t \qquad t \text{ an arbitrary integer}$$
$$x = 23 + 48t \qquad t \text{ an arbitrary integer}$$
$$x = 31 + 48t \qquad t \text{ an arbitrary integer}$$
$$x = 39 + 48t \qquad t \text{ an arbitrary integer}$$
$$x = 47 + 48t \qquad t \text{ an arbitrary integer}$$

Exercise 6

1. How many solutions do each of the following congruences have?
  (a) $7x \equiv 5 \pmod{19}$
  (b) $2x \equiv 12 \pmod 4$
  (c) $9x \equiv 27 \pmod 6$
  (d) $108x \equiv 72 \pmod{99}$
  (e) $57x \equiv 82 \pmod{103}$

Find all the solutions of the following.

2. $3x \equiv 6 \pmod{24}$
3. $3x \equiv 2 \pmod 5$
4. $36x \equiv 8 \pmod{102}$

5. $144x \equiv 216 \pmod{306}$
6. $221x \equiv 111 \pmod{360}$
7. $20x \equiv 7 \pmod{15}$
8. $315x \equiv 11 \pmod{501}$
9. $360x \equiv 3072 \pmod{96}$
10. $75x \equiv 125 \pmod{175}$
11. $39x \equiv 129 \pmod{42}$
12. $16x \equiv 24 \pmod{48}$
13. $350x \equiv 487 \pmod{729}$
14. $95x \equiv 90 \pmod{115}$
15. $36x \equiv 7 \pmod{157}$

8.
DIVISIBILITY AND CONGRUENCES

We can derive rules for divisibility using congruences. We recall that if a number is divisible by 9, then the sum of the digits in its numeral is divisible by 9. To derive this rule using congruences, let us look at the powers of 10:

$$10 \equiv 1 \pmod 9$$
$$10^2 \equiv 10 \cdot 10 \equiv 1 \cdot 1 \pmod 9$$
$$10^3 \equiv 10 \cdot 10 \cdot 10 \equiv 1 \cdot 1 \cdot 1 \equiv 1 \pmod 9$$
$$\vdots$$
$$10^k \equiv 1 \pmod 9$$

Now let

$$N = a_0 10^k + a_1 10^{k-1} + \cdots + a_{k-1} 10 + a_k$$

If N is divisible by 9, it must be congruent to 0 modulo 9. Then

$$a_0 10^k + a_1 10^{k-1} + \cdots + a_{k-1} 10 + a_k \equiv 0 \pmod 9$$
$$a_0 \cdot 1 + a_1 \cdot 1 + \cdots + a_{k-1} \cdot 1 + a_k \equiv 0 \pmod 9$$
$$a_0 + a_1 + \cdots + a_{k-1} + a_k \equiv 0 \pmod 9$$

But $a_0 + a_1 + \cdots + a_k$ is the sum of the digits of the numeral of N. Hence if a number is divisible by 9, then the sum of the digits in its numeral is divisible by 9. We have proved:

THEOREM 4.7: *If a number is divisible by 9, then the sum of the digits in its numeral is divisible by 9.*

Now let us prove

THEOREM 4.8: *If a number is divisible by 8, then the number named by the last three digits in its numeral is divisible by 8.*

Proof: Let

$$N = a_0 10^k + a_1 10^{k-1} + \cdots + a_{k-2} 10^2 + a_{k-1} 10 + a_k$$

Since N is divisible by 8, it is congruent to 0 modulo 8. Hence

$$a_0 10^k + a_1 10^{k-1} + \cdots + a_{k-2} 10^2 + a_{k-1} 10 + a_k \equiv 0 \,(\text{mod } 8)$$

But

$$10^3 \equiv 10 \cdot 10 \cdot 10 \equiv 2 \cdot 2 \cdot 2 \equiv 0 \,(\text{mod } 8)$$
$$10^4 \equiv 0 \,(\text{mod } 8)$$
$$10^5 \equiv 0 \,(\text{mod } 8)$$
$$\vdots$$
$$10^k \equiv 0 \,(\text{mod } 8)$$

Then

$$a_0 10^k + a_1 10^{k-1} + \cdots + a_{k-3} 10^3 + a_{k-2} 10^2$$
$$+ a_{k-1} 10 + a_k \equiv 0 \,(\text{mod } 8)$$
$$a_0 \cdot 0 + a_1 \cdot 0 + \cdots + a_{k-3} \cdot 0 + a_{k-2} 10^2$$
$$+ a_{k-1} 10 + a_k \equiv 0 \,(\text{mod } 8)$$
$$0 + 0 + \cdots + 0 + a_{k-2} 10^2 + a_{k-1} 10 + a_k \equiv 0 \,(\text{mod } 8)$$
$$a_{k-2} 10^2 + a_{k-1} 10 + a_k \equiv 0 \,(\text{mod } 8)$$

But $a_{k-2} 10^2 + a_{k-1} 10 + a_k$ is the number named by the last three digits of the numeral of N.

Exercise 7

Using congruences prove the following.

1. If a number is divisible by 2, then the units digit of its numeral is divisible by 2.

2. If a number is divisible by 5 then the units digit of its numeral is divisible by 5.

3. If a number is divisible by 10, then the units digit of its numeral is divisible by 10.

4. If a number is divisible by 3, then the sum of the digits of its numeral is divisible by 3.

5. If a number is divisible by 4, then the number named by the last two digits of its numeral is divisible by 4.

6. If a number is divisible by 25, then the number named by the last two digits of its numeral is divisible by 25.

7. If $N = a_0 10^k + a_1 10^{k-1} + \cdots + a_{k-1} 10 + a_k$ is divisible by 11, then

$$a_k - a_{k-1} + a_{k-2} - a_{k-2} + \cdots \equiv 0 \pmod{11}.$$

Everiste Galois (1811–1832) at the age of sixteen knew that he was a mathematical genius (his teachers, unfortunately, failed to recognize this fact). The principal object of Galois's research at the age of seventeen was to determine when polynomial equations are solvable by radicals. He demonstrated that in order that an irreducible equation of prime degree be solvable by radicals it is necessary and sufficient that all its roots be rational functions of any two of them. He discovered that an irreducible algebraic equation is solvable by radicals if and only if the symmetric group (see Chapter 6) on the roots is solvable.

Galois invented the term "group" and his theory, called Galois Theory, is one of the main sources of finite field theory and of modern abstract algebra. Galois was educated at the Lycee Louis-de-Grand and Ecole Normale in Paris. He was a rabid republican and the first shots of the revolution of 1830 filled Galois with joy. He was twice imprisoned for his political views. On May 30, 1832, Galois was fatally wounded in a duel and died the next day. The night before the duel took place he wrote a brilliant paper on group theory and addressed it to his friend Auguste Chavalier who preserved it for the world. Galois was buried in the common ditch of the South Cemetery in Paris so that today, there is no trace of his grave. All that remains is his collected work of sixty pages.

SETS, RELATIONS AND FUNCTIONS

5

1.
SETS AND
SET LANGUAGE

In mathematics we use the word **set** to refer to a collection of things. Each individual object in a set is called a **member** or **element** of the set. Sets are usually denoted by capital letters such as A, B, C and so forth and the elements by lower case letters. The phrase "is an element of" is symbolized by $\in$. Thus $a \in B$ is read "a is an element of set B." Negation of set membership is denoted by $\notin$; for example if c is not a member of set B we write $c \notin B$ (read "c is not an element of set B").

In mathematics a set must be **well-defined.** This means that there is a rule whereby set membership or nonmembership can be determined. All the sets in this book will be well-defined sets.

One way to denote a set is by listing all its members. The symbols that represent the elements are enclosed in braces, { }, with a comma separating the elements. For example, the set of the first five letters of the English alphabet may be denoted by

$$\{a,\ b,\ c,\ d,\ e\}$$

Example 1: List the elements of the set of the first four natural numbers.

Solution: $\qquad\qquad \{1,\ 2,\ 3,\ 4\}$

Example 2: List the elements of the set of letters of the English alphabet.

Solution: $\qquad\qquad \{a,\ b,\ c,\ d,\ \ldots,\ z\}$

If a set has many elements we may abbreviate when listing its members. In listing the elements of the set of letters of the alphabet we listed the first four elements and then used three dots to indicate that the letters e through y inclusive have been omitted in the listing. In omitting some of the elements of a set in the listing we must be sure to list enough of the elements so that it is obvious which elements have been omitted.

Example 3: List the elements of the set, N, of natural numbers.

Solution: When we list the members of a nonending set, as the set of natural numbers, we again use three dots and write

$$N = \{1, 2, 3, 4, \ldots\}$$

Note that the set of natural numbers is a nonending set; therefore we cannot list its last element.

Another way of describing a set is by means of a common characteristic possessed by all of its elements. The set of all elements with some property R is denoted by

$$\{x \mid x \text{ has property } R\}$$

This symbol, called **set-builder notation,** is read "the set of all x such that x has property R". Note that the vertical line in set-builder notation is read "such that".

Example 4: Use set-builder notation to denote the set of natural numbers less than 10.

Solution: $\{x \mid x \text{ is a natural number and } x < 10\}$

Example 5: Use set-builder notation to denote the set of integers greater than -4 and less than 16.

Solution: $\{x \mid x \text{ is an integer and } -4 < x < 16\}$

Example 6: Use set-builder notation and denote the set of integers greater than -8

Solution: $\{x \mid x \text{ is an integer and } x > -8\}$

If two sets, A and B, have exactly the same elements they are said to be **equal** and we write $A = B$. Thus

$$P = \{x \mid x \text{ is a natural number and } x < 4\}$$

and

$$Q = \{1, 2, 3\}$$

have exactly the same elements and therefore $P = Q$.

A set that has no elements is called the **empty set** and is denoted by $\emptyset$. The set of pink elephants is an example of the empty set. Other examples of empty sets are

The set of icebergs in the Nile River.

The set of states of the United States with a population of less than 100.

The set of natural numbers greater than 3 and less than 4.

Consider

$$A = \{w, o, m, e, n\}$$

and

$$B = \{d, a, m, s, e, l\}$$

Let us form the set of elements that are members of both set A and set B. This set is called the **intersection** of sets A and B and is denoted by $A \cap B$ (read A intersection B). This set contains the elements m and e:

$$A \cap B = \{m, e\}$$

We see then that the **intersection** of two set A and B, denoted by $A \cap B$, is the set of elements common to the two sets.

Example 7:   Given $P = \{1, 2, 3, 4, 5, 6\}$ and $Q = \{2, 4, 6, 8\}$, find $P \cap Q$.

Solution: $P \cap Q = \{2, 4, 6\}$

Example 8:   Given $R = \{p, e, r, t\}$ and $S = \{u, g, l, y\}$, find $R \cap S$.
Solution: Since R and S have no elements in common their intersection is the empty set. Thus

$$R \cap S = \emptyset$$

If two sets, A and B, have no elements in common, that is if $A \cap B = \emptyset$ then A and B are called **disjoint** sets or **mutually exclusive** sets.

Again consider sets A and B above. Let us form a new set composed of all the elements that are members of A or of B or of both A and B. This set is called the **union** of sets A and B and is denoted by $A \cup B$. We see that

$$A \cup B = \{w, o, m, e, n, d, a, s, l\}$$

The union of two sets A and B, denoted by $A \cup B$, is the set of elements contained in set A or set B (notice here "or" is used in the inclusive sense, see Chapter 1, Section 6).

If A and B are two sets and $A \cap B = A$, then A is called a **subset** of B. This is denoted by $A \subseteq B$ (Read: A is a subset of B). From this definition we see that A is a subset of B if every element of A is also an element of B. Since $A \cap A = A$, we see that *every set is a subset of itself*. Also since $A \cap \emptyset = \emptyset$ we see that by our definition of subset, *the empty set is a subset of every set*.

If all the sets in a particular discussion are subsets of one set, we call this set the **universal set** or the **universe** and denote it by U. For example

$$A = \{a, b, c\}$$
$$B = \{w, x, y, z\}$$
$$C = \{p, q, r, s, t, u, v, w\}$$

are all subsets of

$$U = \{a, b, c, d, \ldots, z\}$$

Suppose $A \subseteq U$. Now let us consider the set of all elements of U that are not elements of A. We call this set the **complement** of A and denote it by A' (read A complement).

Example 9: Given

$$U = \{1, 2, 3, \ldots, 10\}$$
$$A = \{2, 4, 6, 8\}$$
$$B = \{3, 6, 9\}$$
$$C = \{1, 4, 7, 10\}$$

find (a) A'; (b) $(A \cup B)'$; (c) $(A \cap C)'$; (d) $(A \cup C)'$

Solution: (a) $A' = \{1, 3, 5, 7, 9, 10\}$
(b) $A \cup B = \{2, 3, 4, 6, 8, 9\}$, hence
    $(A \cup B)' = \{1, 5, 7, 10\}$
(c) $A \cap C = \{4\}$, hence
    $(A \cap C)' = \{1, 2, 3, 5, 6, 7, 8, 9, 10\}$
(d) $A \cup C = \{1, 2, 4, 6, 7, 8, 10\}$, hence
    $(A \cup C)' = \{3, 5, 9\}$

2.
SETS OF NUMBERS

In this chapter we will be discussing sets of numbers. You are familiar with the set, N, of **natural numbers** (also called the **counting numbers** and the **positive integers**):

$$N = \{1, 2, 3, 4, \ldots\}$$

When zero is annexed to this set, we have the set, W, of **whole numbers** (also called the **nonnegative integers**):

$$W = \{0, 1, 2, 3, \ldots\}$$

In Chapter 2 we studied the set, I, of **integers:**

$$I = \{\ldots, -3, -2, -1, 0, 1, 2, 3, \ldots\}$$

The set, Q, of **rational numbers** is the set of all numbers that can be represented in the form $\frac{a}{b}$, called a **fraction,** where a and b are integers and $b \neq 0$. Using set-builder notation we denote the set of rational numbers by

$$Q = \left\{ \frac{a}{b} \,\middle|\, a \in I, b \in I \text{ and } b \neq 0 \right\}$$

Among the elements of Q are $\frac{1}{2}$, $-\frac{3}{4}$, 9, -8, and 0.

As stated above, rational numbers can be named by common fractions or simply fractions. Another method of naming rational numbers is by decimal fraction notation, called **decimals.** All rational numbers can be named by terminating or repeating nonterminating decimals. For example

$$\frac{1}{4} = 0.25 \qquad \frac{1}{3} = 0.333\ldots = 0.\overline{3}$$

$$\frac{3}{5} = 0.6 \qquad \frac{1}{6} = 0.1666\ldots = 0.1\overline{6}$$

$$\frac{5}{8} = 0.625 \qquad \frac{5}{7} = 0.714285714285\ldots = 0.\overline{714285}$$

$$\frac{11}{16} = 0.6875 \qquad \frac{9}{11} = 0.8181\ldots = 0.\overline{81}$$

The set S of **irrational numbers** is the set of all numbers whose decimal representations are nonterminating and nonrepeating. An irrational number cannot be represented in the form $\frac{a}{b}$ where a and b are integers and $b \neq 0$. Among the elements of S are $\sqrt{2}$, $\sqrt[3]{3}$, $-\sqrt{7}$, and π.

The set R of **real numbers** is the union of the set of rational numbers and the set of irrational numbers. That is

$$R = Q \cup S$$

Example 1: Use set-builder notation to denote the set of real numbers greater than 4.

Solution: $$\{x \mid x \in R \text{ and } x > 4\}$$

Example 2: Use set-builder notation to denote the set of real numbers greater than or equal to $-\sqrt{2}$ and less than or equal to $\sqrt{2}$.

Solution: $$\{x \mid x \in R \text{ and } -\sqrt{2} \le x \le \sqrt{2}\}$$

Exercise 1

1. List the elements of the following sets.
   (a) The set of days of the week.
   (b) The set of months of the year with exactly 30 days.
   (c) The set of seasons of the year.
   (d) The set of days of the week whose names begin with the letter t.
   (e) The set of states of the United States whose names begin with the letter v.
2. List the elements of the following sets.
   (a) $\{x \mid x \text{ is a natural number and } x < 10\}$
   (b) $\{x \mid x \text{ is an integer and } x > -7 \text{ and } x < 12\}$
   (c) $\{x \mid x \text{ is an even natural number}\}$
   (d) $\{x \mid x \text{ is an integer}\}$
3. Use set-builder notation to denote the following sets.
   (a) $\{1, 2, 3, 4, 5, 6, 7\}$
   (b) $\{\text{January, February, March,} \ldots, \text{December}\}$
   (c) $\{a, b, c, d, e, f, g, h\}$
   (d) $\{\ldots, -4, -2, 0, 2, 4, 6, \ldots\}$
4. Given $A = \{0, 1, 2, 3, 4, 5\}$, $B = \{0, 3, 6\}$, and $C = \{1, 3, 5, 7\}$, find
   (a) $A \cup B$ (d) $A \cap C$
   (b) $A \cap B$ (e) $(A \cup B) \cap C$
   (c) $A \cup C$ (f) $(A \cap B) \cup C$
5. Given $P = \{p, \ell, a, s, t, e, r\}$, $Q = \{m, a, s, o, n\}$ and $R = \{b, u, i, \ell, d, e, r\}$, find
   (a) $P \cup Q$ (d) $(P \cap R) \cup Q$
   (b) $R \cap P$ (e) $(R \cap Q) \cup P$
   (c) $(Q \cup P) \cup R$ (f) $P \cap (Q \cup R)$
6. Let $U = \{x \mid x \text{ is a natural number and } x < 15\}$ be the universe and consider the following subsets of U:

$$A = \{1, 3, 5, 7, 9, 11, 13\}$$
$$B = \{2, 4, 6, 8, 10, 12, 14\}$$
$$C = \{3, 6, 9, 12\}$$
$$D = \{4, 8, 12\}$$
$$E = \{1, 4, 7, 10, 13\}$$

Find

(a) $A \cap B$ (e) $B \cap A'$

(b) $C \cup D$ (f) $C' \cup D'$

(c) $(C \cup D)'$ (g) $(D \cup E)' \cap A$

(d) $(A \cap B)' \cup C$ (h) $(A \cap C)' \cup (D \cap E)'$

7. Given the universal set U and $A \subseteq U$, find

(a) $A \cup U$ (d) $A \cup \emptyset$

(b) $A \cap U$ (e) $A \cap A'$

(c) $A \cap \emptyset$ (f) $A \cup A'$

8. If $A \subseteq B$ and $B \subseteq A$, is $A = B$ or is $A \neq B$?

9. Give three examples of the empty set.

In each of the following A and B are subsets of the universe U. List the elements in (a) $A' \cap B'$ and (b) $A' \cup B'$ (Exercises 10–15).

10. $U = \{1, 2, 3, 4, 5\}$; $A = \{4, 3\}$; $B = \{1, 3, 5\}$

11. $U = \{1, 2, 3, \ldots, 10\}$; $A = \{1, 3, 5, 7\}$; $B = \{2, 4, 6, 8, 10\}$

12. $U = \{1, 2, 3, 4, \ldots, 12\}$; $A = \{2, 5, 8, 11\}$; $B = \{3, 5, 7, 9\}$

13. $U = \{1, 2, 3, 4, 5, 6, 7\}$; $A = \{1, 6\}$; $B = \{2, 4, 6\}$

14. $U = \{1, 2, 3, 4, \ldots, 10\}$; $A = \emptyset$; $B = \{2, 5, 9\}$

15. $U = \{1, 2, 3, \ldots, 100\}$; $A = \{1, 3, 5, 7, \ldots, 99\}$;
$B = \{2, 4, 6, \ldots, 98\}$

16. Use set-builder notation to denote the set of real numbers less than $\sqrt{3}$.

17. Use set-builder notation to denote the set of real numbers not equal to $\sqrt{10}$.

18. Use set-builder notation to denote the set of real numbers greater than 6 and less than 25.

19. Use set-builder notation to denote the set of real numbers greater than or equal to $-\frac{1}{3}$ and less than or equal to $\frac{5}{8}$.

20. Use set-builder notation to denote the set of real numbers greater than $-\sqrt{13}$ and less than $\sqrt{13}$.

3.
CARTESIAN PRODUCTS

Let us consider two elements, a and b, of set S. The subset of S containing these two elements may be denoted by $\{a, b\}$ or by $\{b, a\}$; the order in which we list the two elements in this set is immaterial. If, on the other hand, we are interested in a and b, but the order in which we designate the elements is important, we have what is called an **ordered pair** of elements. We designate an ordered pair of elements a and b, in that order, by the symbol (a, b). We call a the **first component** of the ordered pair (a, b) and b the **second component.**

Two ordered pairs (a, b) and (c, d) are defined to be **equal,** denoted by $(a, b) = (c, d)$, if and only if $a = c$ and $b = d$. Thus

$$(3, 4) = (2 + 1, 2 + 2)$$

since $3 = 2 + 1$ and $4 = 2 + 2$. Note that (a, b) and (b, a) are equal only if $a = b$.

Now let us consider the two sets

$$A = \{1, 2, 3\} \qquad \text{and} \qquad B = \{4, 5\}$$

Let us form all possible ordered pairs such that the first component from each ordered pair is an element of A and the second component is an element of B. We pair each element of A with each element of B in a systematic manner using the tree diagram shown in Figure 5.1. We see

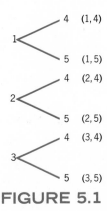

FIGURE 5.1

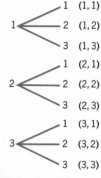

FIGURE 5.2

from Figure 5.1 that it is possible to form six ordered pairs using the elements of A as the first components and the elements of B as the second components. This set of ordered pairs is called the **Cartesian product** of A and B and is denoted by $A \times B$ (read "A cross B"):

$$A \times B = \{(1, 4), (1, 5), (2, 4), (2, 5), (3, 4), (3, 5)\}$$

Now let us form $A \times A$. Using the tree diagram in Figure 5.2 we see that it is possible to form nine ordered pairs using the elements of A as components. Then

$$A \times A = \{(1, 1), (1, 2), (1, 3), (2, 1), (2, 2), (2, 3), (3, 1), (3, 2), (3, 3)\}$$

The reader should construct his own tree diagrams to see that

$$B \times A = \{(4, 1), (5, 1), (4, 2), (4, 3), (5, 2), (5, 3)\}$$

and

$$B \times B = \{(4, 4), (4, 5), (5, 4), (5, 5)\}$$

Example 1: Given $P = \{a, b\}$ and $Q = \{x, y, z\}$, form $P \times Q$ and $P \times P$.

Solution: $P \times Q = \{(a, x), (a, y), (a, z), (b, x), (b, y), (b, z)\}$
$P \times P = \{(a, a), (a, b), (b, a), (b, b)\}$

Example 2: Given $S = \{1, 2, 3, 4, 5\}$ and $T = \{a, b, c, d\}$, how many elements are in (a) $S \times S$? (b) $S \times T$? (c) $T \times T$? (d) $T \times S$?

Solution: (a) 25
(b) 20
(c) 16
(d) 20

4.
GRAPHING CARTESIAN PRODUCTS

We graph a set of numbers using a number line. A **number line** is a line of indefinite extent whose points are associated with the elements of a set of numbers. A number line on which the set of whole numbers has been graphed is shown in Figure 5.3a. Observe that although every whole number is associated with a point on the line, not every point is associated with a whole number. Figure 5.3b shows a number line on which the integers have been graphed. Again we observe that every integer is associated with a point on the line, but not every point is associated with an integer.

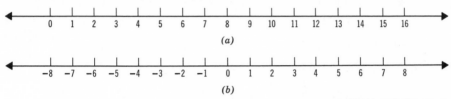

(a)

(b)

FIGURE 5.3

It can be shown, although we shall not do it here, that a one-to-one correspondence can be established between the points on a line and the real numbers. By a **one-to-one correspondence** we mean that with every point on the line we can match one and only one real number and with every real number we can match one and only one point on the line. A point that is associated with a number is called the **graph** of the number. The number associated with a point on the line is called the **coordinate** of the point. When points on a line are thus associated with the real numbers, the line is called the **real number line.**

Example 1: On the real number line graph the set

$$\{x \mid x \text{ is a whole number and } 3 < x < 7\}$$

Solution:

FIGURE 5.4

135

The points on the line that are graphs of the whole numbers greater than 3 and less than 7 are indicated by large dots.

Example 2: On the real number line graph the set

$$\{x \mid x \text{ is a real number and } 3 < x < 7\}$$

Solution:

FIGURE 5.5

The heavy trace is the graph of the given set of real numbers. Every point on the line between the point which is the graph of 3 and the point that is the graph of 7 is the graph of an element of the set. The open circles about the points associated with 3 and 7 indicate that these points are not included in the graph.

Example 3: On the real number line graph the set

$$\{x \mid x \text{ is a real number and } -2 \leq x \leq 5\}$$

Solution:

FIGURE 5.6

The graph of this set is shown in Figure 5.6. The filled in circles at the points which are graphs of -2 and 5, respectively, indicate that these points are included in the graph.

To graph ordered pairs of numbers we use two number lines; one placed horizontally and the other vertically. These two number lines are perpendicular to each other and intersect at the point on each number line that is the graph of zero (Figure 5.7). Each of these two lines is called an **axis.** The horizontal axis is called the **axis of abscissas** or the **x-axis.** The vertical axis is called the **axis of ordinates** or the **y-axis.** The point at which the two axes meet is called the **origin.** The plane (flat surface) in which these two axes lie is called the **Cartesian plane.**

136

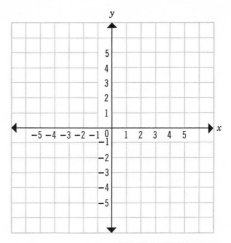

FIGURE 5.7

We now associate every point P in the plane with an ordered pair of real numbers. To do this we draw horizontal and vertical lines through each point on the plane (Figure 5.8). We then examine the coordinates of the points at which these horizontal and vertical lines intersect the two axes.

Consider the point P on the plane in Figure 5.9. Here the vertical line through P intersects the x-axis at the point with coordinate -4. We use -4 as the first component of the ordered pair with P as its graph. The horizontal line through P intersects the y-axis at the point whose coordinate is 2. We use 2 as the second component of the ordered pair with

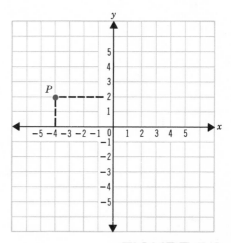

FIGURE 5.8

137

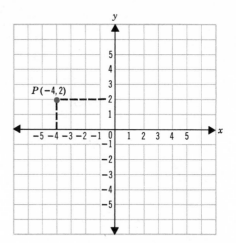

FIGURE 5.9

P as its graph. The **graph** of the ordered pair $(-4, 2)$ is *P*; the **coordinates** of *P* are the components of the ordered pair $(-4, 2)$. When we speak of points on the plane and their coordinates, we usually say "point $(-4, 2)$" to mean the point that is the graph of the ordered pair $(-4, 2)$. When we say "$(-4, 2)$ are the coordinates of point *P*" we mean that -4 is the *x*-coordinate and 2 is the *y*-coordinate of the point *P*.

This method of associating ordered pairs of real numbers with points in the Cartesian plane is an example of a **coordinate system.** Note that

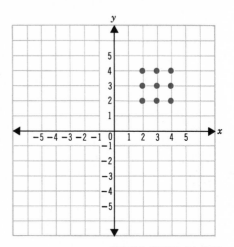

FIGURE 5.10

all points on the x-axis have coordinates with second components 0; all points on the y-axis have coordinates with first components 0. The coordinates of the origin are (0, 0).

Example 1:   Given $A = \{2, 3, 4\}$, graph $A \times A$ on the Cartesian plane.
   Solution: $A \times A = \{(2, 2), (2, 3), (2, 4), (3, 2), (3, 3), (3, 4), (4, 3),$ $(4, 2), (4, 4)\}$. The graph of $A \times A$ is the set of all points on the Cartesian plane whose coordinates are the ordered pairs of $A \times A$. This graph is shown in Figure 5.10.

Example 2:   Graph $I \times I$, that is the graph of the Cartesian product of the set of integers $I = \{\ldots, -2, -1, 0, 1, 2, \ldots\}$.
   Solution: It is impossible to graph every point in $I \times I$ because the graph extends indefinitely in all directions. The incomplete graph in Figure 5.11 indicates the appearance of the graph in a portion of the plane.

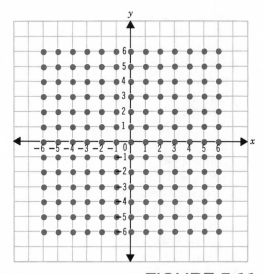

FIGURE 5.11

Example 3: $P = \{x \mid x$ is a real number and $1 \leq x \leq 2\}$. Graph $P \times P$.
   Solution: The graph of $P \times P$ consists of all points on the square together with all the points inside of the square as shown in Figure 5. 12.

139

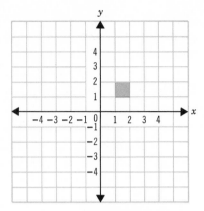

FIGURE 5.12

Exercise 2

1. Given $A = \{-2, 0, 2\}$ and $B = \{1, 2, 3\}$. List the elements of
(a) $A \times A$ (c) $B \times A$
(b) $A \times B$ (d) $B \times B$

2. Given $P = \{-4, 4\}$. List the elements of $P \times P$.

3. Given $S = \{x \mid x \text{ is an integer and } -1 \leq x \leq 1\}$. List the elements of $S \times S$.

4. Given $D = \{d, o, g\}$. List the elements of $D \times D$.

5. If A is a set containing 4 elements and B is a set containing 6 elements, give the number of elements in
(a) $A \times B$ (c) $A \times A$
(b) $B \times B$ (d) $B \times A$

6. If P is a set containing k elements and Q is a set containing n elements, give the number of elements in
(a) $P \times Q$ (c) $Q \times Q$
(b) $Q \times P$ (d) $P \times P$

7. Use a Cartesian plane to graph the following ordered pairs.
(a) $(3, 2)$ (e) $(-\frac{3}{2}, -\frac{1}{2})$
(b) $(0, 5)$ (f) $(-4, 0)$
(c) $(-1, 4)$ (g) $(0, 0)$
(d) $(\frac{1}{2}, -\frac{1}{2})$ (h) $(0, -4)$

8. What are the coordinates of a point midway between $(0, 6)$ and $(0, 10)$?

9. What are the coordinates of a point midway between $(4, 0)$ and $(12, 0)$?

10. What are the coordinates of a point midway between $(-5, 0)$ and $(15, 0)$?

11. What are the coordinates of a point midway between $(0, -12)$ and $(0, 8)$?

12. Given the points $(0, 0)$, $(0, 6)$ and $(4, 0)$, determine a fourth point so that the four points are vertices of a rectangle.

13. Given the points $(-3, 0)$, $(5, 0)$, and $(-3, 4)$, determine a fourth point so that the four points are vertices of a rectangle.

14. Given the points (2, 4), (8, 4) and (5, 7), determine the coordinates of a fourth point so that the points are vertices of a parallelogram.

15. Given $A = \{-3, 2, 4\}$.

(a) List the elements of $A \times A$.

(b) Graph $A \times A$ on the Cartesian plane.

16. Given $P = \{x \mid x$ is an integer and $-1 \le x \le 0\}$.

(a) List the elements of $P \times P$.

(b) Graph $P \times P$ on the Cartesian plane.

17. Given $S = \{x \mid x$ is a real number and $-1 \le x \le 1\}$. Graph $S \times S$ on the Cartesian plane.

18. Given $T = \{x \mid x$ is a real number and $4 \le x \le 6\}$. Graph $T \times T$ on the Cartesian plane.

19. Given $A = \{1, 2, 3, 4\}$

(a) List the elements of $A \times A$.

(b) Graph $A \times A$ on the Cartesian plane.

(c) What is the subset of $A \times A$ satisfying the condition that the sum of the x and y components of the ordered pairs equals 5?

20. Given $R = \{-1, 0, 1, 2\}$

(a) List the elements of $R \times R$

(b) Graph $R \times R$ on the Cartesian plane

(c) What is the subset of $R \times R$ satisfying the condition that the first component of the ordered pair is greater than the second?

5.
RELATIONS

Let us consider two sets A and B. A **relation on A × B** is a subset of the Cartesian product $A \times B$. For example if

$$A = \{2, 3, 4\} \quad \text{and} \quad B = \{5, 7\}$$

then

$$A \times B = \{(2, 5), (2, 7), (3, 5), (3, 7), (4, 5), (4, 7)\}$$

Any subset of $A \times B$ is a relation on $A \times B$. Some examples of relations on $A \times B$ are sets F, G, and H below.

$$F = \{(2, 5), (3, 5), (4, 5)\}$$
$$G = \{(2, 5), (2, 7)\}$$
$$H = \{(4, 5), (2, 7)\}$$

If F is a relation on $A \times A$, we say simply **F is a relation on A.** In this book unless stated otherwise we shall consider relations on R, the set of real numbers. A relation on a set A is a set of ordered pairs both of whose components are elements of set A. If the relation F under discussion is on R the set of real numbers we shall just say the relation F and it will be understood that relation F is a relation on R the set of real numbers.

Example 1:   Given $A = \{1, 2, 3, 4, 5\}$. List the elements of the relation F on A:

$$F = \{(x, y)|(x, y) \in A \times A \text{ and } x + y = 5\}$$

Solution: We are looking for a particular subset of $A \times A$. The elements of this subset are ordered pairs of the form (x, y) where x and y are elements of A and their sum, $x + y$, equals 5. Then

$$F = \{(1, 4), (2, 3), (3, 2), (4, 1)\}$$

Example 2:   Given $A = \{x | x \text{ is an integer and } -3 < x < 3\}$ and the relation K on A:

$$K = \{(a, b)|(a, b) \in A \times A \text{ and } a = b\}$$

Graph relation K.

Solution: The elements of K are all the ordered pairs of the form (a, b) where a and b are integers greater than -3 and less than 3 and $a = b$. The graph of K is shown in Figure 5.13.

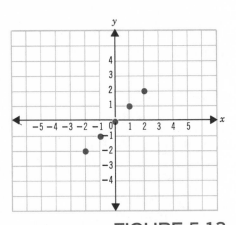

FIGURE 5.13

Example 3: Given the set of real numbers R. Graph the relation F on R:

$$F = \{(x, y)\,|\,(x, y) \in R \times R,\ 1 < x < 2 \text{ and } -2 < y < 2\}$$

Solution:

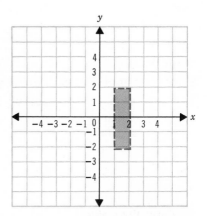

FIGURE 5.14

The graph of F consists of all points whose coordinates are ordered pairs (x, y) such that both x and y are real numbers and x is greater than 1 and less than 2 and y is greater than -2 and less than 2. The graph of F is shown in Figure 5.14. All points inside of the rectangle are points of the graph. The sides of the rectangle are not part of the graph. To indicate that the sides of the rectangle are not included in the graph they are shown by dotted lines.

A relation, then, is a set of ordered pairs. The **domain** of a relation is the set of all first components of the ordered pairs of the relation. The **range** of a relation is the set of all second components of the ordered pairs of the relation. Consider the relation

$$R = \{(1, 2), (3, 5), (4, 1), (0, 1)\}$$

The domain of R is $\{0, 1, 3, 4\}$; the range of R is $\{1, 2, 5\}$.

Example 4: What is the domain of the relation F in Example 3? What is its range?

Solution: The domain of F is $\{x\,|\,x$ is a real number and $1 < x < 2\}$
The range of F is $\{y\,|\,y$ is a real number and $-2 < y < 2\}$.

143

6.
FUNCTIONS

We may think of a relation on $A \times B$ as a "matching" or "pairing" operation in which an element of the domain is matched or paired with one or more elements of the range. When relations are interpreted in this way they are called **mappings.** Since a relation is a set of ordered pairs, we think of the pairing as starting in the domain and proceeding to the range.

Let us consider the relation

$$F = \{(0,\ 1),\ (1,\ 2),\ (3,\ 3),\ (0,\ 2)\}$$

We can picture this matching of the elements of the domain of F with the elements of the range of F in the diagram in Figure 5.15, called a **mapping diagram.** This relation pairs 0 with 1 and 2, 1 with 2 and 3 with 3.

Now let us consider a second relation, H,

$$H = \{(0,\ 1),\ (1,\ 2),\ (2,\ 3),\ (3,\ 4)\}$$

The mapping diagram for relation H is shown in Figure 5.16.

Relation H, unlike relation F, matches each element in the domain with one and only one element in the range. Relations with this property are called functions. We see, then, that a **function** is a relation in which no two ordered pairs have the same first component and different second components. A function matches each element in its domain with one and only one element in its range. We observe that *every function is a relation but every relation is not a function.*

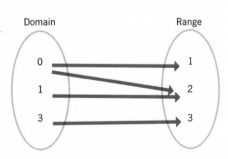

FIGURE 5.15

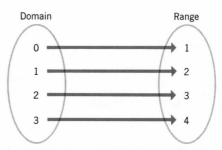

FIGURE 5.16

Example 1: Is the relation $\{(-2, 4), (1, 2), (4, 0)\}$ a function?

Solution: Yes, because no two of its ordered pairs have the same first component and different second components.

Example 2: What are the domain and the range of the relation

$$\{(0, -1), (0, -2), (0, -3), (1, 2), (3, -3)\}$$

Is this relation a function?

Solution: Domain: $\{0, 1, 3\}$
Range: $\{-1, -2, -3, 2\}$
The relation is not a function since $(0, -1)$, $(0, -2)$ and $(0, -3)$ belonging to the relation have the same first component and different second components.

Example 3: Construct a mapping diagram for the relation in Example 2 above.

Solution: The mapping diagram is given in Figure 5.17. As with relations, when we speak of a function, f, on the set R of real numbers we shall simply say "the function f" and it will be understood we mean the function f on R.

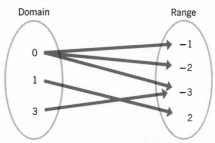

FIGURE 5.17

145

1. Given $A = \{1, 2, 3, 4, 5\}$. List the elements of the relation R on A:

$$R = \{(x, y) \mid (x, y) \in A \times A \text{ and } xy = 4\}$$

2. Consider the relation F on the set W of whole numbers:

$$F = \{(x, y) \mid (x, y) \in W \times W \text{ and } x + y = 8\}$$

List the elements of F.

3. Draw a mapping diagram for each of the following relations.
   (a) $\{(0, 1), (1, 4), (4, 2)\}$
   (b) $\{(-2, 4), (-1, 1), (0, 0), (1, 1), (2, 4)\}$
   (c) $\{(\frac{9}{4}, \frac{3}{2}), (\frac{25}{16}, \frac{5}{4}), (0, 0), (9, 3)\}$

4. Tell whether each of the following relations is a function.
   (a) $\{(-3, 2), (-4, 1), (-2, 0), (1, 3)\}$
   (b) $\{(4, 2), (4, -2), (2, 2), (2, -2), (0, 0)\}$
   (c) $\{(\frac{1}{2}, \frac{3}{4}), (\frac{3}{4}, \frac{1}{2}), (0, 0)\}$
   (d) $\{(0, 1), (2, 4), (3, 0), (2, -1), (0, 0)\}$

5. Give the domain and range of the following relations.
   (a) $\{(0, 6), (1, 7), (2, 8), (3, 9), (4, 10)\}$
   (b) $\{(3, 4), (-2, 5), (4, 3), (-2, 7)\}$
   (c) $\{(-2, -8), (-1, -1), (0, 0), (1, 1), (2, 8)\}$
   (d) $\{(-2, 4), (-1, 1), (1, 1), (2, 4), (4, 16)\}$

6. Given $A = \{0, 1, 3\}$ and $B = \{-1, -2, 0\}$. (a) Is $A \times A$ a function? (b) Is $B \times B$ a function?

7. Let $A = \{0, 1, 2, 3, 4, 5\}$. Given the relation R on A:

$$R = \{(x, y) \mid (x, y) \in A \times A \text{ and } x > y\}$$

Graph R.

8. Given the relation F on the set of real numbers R:

$$F = \{(x, y) \mid (x, y) \in R \times R \text{ and } 0 \leq x \leq 2 \text{ and } -1 \leq y \leq 2\}$$

Is F a function?

9. Graph relation F in Exercise 8.

10. What is the domain and relation of relation F in Exercise 8?

7.
FUNCTION NOTATION

A function may be thought of as a rule that associates every element of the domain D with one and only one element of the range A. It is convenient to think of a function as a machine into

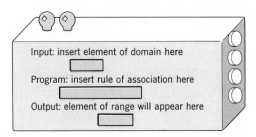

FIGURE 5.18

which we feed elements of the domain D. We also indicate the range A and the rule of association. The machine processes every element of the domain with some element of the indicated range. Such a function machine is pictured in Figure 5.18. We place an element x of the domain in the machine as shown. We then indicate the range of the function. The machine is then set to perform the necessary operations to produce the element of the range associated with the element x put in the machine. It is then turned on and for each element x put in, it produces the paired element of the range A. For example, suppose we program our function machine to multiply each element of the domain by 3 and then subtract 1. Like any calculator or computer, our function machine cannot think; it merely takes the numbers we feed into it and cranks out the associated numbers of the range. If for some element in the domain the associated number is not an element of the range, the machine rings a bell and flashes a red light. Thus after our machine is programmed it associates with the number x of the domain the number $3x - 1$. If 4 is put into the input, the machine processes $3 \cdot 4 - 1 = 11$.

Every function is a set of ordered pairs. Each ordered pair (x, y) of the function consists of a first component x which is an element of the domain and a second component y which is an element of the range.

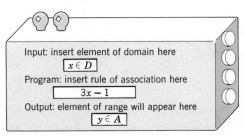

FIGURE 5.19

147

When the domain of a function is a finite set (that is, it has a limited number of elements), it is possible to list every ordered pair that belongs to the function. For example, if the domain of a function is $D = \{0, 1, 2, 3, 4, 5, 6\}$ and the rule of association is "multiply each number by 4 and add 1" we find the elements of the range are

$$4 \cdot 0 + 1 = 1 \qquad 4 \cdot 4 + 1 = 17$$
$$4 \cdot 1 + 1 = 5 \qquad 4 \cdot 5 + 1 = 21$$
$$4 \cdot 2 + 1 = 9 \qquad 4 \cdot 6 + 1 = 25$$
$$4 \cdot 3 + 1 = 13$$

The set of ordered pairs of this function is

$$\{(0, 1), (1, 5), (2, 9), (3, 13), (4, 17), (5, 21), (6, 25)\}$$

Functions with finite domains are seldom encountered. Usually the functions studied in mathematics have domains which are infinite sets such as the set of natural numbers, the set of whole numbers, the set of integers, the set of real numbers, and so forth. Now let us look at a function whose domain is the set of real numbers, and the rule of association is "the square of each number." Some ordered pairs belonging to this function are

$$\left(-\frac{1}{2}, \frac{1}{4}\right) \qquad \left(\frac{3}{4}, \frac{9}{16}\right) \qquad (\sqrt{2}, 2) \qquad (1.5, 2.25)$$

How can we designate this function? We see that the domain is the set of real numbers and the ordered pairs of the function are of the form (x, y) where $y = x^2$. We denote this function by

$$\{(x, y) \mid y = x^2 \text{ and } x \in R\}$$

or, when the domain is understood to be the set of real numbers, as

$$\{(x, y) \mid y = x^2\}$$

In this chapter, unless stated otherwise *the domain of all functions will be understood to be the set of real numbers except for those real numbers which result in division by zero or taking an even power of a negative number.* Thus the domain of the function

$$f = \left\{(x, y) \mid y = \frac{1}{x}\right\}$$

is the set of all real numbers except the real number 0. Why?

We usually name functions by the use of letters f, g, h, $\ldots$, F,

G, H, . . . For example the function f may be the set of ordered pairs (x, y) for which $y = -x$. We write this

$$f = \{(x, y)\,|\,y = -x\}$$

Frequently if x is an element of the domain of a function f instead of using y to denote the element of the range associated with x we use the symbol $f(x)$. Then each ordered pair of the function f is denoted by $(x, f(x))$. The symbol $f(x)$ is read "the value of f at x" or "f at x." Thus $f(x)$ is the name of an element in the range of the function and is not the product of f and x. Using this notation we may denote the function

$$f = \{(x, y)\,|\,y = -x\}$$

by

$$f = \{(x, f(x))\,|\,f(x) = -x\}$$

We call the second component of an ordered pair belonging to a function a **value** of the function. For example for the function above the value of f at 1, denoted by $f(1)$, is -1; the value of f at 2, denoted by $f(2)$, is -2; the value of f at -4, denoted by $f(-4)$, is 4.

Example 1: Given $f = \{(x, f(x))\,|\,f(x) = x^2 + 1\}$, find (a) $f(1)$; (b) $f(-2)$; (c) $f(-1)$; (d) $f(x + h)$.

Solution: (a) $f(1) = 1^2 + 1 = 2$
(b) $f(-2) = (-2)^2 + 1 = 5$
(c) $f(-1) = (-1)^2 + 1 = 2$
(d) $f(x + h) = (x + h)^2 + 1 = x^2 + 2xh + h^2 + 1$

Example 2: Given $f = \{(x, f(x))\,|\,f(x) = x + 3\}$ and $g = \{(x, g(x))\,|\,g(x) = x^2\}$, find (a) $f(-3)$; (b) $g(f(2))$; (c) $\dfrac{f(3)}{g(4)}$

Solution: (a) $f(-3) = -3 + 3 = 0$
(b) $f(2) = 2 + 3 = 5$
    $g(f(2)) = g(5) = 5^2 = 25$
(c) $f(3) = 3 + 3 = 6$
    $g(4) = 4^2 = 16$
    $\dfrac{f(3)}{g(4)} = \dfrac{6}{16} = \dfrac{3}{8}$

Example 3: State the domain and range of

$$f = \left\{ (x, f(x)) \mid f(x) = \frac{2}{x-3} \right\}$$

Solution: The domain is the set of all real numbers except 3 since when $x = 3$, $f(x) = \frac{2}{3-3} = \frac{2}{0}$ and division by 0 is not defined. The domain of the function is $D = \{x \mid x \in R$ and $x \neq 3\}$. To find the range we solve $f(x) = \frac{2}{x-3}$ for x and seek values of $f(x)$ for which x is real:

$$x = \frac{2}{f(x)} + 3$$

We see that the range is $\{f(x) \mid f(x) \neq 0\}$.

Example 4: State the domain and range of the function

$$f = \{(x, y) \mid y = x^2 - 4\}$$

Solution: Since all real values of x give real values of y the domain of the function is

$$\{x \mid x \in R\}$$

To find the domain we solve $y = x^2 - 4$ for x:

$$x^2 = y + 4$$
$$x = \sqrt{y+4} \qquad \text{or} \qquad x = -\sqrt{y+4}$$

We now seek values of y for which x is a real number. Thus we want the solution set of the inequality

$$y + 4 \geq 0$$

The inequality is equivalent to

$$y \geq -4$$

and the range of the function is

$$\{y \mid y \geq -4\}$$

8.
GRAPHS OF FUNCTIONS

The **graph** of a function is the graph of the set of all ordered pairs belonging to the function.

Example 1: Graph the function $\{(-2, 0), (-1, 1), (0, 2), (1, 3), (2, 4)\}$
  Solution:

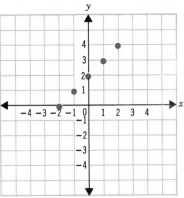

FIGURE 5.20

Example 2: Graph the function $\{(-2, -5), (2, 4), (3, 0), (4, -1)\}$.
  Solution:

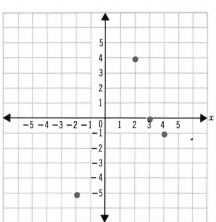

FIGURE 5.21

151

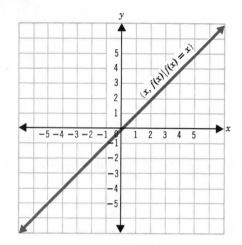

FIGURE 5.22

Notice in each of the examples above the domain was a finite set of real numbers. Now let us consider the function

$$f = \{(x, f(x)) \mid f(x) = x\}$$

The graph of f is the set of all points (x, y) on the Cartesian plane such that x is a real number and $y = f(x)$. Since $f(x) = x$, we see that all ordered pairs of this function are of the form (a, a) where a is a real number. Some ordered pairs of this function are

$$(-3, -3) \; (-2, -2) \; (-1, -1) \; (0, 0) \; (1, 1) \; (2, 2) \; (3, 3)$$

Graphing these ordered pairs we see that the graph of f appears to be a straight line as shown in Figure 5.22.

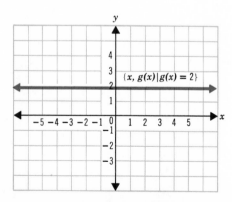

FIGURE 5.23

Example 3:   Graph $g = \{(x, g(x))\,|\,g(x) = 2\}$

    Solution: We see that every ordered pair of this function is of the form $(x, 2)$ where $x \in R$. The graph of g is shown in Figure 5.23.

A function such as g is called a **constant function.** All ordered pairs of a constant function are of the form (x, k) where $x \in R$ and k is a fixed real number.

Exercise 4

1. Given $f = \{(x, f(x))\,|\,f(x) = x^3\}$, find
   - (a) $f(-1)$
   - (b) $f(0)$
   - (c) $f(-3)$
   - (d) $f(2)$
   - (e) $f(4)$
   - (f) $f(-2)$

2. Given $f = \left\{(x, f(x))\,|\,f(x) = \dfrac{x+3}{x^2} \text{ and } x \neq 0\right\}$, find

   - (a) $f(1)$
   - (b) $f(-1)$
   - (c) $f(-3)$
   - (d) $f(2x)$
   - (e) $f(h)$
   - (f) $f(h + x)$

3. Given $f = \{(x, f(x))\,|\,f(x) = \tfrac{1}{2}x + 1\}$ and $h = \{(x, h(x))\,|\,h(x) = x^2 + 5\}$ find
   - (a) $f(h(1))$
   - (b) $h(f(6))$
   - (c) $\dfrac{f(h(4))}{h(f(2))}$
   - (d) $f(3) + h(-1)$
   - (e) $h(-2) + f(-4)$
   - (f) $\dfrac{f(0)}{f(h(-3))}$

4. Given $t = \{(x, t(x))\,|\,t(x) = x^2 + 2x + 1\}$ and $s = \{(x, s(x))\,|\,s(x) = 2x - 3\}$ find
   - (a) $t(-1)$
   - (b) $t(3) + s(5)$
   - (c) $\dfrac{t(0)}{s(0)}$
   - (d) $s(t(1)) + t(s(1))$
   - (e) $s(2x) + t(x)$
   - (f) $t(x + h)$

5. Let $f = \{(x, f(x))\,|\,f(x) = \tfrac{1}{x} \text{ and } x \neq 0\}$. Find the element of the domain so that $f(x) = \tfrac{2}{3}$.

6. Let $h = \{(x, h(x))\,|\,h(x) = x^2 - 1\}$. Find all the elements of the domain so that the value of the function is
   - (a) 8
   - (b) 0
   - (c) 15
   - (d) 1
   - (e) 4
   - (f) 28

7. Let $t = \{(x, t(x))\,|\,t(x) = x^2 - 2x + 3\}$, find
   - (a) $t(x + 3)$
   - (b) $t(x - 2)$
   - (c) $t(x + h)$
   - (d) $\dfrac{t(x + h) - t(x)}{h}$

8. What elements of the domain, the set of real numbers, must be excluded in order to eliminate division by zero?

(a) $s = \left\{(x, s(x)) \mid s(x) = \dfrac{1}{x}\right\}$

(b) $g = \left\{(x, g(x)) \mid g(x) = \dfrac{4}{x+1}\right\}$

(c) $h = \left\{(x, h(x)) \mid h(x) = \dfrac{3}{x-5}\right\}$

(d) $f = \left\{(x, f(x)) \mid f(x) = \dfrac{3x}{x^2 - 9}\right\}$

Graph the following functions.

9. $\{(-5, 3), (-3, 7), (-1, -2), (0, 3), (1, 2), (3, -2)\}$

10. $\{(-3, 2), (-4, 1), (-1, 5), (1, 5), (2, -4), (3, 3)\}$

11. $\{(-4, 2), (-3, \frac{1}{2}), (-2, 0), (-1, 3), (0, -2), (1, 4), (2, -5)\}$

12. $\{(x, f(x)) \mid f(x) = 2x\}$

13. $\{(x, f(x)) \mid f(x) = -x\}$

14. $\{(x, f(x)) \mid f(x) = -3\}$

15. $\{(x, f(x)) \mid f(x) = 4\}$

9.
LINEAR FUNCTIONS

Any function of the form

$$f = \{(x, y) \mid y = mx + b \text{ and } m \neq 0\}$$

is a **linear function.** Examples of linear functions are

$$f = \{(x, y) \mid y = 2x + 3\}$$
$$g = \{(x, y) \mid y = \tfrac{1}{2}x - 2\}$$
$$h = \{(x, y) \mid y = x + \tfrac{3}{4}\}$$

The graph of a linear function with domain the set R of real numbers is a straight line. Since two points determine a straight line, we need to find only two ordered pairs of a linear function in order to draw its graph. In practice, the two ordered pairs most easily found are those with first or second components zero, that is, ordered pairs of the form $(x, 0)$ and $(0, y)$.

Replacing y by 0 in $y = mx + b$ gives $0 = mx + b$ and $x = -\frac{b}{m}$. Since $m \neq 0$, $(-\frac{b}{m}, 0)$ is an ordered pair of the function. Similarly, if x is replaced by 0 in $y = mx + b$, we have $y = 0 + b = b$ and $(0, b)$ is another ordered pair of this function. Since the points with $(0, b)$ and $(-\frac{b}{m}, 0)$ as coordinates are points where the graph crosses the y-axis and x-axis respectively, they are easily located on the Cartesian plane. The number $-\frac{b}{m}$ is called the **x-intercept** of the graph and the number b is called the **y-intercept.**

Example 1: Given $f = \{(x, y) | y = 3x - 4\}$, find the x- and y-intercepts of the function and graph.

Solution: Here $m = 3$ and $b = -4$, hence the x-intercept is $\frac{4}{3}$ and the y-intercept is -4. The graph of the function is shown in Figure 5.24.

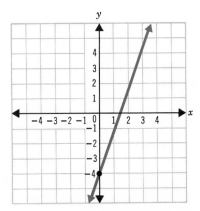

FIGURE 5.24

It may happen that the graph of a linear function intercepts the axes at the origin. In this case, the intercepts are not different numbers and hence do not represent different points. When this happens, another ordered pair of the function must be determined in order to draw the line that is the graph of the function.

Example 2: Given $g = \{(x, y) | y = 2x\}$, find the x- and y-intercepts and graph the function.

Solution: Here $m = 2$ and $b = 0$. The x-intercept is $\frac{0}{2} = 0$. The y-intercept is also 0. We see then that the graph of the function passes through the origin. In order to draw the graph we must determine another ordered pair of the function. If

155

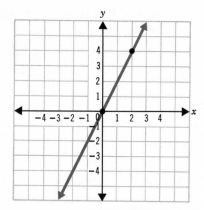

FIGURE 5.25

$x = 2$, then $y = 2 \cdot 2 = 4$, so $(2, 4)$ is an ordered pair of the function. The graph of the function is shown in Figure 5.25.

We now define the **slope** of the graph of a linear function. The slope of the graph of the linear function

$$f = \{(x, y) | y = mx + b\}$$

is defined to be m. The slope of a line is the ratio of the rise to the run of any segment of the line. (Figure 5.26) Notice, if P_1 and P_2 are two distinct points on the line and have coordinates (x_1, y_1) and (x_2, y_2) respectively, then the ratio of the rise to the run of segment $\overline{P_1 P_2}$, that

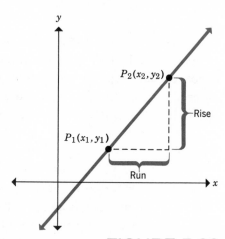

FIGURE 5.26

is the slope of the function whose graph is the line in the figure, is

$$m = \frac{y_2 - y_1}{x_2 - x_1}, \; x_2 \neq x_1$$

Example 3: Given $g = \{(x, y) \mid y = \frac{1}{3}x + 6\}$, find the slope of the graph of the function.

Solution: The slope is $\frac{1}{3}$.

Example 4: Given $(4, -1)$ and $(-2, -3)$, the coordinates of two points. What is the slope of the line determined by these two points?

Solution: Using the formula for the slope of a line with $(x_1, y_1) = (4, -1)$ and $(x_2, y_2) = (-2, -3)$ we have

$$m = \frac{-3 - (-1)}{-2 - 4}$$

$$= \frac{-3 + 1}{-2 - 4}$$

$$= \frac{-2}{-6}$$

$$= \frac{1}{3}$$

Exercise 5

Graph the following linear functions. Give the x- and y-intercepts of each graph (Exercises 1–10).

  1. $\{(x, y) \mid y = 4x + 8\}$
  2. $\{(x, y) \mid y = 2x - 4\}$
  3. $\{(x, y) \mid y = \frac{1}{3}x + 6\}$
  4. $\{(x, y) \mid y = \frac{1}{2}x - 1\}$
  5. $\{(x, y) \mid y = -3x + 6\}$
  6. $\{(x, y) \mid y = -\frac{3}{4}x + 9\}$
  7. $\{(x, y) \mid y = -\frac{1}{5}x - 2\}$
  8. $\{(x, y) \mid y = 4x\}$
  9. $\{(x, y) \mid y = -2x\}$
  10. $\{(x, y) \mid y = \frac{3}{2}x - 2\}$

Give the slope of the graph of each of the following (Exercises 11–20).

  11. $\{(x, y) \mid y = 3x - 4\}$
  12. $\{(x, y) \mid y = 5x - 10\}$
  13. $\{(x, y) \mid y = \frac{3}{5}x - 3\}$

14. $\{(x, y)\,|\,y = -3x + 9\}$
15. $\{(x, y)\,|\,y = -\frac{2}{3}x\}$
16. $\{(x, y)\,|\,y = -\frac{5}{8}x - 4\}$
17. $\{(x, y)\,|\,y = -3x + 3\}$
18. $\{(x, y)\,|\,y = 7x + \frac{1}{2}\}$
19. $\{(x, y)\,|\,y = \frac{5}{2}x + \frac{3}{4}\}$
20. $\{(x, y)\,|\,y = -3x + \frac{7}{16}\}$

Give the slope of the line determined by the points whose coordinates are given.

21. $(3, -2); (2, 4)$
22. $(-1, 4); (-2, 2)$
23. $(0, 3); (3, -2)$
24. $(1, 6); (-1, 2)$
25. $(\frac{1}{2}, -\frac{3}{2}); (\frac{3}{4}, \frac{1}{4})$

10.
THE QUADRATIC FUNCTION

The function

$$f = \{(x, y)\,|\,y = ax^2 + bx + c\}$$

with a, b, and c real numbers and $a \neq 0$ is called a **quadratic function.**
Examples of some quadratic functions are

$$f = \{(x, y)\,|\,y = x^2\}$$
$$g = \{(x, y)\,|\,y = 2x^2 + 3x\}$$
$$h = \{(x, y)\,|\,y = 3x^2 - 4x + 5\}$$
$$k = \{(x, y)\,|\,y = -x^2 + 4\}$$

Now let us graph the quadratic function

$$f = \{(x, y)\,|\,y = x^2\}$$

The graph of this function consists of all points (x, y) in the Cartesian plane whose coordinates are ordered pairs of the function, that is, all ordered pairs that satisfy the equation $y = x^2$. Some ordered pairs that belong to this quadratic function are

$$(-3, 9), (-2, 4), (-1, 1), (0, 0), (1, 1), (2, 4), (3, 9)$$

These ordered pairs are plotted in Figure 5.27. These points obviously do not lie on a straight line. If we were to obtain many more ordered pairs of this function and graph them, we could see that each point would lie on the graph in Figure 5.28. This curve is called a **parabola.** It extends indefinitely upward to the left and right as indicated by the arrows in the figure.

We can obtain a great deal of information about this quadratic function by inspecting its graph. Notice that the graph has a lowest point called the **vertex** of the parabola. It is symmetric about a vertical line through the vertex. At the vertex the function has its **minimum value.** The y coordinate of the vertex is the minimum value of the function. Since the parabola opens upward we say that it is **concave upward.** From the graph we see that the vertex of our parabola is (0, 0) and the minimum value of the quadratic function is 0.

It can be proved, although we shall accept it without proof, that the graph of any quadratic function

$$f = \{(x, y) \mid y = ax^2 + bx + c\}$$

where a, b, and c are real numbers and $a \neq 0$ is a **parabola** and that the parabola opens upward if $a > 0$ and opens downward if $a < 0$. If the

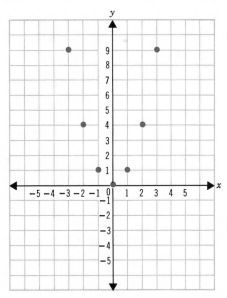

FIGURE 5.27

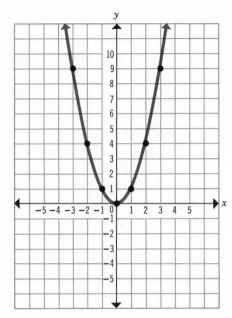

FIGURE 5.28

parabola opens upward it is said to be **concave upward;** if it opens downward it is said to be **concave downward.** If a parabola opens upward, it has a lowest point called the **vertex** of the parabola. The function has its **minimum value** at this point. If a parabola opens downward it has a highest point called the **vertex** of the parabola. The function has its **maximum value** at this point.

Example 1:   Graph the quadratic function

$$f = \{(x, y) \mid y = x^2 - 2x - 3\}$$

Solution: We find some ordered pairs belonging to this function by assigning values to x in the equation $y = x^2 - 2x + 8$ and finding the values of y associated with the assigned values of x. Some ordered pairs of this function are

$$(-2, 5), (-1, 0), (0, -3), (1, -4), (2, -3), (3, 0), (4, 5)$$

Plotting these points and drawing a smooth curve connecting them, we have the graph in Figure 5.29. The vertex of the parabola is (1, −4). The parabola opens upward. The point at which the function has its minimum value is (1, −4). The minimum value of the function is −4.

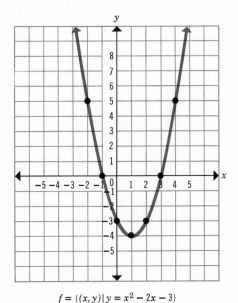

$$f = \{(x, y)\,|\,y = x^2 - 2x - 3\}$$

FIGURE 5.29

Example 2: Graph the quadratic function

$$f = \{(x, y)\,|\,y = -x^2 + 2x - 1\}$$

Solution: Since $a = -1$, which is less than zero, the graph of this function is a parabola that is concave downward. Some ordered pairs of the function are

$$(-2, -9),\ (-1, -4),\ (0, -1),\ (1, 0),\ (2, -1),$$
$$(3, -4),\ (4, -9)$$

Plotting these points and drawing a smooth curve connecting them, we have the graph in Figure 5.30. The vertex of the parabola is (1, 0). The parabola is concave downward. The point at which the function has its maximum value is (1, 0). The maximum value of the function is 0.

161

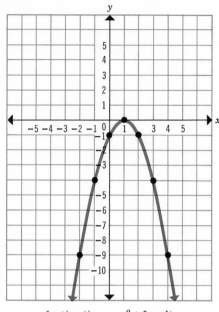

$$f = \{(x, y) \mid y = -x^2 + 2x - 1\}$$

FIGURE 5.30

Exercise 6

Graph the following quadratic functions (Exercises 1–9).

1. $f = \{(x, y) \mid y = x^2 - 4x + 1\}$
2. $f = \{(x, y) \mid y = x^2 + 6x + 8\}$
3. $f = \{(x, y) \mid y = x^2 + x + 1\}$
4. $f = \{(x, y) \mid y = x^2 - 2x + 4\}$
5. $f = \{(x, y) \mid y = -4x^2\}$
6. $f = \{(x, y) \mid y = x^2 + 2\}$
7. $f = \{(x, y) \mid y = x^2 - 2x + 3\}$
8. $f = \{(x, y) \mid y = -x^2 - 1\}$
9. $f = \{(x, y) \mid y = -x^2 + 1\}$
10. Identify the following as linear or quadratic functions.
    (a) $f = \{(x, y) \mid y = 2x + 3\}$
    (b) $f = \{(x, y) \mid y = x^2 + x + 2\}$
    (c) $f = \{(x, y) \mid y = -x + 6\}$
    (d) $f = \{(x, y) \mid y = -x^2 - 3x + 4\}$
    (e) $f = \{(x, y) \mid y = 3 + x^2\}$

11. Tell whether the graphs of the following quadratic functions are concave upward or concave downward.

   (a) $f = \{(x, y) \mid y = x^2 + 3\}$
   (b) $f = \{(x, y) \mid y = -2x^2 + 3x + 6\}$
   (c) $f = \{(x, y) \mid y = 8 - x^2\}$
   (d) $f = \{(x, y) \mid y = 2x^2 + 5x + 9\}$
   (e) $f = \{(x, y) \mid y = 3x^2 + 6x + 8\}$

12. Tell whether the following quadratic functions have a maximum value or a minimum value.

   (a) $f = \{(x, y) \mid y = 4x^2\}$
   (b) $f = \{(x, y) \mid y = -x^2\}$
   (c) $f = \{(x, y) \mid y = 4x^2 - 3x + 8\}$
   (d) $f = \{(x, y) \mid y = 9 + x^2\}$
   (e) $f = \{(x, y) \mid y = 9 - x^2\}$

13. What is the graph of a quadratic function called?

14. If a quadratic function has a maximum value does its graph open upward or downward?

15. If a quadratic function has a minimum value does its graph open upward or downward?

16. If the graph of the quadratic function $\{(x, y) \mid y = ax^2 + bx + c\}$ is concave upward is $a > 0$ or is $a < 0$?

Emmy Noether (1882–1935) was the most significant creative abstract algebraist of our own times. Fortunately she
lived at a time when there was no objection to women pursuing a higher education in mathematics, and she received
her doctorate degree from the University of Erlangen where
her father was a professor of mathematics. In 1916 she
moved to Göttingen where she worked at the university
with the famous mathematician David Hilbert (1862–1943).
Unfortunately, tradition prevented her from obtaining a
professorship at the university because of her sex. However, she gave lectures announced under Hilbert's name
and these lectures gave evidence of her great creative
power in abstract algebra.

In 1933, the Nazis expelled Miss Noether from Germany
because she was Jewish. She then came to the United
States and taught at Bryn Mawr College until her death in
1935.

GROUPS
AND
FIELDS

6

ABSTRACT
MATHEMATICAL SYSTEMS

A **mathematical system** is any nonempty set S of elements, $S = \{a, b, c, \ldots\}$, together with one or more binary operations defined on the elements of the set, and a set of **axioms** or **postulates.**

A **binary operation** on a set S is a rule that associates with each ordered pair (a, b) of elements of S a uniquely defined element of S. Ordinary addition and multiplication performed on the set of integers are binary operations: to each ordered pair of integers a and b are associated unique integers called, respectively, the **sum** of a and b, denoted by $a + b$, and the **product** of a and b, denoted by ab, $a \cdot b$ or $a \times b$. For example, if the operation is addition:

$$(2, 3) \text{ is associated with } 2 + 3 = 5$$
$$(3, 6) \text{ is associated with } 3 + 6 = 9$$
$$(6, 3) \text{ is associated with } 6 + 3 = 9$$

If the operation is multiplication:

$$(2, 3) \text{ is associated with } 2 \times 3 = 6$$
$$(3, 6) \text{ is associated with } 3 \times 6 = 18$$
$$(6, 3) \text{ is associated with } 6 \times 3 = 18$$

Example 1: A binary operation $*$ on the set of whole numbers is described as associating with each ordered pair (a, b) of whole numbers the whole number found by multiplying the first component of the ordered pair by 2 and adding the second component. Thus with (a, b) is associated the whole number $2a + b$. Find the whole number associated with the ordered pair $(3, 5)$ by the operation $*$.

Solution: $3 * 5 = (2 \cdot 3) + 5 = 6 + 5 = 11.$

Example 2: A binary operation $*$ on the set of integers is described as associating with each ordered pair (a, b) of integers the integer found by multiplying the first component, a, of the ordered pair by 3 and subtracting twice the second component. Thus with (a, b) is associated $3a - 2b$. Find the integer associated with the ordered pair $(-2, 5)$ by the operation $*$. That is, find $-2 * 5$.

Solution: $-2 * 5 = (3)(-2) - (2)(5) = -6 - 10 = -16.$

In the study of any mathematical system, the concept of relation must be included. Elements of a set are usually thought of in relation to one another. For example, if we consider two integers a and b, they may be "related" in any one of the following ways:

$a = b$ (*a* equals *b*)
$a < b$ (*a* is less than *b*)
$a \mid b$ (*a* divides *b*)
$a \equiv b \pmod{m}$ (*a* is congruent to *b* modulo *m*)
$a \geq b$ (*a* is greater than or equal to *b*)

Each of the symbols, $=$, $<$, $\mid$, $\equiv$, and $\geq$, denotes a verb phrase that expresses a certain binary relation between the integers a and b.

The above relations are mathematical relations. We are certainly familiar with many nonmathematical relations. For example, "is the sister of" is a binary relation between women; "has the same horsepower as" is a binary relation between automobiles; and "has a longer life span than" is a binary relation between animals.

We shall use the symbol R to denote any relation. A relation R is a **binary relation** on a given set S of elements if given two elements a and b of S, *in that order,* a is related to b, denoted by aRb (read: a is related to b) or a is not related to b, denoted by $a\not{R}b$ (read: a is not related to b).

Especially important in mathematics are relations, R, on a set S that satisfy the following properties.

1. REFLEXIVE PROPERTY: aRa for all elements a of S.

2. SYMMETRIC PROPERTY: If aRb, then bRa for all elements a and b of S.

3. TRANSITIVE PROPERTY: If aRb and bRc, then aRc for all elements a, b, and c of S.

Relations which satisfy the reflexive, symmetric, and transitive properties are called **equivalence relations.** The most familiar example in mathematics of an equivalence relation is the ordinary equality relation, denoted by $=$. Let us consider the set I of integers. Certainly $a = a$ for every integer. If a and b are any two integers and $a = b$, then $b = a$. For any integers a, b, and c, if $a = b$ and $b = c$ then $a = c$.

Example 3: Is the binary relation "lives in the same city as" on the set of all people an equivalence relation?

Solution: aRa: a lives in the same city as a.

If aRb, then bRa: If a lives in the same city as b, then b lives in the same city as a.

If aRb and bRc, then aRc: If a lives in the same city as b and b lives in the same city as c, then a lives in the same city as c.

The relation is an equivalence relation.

Example 4: Is the binary relation "sells at a lower price than" on the set of all glamour stocks an equivalence relation?

Solution: $a\cancel{R}a$: a does not sell at a lower price than a (itself).

If aRb, then $b\cancel{R}a$: If a sells at a lower price than b, then b does not sell at a lower price than a.

If aRb and bRc, then aRc: If a sells at a lower price than b and b sells at a lower price than c, then a sells at a lower price than c.

The relation is not an equivalence relation since it is neither reflexive nor symmetric.

In the study of mathematical systems we shall be especially interested in the equality relation. In addition to being an equivalence relation and satisfying the reflexive, symmetric, and transitive properties, the equality relation also satisfies the following properties.

E-1. **(Substitution Property):** *For all a and b, if $a = b$, then in any statement involving a, b may be substituted for a without changing the truth or falsity of the statement.*

E-2. **(Addition Property):** *For all a, b, and c, if $a = b$, then $a + c = b + c$ and $c + a = c + b$.*

E-3. **(Multiplication Property):** *For all a, b, and c, if $a = b$, then $ac = bc$ and $ca = cb$.*

It is quite easy to prove the addition (E-2) and multiplication (E-3) properties of the equality relation using the substitution property, but we shall accept these two properties without proof.

Example 5: What properties of the equality relation are illustrated by the following? (The variables represent real numbers.)
(a) If $a + 2 = 7$, then $a = 5$
(b) If $x = 6$ and $y + x = 9$, then $y + 6 = 9$
(c) If $6n = 24$, then $n = 4$

Solution: (a) Addition property; adding -2 to each member of $a + 2 = 7$.
(b) Substitution property; substituting 6 for x in $y + x = 9$.
(c) Multiplication property; multiplying each member of $6n = 24$ by $\frac{1}{6}$.

Exercise 1

1. Give the unique positive integer associated with each ordered pair below if the operation is (1) addition, (2) multiplication.
(a) (116, 349)   (c) (241, 39)
(b) (74, 968)   (d) (87, 56)

2. Which of the following relations are equivalence relations for the given sets?
(a) "Is similar to" for the set of all triangles.
(b) "Is the brother of" for the set of all men.
(c) "Has the same initials as" for the set of all people.

3. Which of the following relations are equivalence relations for the given sets?
(a) "Is less than" for the set of whole numbers.
(b) "Is the sister of" for the set of all people.
(c) "Has the same number of children as" for the set of all married couples.

4. Tell which of the three properties of an equivalence relation is satisfied by each of the relations described in problem 2.

5. Tell which of the three properties of an equivalence relation are satisfied by each of the relations described in problem 3.

6. Which of the adjectives "reflexive," "symmetric," "transitive" is applicable to the following relations?
(a) "Is the father of" for the set of all people.
(b) "Is East of" for cities in the United States.
(c) "Lives within three miles of" for the set of all people.
(d) "Is the wife of" for the set of all people.

7. A binary operation * on the set of integers is defined as associating with each ordered pair (a, b) of integers the integer found by multiplying the first component a of the ordered pair by 5 and adding the second component. Thus with (a, b) is associated the integer $5a + b$. Find the integers associated with the following ordered pairs by the binary operation $*$.

(a) $(3, 6)$ (d) $(-2, -4)$
(b) $(-1, 6)$ (e) $(4, -3)$
(c) $(0, -7)$ (f) $(-2, 0)$

8. A binary operation $*$ on the set of integers is defined as associating with each ordered pair (a, b) of integers the remainder when the product of the two components of the ordered pair is divided by three. Thus $5 * 3 = 0$ since $5 \times 3 = 15$ and 15 divided by 3 gives a quotient of 5 and a remainder of 0. Find

(a) $3 * 6$ (d) $(-5) * (-6)$
(b) $4 * 8$ (e) $(-7) * 3$
(c) $8 * 4$ (f) $0 * (-9)$

9. Justify each statement below by a property of the equality relation on the set of real numbers.

(a) If $a = 7$, then $7 = a$.
(b) If $x + 9 = 27$, then $x = 18$.
(c) If $x + y = k$ and $k = 36$, then $x + y = 36$.
(d) If $25a = 625$, then $a = 125$.

10. Determine which ones of the three properties "reflexive", "symmetric", and "transitive" apply to the following relations on the set of integers.

(a) "Is less than."
(b) "Is greater than."
(c) "Divides."
(d) "Is a multiple of."

2.
MATHEMATICAL FIELDS

One of the most familiar mathematical systems is a field. A **field,** is a set, F, of elements, $\{a, b, c, d, \ldots\}$, together with two binary operations called **addition** and **multiplication,** denoted respectively by $+$ and $\times$,† that obey the following postulates or axioms.

F-1. **(Commutative Property of Addition):** *If a and b are in F, then*

$$a + b = b + a$$

† Multiplication is also denoted by a raised dot and placing two numbers adjacent to each other.

F-2. **(Associative Property of Addition):** *If a, b, and c are in F, then*

$$(a + b) + c = a + (b + c)$$

F-3. **(Additive Identity Axiom):** *There exists in F an element, 0, called the* **additive identity,** *such that for all a in F,*

$$a + 0 = 0 + a = a$$

F-4. **(Additive Inverse Axiom):** *For every element a in F, there exists an element of F denoted by* $-a$ *and called the* **additive inverse** *or* **opposite** *of a, such that*

$$a + (-a) = (-a) + a = 0$$

F-5. **(Commutative Property of Multiplication):** *If a and b are in F, then*

$$ab = ba$$

F-6. **(Associative Property of Multiplication):** *If a, b, and c are in F, then*

$$(ab)c = a(bc)$$

F-7. **(Multiplicative Identity Axiom):** *There exists in F an element, 1, called the* **multiplicative identity,** *such that for all a in F,*

$$a \cdot 1 = 1 \cdot a = a$$

F-8. **(Multiplicative Inverse Axiom):** *For every element a $\neq$ 0 in F, there exists an element denoted by a^{-1} in F, called the* **multiplicative inverse** *of F, such that*

$$a \cdot a^{-1} = a^{-1} \cdot a = 1$$

F-9. **(Distributive Property):** *If a, b, and c are elements in F, then*

$$a(b + c) = ab + ac$$

These postulates fall into three classes:

1. Four postulates describing the behavior of the elements under addition.

2. Four postulates describing the behavior of the elements under multiplication.

3. One postulate connecting addition and multiplication.

Example 1: Consider the set, $W = \{0, 1, 2, 3, \ldots\}$, of whole numbers and the operations of addition and multiplication. Check the field postulates for this system. State the properties, if any, that do not hold for the whole number system. Is the whole number system a field?

Solution: The whole number system is not a field since postulates F-4 and F-8 are not satisfied.

Example 2: Consider the set, $I = \{\ldots -2, -1, 0, 1, 2, \ldots\}$, of integers and the operations of addition and multiplication. Check the field postulates for this system. State the properties, if any, that do not hold for the system of integers. Is the system of integers a field?

Solution: The system of integers is not a field since postulate F-8 is not satisfied. With the exception of $+1$ and -1, no integers have multiplicative inverses.

Example 3: Consider the set of rational numbers. A **rational number** is a number that may be represented as the quotient of two integers a and b, denoted by $\frac{a}{b}$, $b \neq 0$. Two rational numbers $\frac{a}{b}$ and $\frac{c}{d}$ are defined to be equal, denoted by $\frac{a}{b} = \frac{c}{d}$, if and only if $ad = bc$. We define addition and multiplication of rational numbers in the following manner:

$$\frac{a}{b} + \frac{c}{d} = \frac{ad + bc}{bd} \qquad \frac{a}{b} \cdot \frac{c}{d} = \frac{ac}{bd}$$

Is the set of rational numbers together with addition and multiplication defined above a field?

Solution: All of the field postulates can easily be checked to show that the rational number system is a field, called the **rational number field.** We shall check the commutative property of addition. We compute

$$\frac{a}{b} + \frac{c}{d} = \frac{ad + bc}{bd}$$

and

$$\frac{c}{d} + \frac{a}{b} = \frac{cb + da}{db}$$

These two expressions are equal since the commutative property holds for addition and multiplication of integers:†

$$\frac{ad + bc}{bd} = \frac{da + cb}{db}$$ Commutative property of multiplication of integers

$$= \frac{cb + da}{db}$$ Commutative property of addition of integers

$$= \frac{c}{d} + \frac{a}{b}$$ Definition of addition of rational numbers

Now we check that every rational number except zero has a multiplicative inverse. We note that the multiplicative inverse (**reciprocal**) of $\frac{a}{b}$, $a \neq 0$, is $\frac{b}{a}$, since

$$\frac{a}{b} \cdot \frac{b}{a} = \frac{b}{a} \cdot \frac{a}{b} = \frac{1}{1}$$

The reader should check the other field postulates.

Example 4:   Let us consider a system consisting of the set S of two distinct elements, even and odd:

$$S = \{\text{even, odd}\}$$

and two binary operations, addition, denoted by $+$, and multiplication, denoted by $\times$, defined as follows:

odd + odd = even	odd × odd = odd
odd + even = odd	odd × even = even
even + odd = odd	even × odd = even
even + even = even	even × even = even

Is this system a field?

Solution: We can more easily verify that this system is a field if we make an operation table (Table 6.1) showing the operations defined above.

Examining the operation tables we see that

$$\text{even} + \text{odd} = \text{odd} + \text{even}$$

† See Chapter 2, Section 1.

173

TABLE 6.1
Operation tables for even-odd

+	even	odd		×	even	odd
even	even	odd		even	even	even
odd	odd	even		odd	even	odd

and

$$\text{even} \times \text{odd} = \text{odd} \times \text{even}$$

hence, both operations are commutative. The identity for addition is even; the identity for multiplication is odd. Each element is its own additive inverse. The multiplicative inverse of odd is odd (even is the additive identity and hence is not required to have a multiplicative inverse.) To check the associative property of addition, we must replace each of a, b, and c, in turn by even and odd in

$$(a + b) + c = a + (b + c)$$

and see whether in each case a true statement results. We see that in the Even-Odd system there are $2 \times 2 \times 2 = 2^3 = 8$ cases to test. If we check each of the eight cases, we find that $(a + b) + c = a + (b + c)$ is a true statement and addition is associative.

In a similar manner we check to see whether or not the associative property of multiplication and the distributive property are satisfied. In each case we find that true statements result and multiplication is associative and the distributive property holds. Since all of the field postulates are satisfied, this system is a field.

3.
FIELD THEOREMS

We can prove some familiar theorems for the elements of a field using the field postulates. Most of us are familiar with these theorems from our study of algebra in which we were introduced to the field of real numbers.

THEOREM 6.1: *If a is an element of a field F, then $a \cdot 0 = 0$.*

    Proof:

$$a \cdot 0 = a \cdot 0 + 0 \qquad \text{Additive identity axiom}$$
$$0 + 0 = 0 \qquad \text{Additive identity axiom}$$
$$a(0 + 0) = a \cdot 0 + 0 \qquad \text{Substitution property of equality}$$
$$\text{(substituting } 0 + 0 \text{ for } 0)$$
$$a \cdot 0 + a \cdot 0 = a \cdot 0 + 0 \qquad \text{Distributive property}$$

Since every element of F has an additive inverse, $a \cdot 0$ has an additive inverse $-(a \cdot 0)$. Then

$$-(a \cdot 0) + [a \cdot 0 + a \cdot 0] = -(a \cdot 0) + [a \cdot 0 + 0] \qquad \text{Addition property of equality}$$
$$[-(a \cdot 0) + a \cdot 0] + a \cdot 0 = [-(a \cdot 0) + a \cdot 0] + 0 \qquad \text{Associative property of addition}$$
$$0 + a \cdot 0 = 0 + 0 \qquad \text{Additive inverse axiom}$$
$$a \cdot 0 = 0 \qquad \text{Additive identity axiom}$$

Theorem 6.1 is sometimes called the **multiplication property of zero.**

THEOREM 6.2: *If a is an element of a field F, then $(-1)a = -a$.*

    Proof:

$$a + (-a) = 0 \qquad \text{Additive inverse axiom}$$
$$= a \cdot 0 \qquad \text{Theorem 6.1}$$
$$= 0 \cdot a \qquad \text{Commutative property of multiplication}$$
$$= [1 + (-1)]a \qquad \text{Substitution property of equality (substituting } 1 + (-1) \text{ for } 0) \text{ and the additive inverse axiom}$$
$$= 1 \cdot a + (-1) \cdot a \qquad \text{Distributive property}$$

We now have

$$a + (-a) = 1 \cdot a + (-1) \cdot a$$

Adding $-a$ to each member of the above equation we obtain

$$-a + [a + (-a)] = -a + [1 \cdot a + (-1)a] \qquad \text{Addition property of}$$
equality
$$(-a + a) + (-a) = (-a + 1 \cdot a) + (-1)a \qquad \text{Associative property of}$$
addition
$$(-a + a) + (-a) = (-a + a) + (-1)a \qquad \text{Multiplicative identity}$$
axiom
$$0 + (-a) = 0 + (-1)a \qquad \text{Additive inverse axiom}$$
$$-a = (-1)a \qquad \text{Additive identity axiom}$$

THEOREM 6.3: *For all elements a and b of a field F, if $ab = 0$, then $a = 0$ or $b = 0$.*

Proof: Either $a = 0$ or $a \neq 0$. If $a = 0$, then the theorem is true. If $a \neq 0$, then it has a multiplicative inverse a^{-1} in F by the multiplicative inverse axiom. Then

$ab = 0$	Given
$a^{-1}(ab) = a^{-1} \cdot 0$	Multiplication property of equality
$(a^{-1} \cdot a)b = a^{-1} \cdot 0$	Associative property of multiplication
$(a^{-1} \cdot a)b = 0$	Theorem 6.1
$1 \cdot b = 0$	Multiplicative inverse axiom
$b = 0$	Multiplicative identity axiom

We see from the above that if $ab = 0$, then $a = 0$ or $b = 0$.

In the field of real numbers, for example, if the product of two factors is 0, then Theorem 6.3 assures us that one or the other of them is 0. For example, if x represents a real number and $(x + 2)(x - 4) = 0$, by Theorem 6.3 we know that $x + 2 = 0$ or $x - 4 = 0$.

THEOREM 6.4: *If x is an element of a field F, then $-(-x) = x$.*

Proof:
$$-(-x) + (-x) = 0$$
$$x + (-x) = 0$$
$$-(-x) + (-x) = x + (-x)$$
$$[-(-x) + (-x)] + x = [x + (-x)] + x$$
$$-(-x) + [(-x) + x] = x + [(-x) + x]$$
$$-(-x) + 0 = x + 0$$
$$-(-x) = x$$

The reasons for the steps in Theorem 6.4 have been omitted. The reader should verify each step in the proof with a field axiom, a theorem, or a property of the equality relation.

Theorem 6.4 tells us that the opposite of the opposite of an element of a field F is the element. Thus in the field of real numbers $-(-3) = 3$.

Exercise 2

1. In the set of rational numbers what is the multiplicative inverse of each of the following?

(a) $\dfrac{3}{5}$

(b) $\dfrac{3}{4}$

(c) $\dfrac{7}{-6}$

(d) $\dfrac{-6}{7}$

(e) $\dfrac{-7}{15}$

(f) $\dfrac{19}{12}$

2. In the set of rational numbers what is the additive inverse of each of the following?

(a) $\dfrac{1}{2}$

(b) $\dfrac{3}{4}$

(c) $-\dfrac{1}{4}$

(d) $-\dfrac{2}{3}$

(e) $-\dfrac{1}{2}$

(f) $\dfrac{7}{9}$

3. State whether or not the given sets together with the binary operations of addition and multiplication defined on the elements of the set are fields. If the systems are not a field, state at least one field postulate that is not satisfied.
   (a) The set of all positive rational numbers.
   (b) The set of all even positive integers.
   (c) The set of all odd positive integers.
   (d) The set of all negative integers.
   (e) The set of all real numbers.
   (f) $\{0, 1, 2, -2, -1\}$
   (g) The set of all integral multiples of 5.
   (h) The set of all integral multiples of 3.

4. Verify that multiplication of rational numbers is commutative.

5. Verify that addition of rational numbers is associative.

6. Verify that $\frac{0}{b}$, $b \neq 0$, is the additive identity for the set of rational numbers.

7. Verify that multiplication of rational numbers is associative.

8. Verify that $\frac{b}{b}$, $b \neq 0$, is the multiplicative identity for the set of rational numbers.

9. Using Table 6.1, verify the following.
   (a) even + (odd + odd) = (even + odd) + odd
   (b) even × (odd + odd) = (even × odd) + (even × odd)

10. Verify that even is the additive identity for the even-odd field.

11. Verify that odd is the multiplicative identity for the even-odd field.

12. Consider the set I of integers and addition (mod 7) and multiplication (mod 7). Check the field postulates for this system. State the properties, if there are any, that do not hold for the system. Is this system a field?

13. Consider the set I of integers and addition (mod 8) and multiplication (mod 8). Check the field postulates for this system. State the properties, if there are any, that do not hold for the system. Is this system a field?

14. Consider the set I of integers and addition (mod 12) and multiplication (mod 12). Check the field postulates for this system. State the properties, if there are any, that do not hold for the system. Is this system a field?

15. For t, y, and z elements of a field, prove that if $ty = tz$ and $t \neq 0$, then $y = z$.

16. For t, y, and z elements of a field, prove that if $t + y = t + z$, then $y = z$.

17. For t and y are elements of a field, prove $t(-y) = -(ty)$.

18. Prove that there is only one additive identity in a field. (*Hint:* Assume that there are two and prove that they are equal.)

19. Prove that there is only one multiplicative identity in a field. (*Hint:* Assume there are two and prove that they are equal.)

20. We define the quotient of a and b, denoted by $\frac{a}{b}$, a and b elements of a field and $b \neq 0$ as $\frac{a}{b} = a(b^{-1})$. Prove that division, except by zero is always possible in a field. That is, if S is a field, S is closed with respect to division except by zero.

21. Prove that if x and y are elements of a field $(-x)(-y) = xy$.

4.
A FINITE FIELD

Let us consider the modulo-seven system discussed in Chapter 4, Section 1. This system with the binary operations of addition (mod 7) and multiplication (mod 7) is a field which we shall call F_7. For convenience the addition (mod 7) and multiplication (mod 7) tables are given in Table 6.2.

TABLE 6.2
Operation tables for F_7

+	0	1	2	3	4	5	6
0	0	1	2	3	4	5	6
1	1	2	3	4	5	6	0
2	2	3	4	5	6	0	1
3	3	4	5	6	0	1	2
4	4	5	6	0	1	2	3
5	5	6	0	1	2	3	4
6	6	0	1	2	3	4	5

×	0	1	2	3	4	5	6
0	0	0	0	0	0	0	0
1	0	1	2	3	4	5	6
2	0	2	4	6	1	3	5
3	0	3	6	2	5	1	4
4	0	4	1	5	2	6	3
5	0	5	3	1	6	4	2
6	0	6	5	4	3	2	1

That this system satisfies the eleven field axioms was demonstrated in Chapter 4, Section 1. This field is an example of a finite field. A **finite field** is a field with a finite number of elements in it.

In stating the field axioms, the additive inverse of an element a was denoted by the symbol $-a$. We must be careful not to interpret this symbol as "negative a". In the field of real numbers, the additive inverse of a positive real number is indeed the negative of that real number. For example, the additive inverse of $\sqrt{2}$ is $-\sqrt{2}$. We must keep in mind that the additive inverse of an element a of a field is that element which when added to a gives the sum 0, the additive identity. We see in F_7 that:

The additive inverse of 0 is 0 which we may denote by -0
The additive inverse of 1 is 6 which we may denote by -1
The additive inverse of 2 is 5 which we may denote by -2
The additive inverse of 3 is 4 which we may denote by -3
The additive inverse of 4 is 3 which we may denote by -4
The additive inverse of 5 is 2 which we may denote by -5
The additive inverse of 6 is 1 which we may denote by -6

We see from the above, that although there are no negative numbers in F_7, each element has an additive inverse that we denote by the symbol $-a$ (read the opposite of a). Thus since the additive inverse of 2 is 5 we may write

$$-2 = 5 \ (\text{mod } 7)$$

All of the theorems proved in Section 3 about fields in general are true for the finite field F_7. Theorem 6.1 certainly is true since for any element

a of F_7, $a \cdot 0 = 0 \pmod 7$ (this can be checked by observing Table 6.2b).

Theorem 6.2 states that the product of the additive inverse of 1 and any element a of F_7 is the additive inverse of a. Since there are only a finite number of elements in F_7, we can verify the truth of this theorem by replacing a in $(-1)a = -a$ by each of the elements of F_7 in turn and observing that in each case we get a true statement:

$$(-1) \cdot 0 = 6 \cdot 0 = 0 = -0$$
$$(-1) \cdot 1 = 6 \cdot 1 = 6 = -1$$
$$(-1) \cdot 2 = 6 \cdot 2 = 5 = -2$$
$$(-1) \cdot 3 = 6 \cdot 3 = 4 = -3$$
$$(-1) \cdot 4 = 6 \cdot 4 = 3 = -4$$
$$(-1) \cdot 5 = 6 \cdot 5 = 2 = -5$$
$$(-1) \cdot 6 = 6 \cdot 6 = 1 = -6$$

Now let us check to assure ourselves that Theorem 6.4 is true for F_7. Theorem 6.4 states that the additive inverse of the additive inverse of any element of F_7 is that element. Replacing x in $-(-x) = x$ by each of the elements of F_7 in turn we obtain:

$$-(-0) = -0 = 0$$
$$-(-1) = -6 = 1$$
$$-(-2) = -5 = 2$$
$$-(-3) = -4 = 3$$
$$-(-4) = -3 = 4$$
$$-(-5) = -2 = 5$$
$$-(-6) = -1 = 6$$

We see that in each case a true statement results and Theorem 6.4 is satisfied.

We are familiar with the operations of subtraction and division in the real number field. We now define these operations in the F_7 field. If a and b are elements of F_7, their **difference** denoted by $a - b \pmod 7$ is defined as the sum of a and the additive inverse of b. The operation of finding the difference is called **subtraction (mod 7)**. Then

$$a - b = a + (-b) \text{ (mod 7)}$$

Example 1: Perform the indicated operations:

$$(3 - 5) + (2 - 6) \pmod 7$$

Solution: $(3 - 5) + (2 - 6) = [3 + (-5)] + [2 + (-6)] \,(\text{mod } 7)$
$$= (3 + 2) + (2 + 1) \,(\text{mod } 7)$$
$$= 5 + 3 \,(\text{mod } 7)$$
$$= 1 \,(\text{mod } 7)$$

As in the field of real numbers when we define division in the F_7 field we exclude division by zero, the additive identity. If a and b are elements of F_7 and $b \neq 0$, their **quotient** denoted by $\frac{a}{b}$ (mod 7) is defined to be the product of a and the multiplicative inverse of b. The operation of finding the quotient is called **division (mod 7).** Then for $b \neq 0$,

$$\frac{a}{b} = a(b^{-1}) \,(\textbf{mod } \textbf{7})$$

Example 2: Perform the indicated operations:

$$\frac{3}{2} \cdot \frac{4}{5} \,(\text{mod } 7)$$

Solution: $\dfrac{3}{2} \cdot \dfrac{4}{5} = 3(2^{-1}) + 4(5^{-1}) \,(\text{mod } 7)$

$$= (3 \cdot 4) + (4 \cdot 3) \,(\text{mod } 7)$$
$$= 5 + 5 \,(\text{mod } 7)$$
$$= 3 \,(\text{mod } 7)$$

Exercise 3

1. Name the multiplicative inverse of the following elements of F_7.
   (a) 2   (d) 1
   (b) 5   (e) 6
   (c) 4   (f) 3

2. Perform the indicated operations in F_7.
   (a) $3 + 2 + 4 \,(\text{mod } 7)$ (e) $4(3 + 2) - (5 + 3) \,(\text{mod } 7)$
   (b) $(3 - 4) + (5 - 2) \,(\text{mod } 7)$ (f) $(6 - 3) \cdot (4 + 6) \,(\text{mod } 7)$
   (c) $(5 + 3) \cdot (4 + 6) \,(\text{mod } 7)$ (g) $(6 \cdot 3) - (4 \cdot 2) \,(\text{mod } 7)$
   (d) $(6 \cdot 3) - (4 \cdot 5) \,(\text{mod } 7)$ (h) $6(4 - 3 + 5) \,(\text{mod } 7)$

3. Perform the indicated operations in F_7.

   (a) $\dfrac{4}{3} + \dfrac{2}{4} \,(\text{mod } 7)$

   (b) $\left(\dfrac{1}{6} \cdot \dfrac{3}{4}\right) - \left(\dfrac{4}{3} \cdot \dfrac{2}{5}\right) \,(\text{mod } 7)$

   (c) $\dfrac{3(4 - 5)}{2(5 + 4)} \,(\text{mod } 7)$

   (d) $\dfrac{2(3 + 4)}{5(3 - 2)} \,(\text{mod } 7)$

   (e) $\dfrac{\frac{3}{4} + \frac{2}{3}}{\frac{1}{5} - \frac{1}{6}} \,(\text{mod } 7)$

   (f) $\dfrac{\frac{4}{3} - \frac{3}{4}}{\frac{5}{2} + \frac{3}{4}} \,(\text{mod } 7)$

4. Verify that in the field F_7, $(-3) \cdot (-4) = 3 \cdot 4$.

5. Verify that in the field F_7, $6 \cdot (-3) = -(6 \cdot 3)$.

6. Verify that in the field F_7, $\dfrac{1}{\frac{3}{4}} = \dfrac{4}{3}$.

7. Verify that in the field F_7, $\left(\dfrac{2}{6}\right)^{-1} = \dfrac{6}{2}$.

8. Verify that in the field F_7, $\dfrac{3}{2} + \dfrac{4}{3} = \dfrac{3 \cdot 3 + 4 \cdot 2}{3 \cdot 2}$.

9. Verify that in the field F_7, $\dfrac{2}{5} - \dfrac{4}{2} = \dfrac{2 \cdot 2 - 5 \cdot 4}{5 \cdot 2}$.

10. Verify that in the field F_7, $(-5) \cdot (-2) = 5 \cdot 2$.

11. Verify that in the field F_7, $\dfrac{\frac{3}{4}}{\frac{4}{5}} = 4$.

12. Consider the modulo-five system (see Chapter 4, Exercise 1, problems 7 and 8). Check the field postulates for this system. State the properties, if any, that do not hold for this system. Is this system a field?

13. Consider the modulo-eight system (see Chapter 4, Exercise 1, problem 10). Check the field postulates for this system. State the properties if any, that do not hold for this system. Is this system a field?

14. In the field F_5 name (1) the additive inverse and (2) the multiplicative inverse of each of the following elements.

(a) 3

(b) 2

(c) 4

(d) 1

15. In the field F_5 subtraction and division are defined in the same way as we defined these operations in the field F_7. Give a definition for subtraction (mod 5) and for division (mod 5) and compute the following.

(a) $3 - 2 \pmod 5$

(b) $\dfrac{4}{3} \pmod 5$

(c) $\dfrac{2}{3} - \dfrac{4}{2} \pmod 5$

(d) $3\left(\dfrac{4}{3} + \dfrac{2}{4}\right) \pmod 5$

(e) $\left(\dfrac{1}{2} \cdot \dfrac{3}{4}\right) - \left(\dfrac{2}{3} + \dfrac{3}{2}\right) \pmod 5$

(f) $\dfrac{\frac{3}{4} + \frac{2}{3}}{\frac{4}{3} + \frac{1}{2}} \pmod 5$

5.
FINDING SOLUTION SETS OF LINEAR EQUATIONS OVER F_7

An equation of the form

(1) $$ax + b = c$$

where a, b, and c are elements of the field F_7 and $a \neq 0$ is called a **linear equation in one variable over $\mathbf{F_7}$**. When the variable x in Equation 1 is replaced by an element of F_7 the resulting equation is either a true or a false statement. If the resulting equation is true for a given replacement for x, that replacement is called a **solution** of the equation. The set of all solutions is called the **solution set** of the equation over F_7.

Let us consider the equation over F_7

(2) $$3x + 4 = 2$$

Since F_7 has a finite number of elements, one way to find the solution set is to replace x, in turn, by every element of the field:

$$3 \cdot 0 + 4 = 4 \neq 2$$
$$3 \cdot 1 + 4 = 0 \neq 2$$
$$3 \cdot 2 + 4 = 3 \neq 2$$
$$3 \cdot 3 + 4 = 6 \neq 2$$
$$3 \cdot 4 + 4 = 2$$
$$3 \cdot 5 + 4 = 5 \neq 2$$
$$3 \cdot 6 + 4 = 1 \neq 2$$

The solution set is $\{4\}$.

Example 1: Find the solution set over F_7 of $5x = 3$.

Solution: We can find the solution set of the given equation by examing Table 6.2*b*. Since $5 \cdot 2 = 3$, the solution set of the given equation is $\{2\}$.

We can prove that linear equations in one variable over the field F_7 have one and only one solution.

183

THEOREM 6.5: *The equation $ax + b = c$ where a, b, and c are elements of the field F_7 and $a \neq 0$, has one and only one solution over F_7, namely $a^{-1}[c + (-b)]$.*

Proof: Since b is an element of F_7, its additive inverse, $-b$, is an element of F. Adding $-b$ to each member of $ax + b = c$ gives

$$(ax + b) + (-b) = c + (-b) \qquad \text{Addition property of equality}$$
$$ax + [b + (-b)] = c + (-b) \qquad \text{Associative property of addition}$$
$$ax + 0 = c + (-b) \qquad \text{Additive inverse axiom}$$
$$ax = c + (-b) \qquad \text{Additive identity axiom}$$

Since $a \neq 0$, it has a multiplicative inverse a^{-1}. Multiplying each member of $ax = c + (-b)$ by a^{-1} we obtain

$$a^{-1}(ax) = a^{-1}[c + (-b)] \qquad \text{Multiplication property of equality}$$
$$(a^{-1} \cdot a)x = a^{-1}[c + (-b)] \qquad \text{Associative property of multiplication}$$
$$1 \cdot x = a^{-1}[c + (-b)] \qquad \text{Multiplicative inverse axiom}$$
$$x = a^{-1}[c + (-b)] \qquad \text{Multiplicative identity axiom}$$

We have shown that $a^{-1}[c + (-b)]$ is a solution of the given equation. To prove that there is one and only one solution let us suppose that there are two solutions, s_1 and s_2. Then

$$as_1 + b = c$$
$$as_2 + b = c$$

Applying the substitution property of equality we obtain

$$as_1 + b = as_2 + b$$

Adding the additive inverse, $-b$, of b to each member of the above equation we obtain

$$(as_1 + b) + (-b) = (as_2 + b) + (-b)$$
$$as_1 + [b + (-b)] = as_2 + [b + (-b)] \qquad \text{Why?}$$
$$as_1 = as_2 \qquad \text{Why?}$$

Since $a \neq 0$, it has a multiplicative inverse a^{-1}. Multi-

5. SOLUTION SETS OF LINEAR EQUATIONS

plying each member of $as_1 = as_2$ by a^{-1} we obtain

$$a^{-1}(as_1) = a^{-1}(as_2)$$
$$(a^{-1} \cdot a)s_1 = (a^{-1} \cdot a)s_2 \qquad \text{Why?}$$
$$1 \cdot s_1 = 1 \cdot s_2 \qquad \text{Why?}$$
$$s_1 = s_2 \qquad \text{Why?}$$

We see then that the two solutions s_1 and s_2 are indeed the same. We therefore conclude that the equation $ax + b = c$ has one and only one solution over F_7, namely $a^{-1}[c + (-b)]$. The solution set of $ax + b = c$ over F_7 is $\{a^{-1}[c + (-b)]\}$.

Example 2: Find the solution set over F_7 of $5x + 3 = 4$.

Solution: The given equation is a linear equation in one variable, with $a = 5$, $b = 3$ and $c = 4$. The multiplicative inverse of 5 is 3 since $5 \cdot 3 = 1$; the additive inverse of 3 is 4 since $3 + 4 = 0$. Applying Theorem 6.5 we find that the solution of the given equation is

$$5^{-1}[4 + (-3)] = 3[4 + 4] = 3 \cdot 1 = 3$$

The solution set is $\{3\}$.

Example 3: Find the solution over F_7 of $3x - 5 = 2$.

Solution: This equation is not in the form $ax + b = c$, called the **standard form** of the equation. Before applying Theorem 6.5 we put the given equation in standard form:

$$3x - 5 = 3x + (-5) \qquad \text{Definition of subtraction}$$
$$= 3x + 2 \qquad \text{Additive inverse of 5 is 2}$$

Then the standard form of the equation is $3x + 2 = 2$. Here $a = 3$, $b = 2$, and $c = 2$. Applying Theorem 6.5 we find the solution is

$$3^{-1}[2 + (-2)] = 5(2 + 5)$$
$$= 5 \cdot 0 = 0$$

The solution set is $\{0\}$.

1. Find the solution sets over the field F_7.

(a) $2x = 5$

(b) $3x = 1$

(c) $5x + 2 = 6$

(d) $3x - 1 = 5$

(e) $x + 6 = 6$

(f) $6x - 4 = 1$

2. Find the solution sets over the field F_7.

(a) $x - 6 = 4$

(b) $3x - 2 = 1$

(c) $4 = 5x + 6$

(d) $3 - 2x = 4$

(e) $1 - 5x = 0$

(f) $2x - 6 = 3$

3. In F_7 we define a^n, where a is an element of F_7 and n is a positive integer, as the product of n factors each of which is a. Thus $3^4 = 3 \cdot 3 \cdot 3 \cdot 3 = 4$. Compute the following.

(a) 2^3

(b) 6^5

(c) 4^4

(d) 5^6

4. Compute the following. (See problem 3 above)

(a) 2^6

(b) 3^6

(c) 1^6

(d) 4^6

(e) 5^6

(f) 6^6

5. Observing the results found in problem 4, make a conjecture as to the value of a^6 where a is an element of the field F_7 and $a \neq 0$.

6. Prove the following theorem: The equation $ax + b = c$ where a, b, and c are elements of the field F_5 and $a \neq 0$ has one and only one solution, namely $a^{-1}[c + (-b)]$.

7. Find the solution sets over the field F_5.

(a) $3x = 4$

(b) $2x + 3 = 2$

(c) $4x - 1 = 3$

(d) $2x - 3 = 1$

(e) $4x + 3 = 1$

(f) $2x - 4 = 1$

8. Find the solution sets over the field F_5.

(a) $x - 3 = 4$

(b) $2x - 4 = 1$

(c) $3 - 3x = 4$

(d) $3x - 4 = 2$

(e) $4 - 2x = 3$

(f) $1 - 3x = 2$

9. It can be proved that any modular system whose modulus is a prime is a field. Theorem 6.5 holds true for any modular system that is a field. Demonstrate that Theorem 6.5 does not hold for the modulo-eight system by showing that

(a) $2x + 1 = 6 \pmod 8$ has no solutions

(b) $2x + 3 = 1 \pmod 8$ has two solutions.

6.
POLYNOMIALS
OVER THE FIELD F_7

An algebraic expression of the form

$$a_0x^n + a_1x^{n-1} + \cdots + a_{n-1}x + a_n$$

where $a_0, a_1, \ldots, a_{n-1}, a_n$ are elements of a field F, $a \neq 0$, and n a nonnegative integer is called a **polynomial of degree n over the field F.** A polynomial of degree n over the finite field F_7 is an algebraic expression of the above form where $a_0, a_1, \ldots, a_n$ are element of F_7 and $a_0 \neq 0$. Some examples of polynomials over F_7 are

(1) $\qquad\qquad\qquad 3x^3 + 2x^2 + 5x + 3$
(2) $\qquad\qquad\qquad 6x^6 - 1$
(3) $\qquad\qquad\qquad x^5 + 2x^3 - 3x + 4$
(4) $\qquad\qquad\qquad 4$

Polynomial (1) is of degree 3; polynomial (2) is of degree 6; polynomial (3) is of degree 5; and polynomial (4) is of degree 0.

Polynomials are usually represented by the symbol $P(x)$. When the variable x in $P(x)$ over a field F is replaced by an element of the field, the polynomial represents an element of the field called the **value** of the polynomial. For example, when x is replaced by 2 in polynomial (1) above we have

$$3 \cdot 2^3 + 2 \cdot 2^2 + 5 \cdot 2 + 3 = 3 \cdot 1 + 2 \cdot 4 + 3 + 3$$
$$= 3 + 1 + 3 + 3 = 3$$

The value of $P(x) = 3x^3 + 2x^2 + 5x + 3$ over F_7 when $x = 2$ is 3. We write $P(2) = 3$. We read this symbol "the value of P at 2 equals 3".

We see then that the value of $P(x) = 6x^6 - 1$ over F_7 when $x = 3$ is

$$P(3) = 6 \cdot 3^6 - 1$$
$$= 6 \cdot 1 - 1$$
$$= 6 - 1$$
$$= 6 + 6$$
$$= 5$$

The value or values of x which make $P(x) = 0$ are called the **zeros** of the polynomial $P(x)$.

Example 1: If $P(x) = x^2 - 3x + 2$ is a polynomial over F_7, find (a) $P(0)$; (b) $P(1)$; and (c) $P(4)$.

Solution: (a) $P(0) = 0^2 - 3 \cdot 0 + 2 = 2$

(b) $P(1) = 1^2 - 3 \cdot 1 + 2$
$= 1 + 4 \cdot 1 + 2 = 0$

(c) $P(4) = 4^2 - 3 \cdot 4 + 2$
$= 2 + 4 \cdot 4 + 2$
$= 2 + 2 + 2$
$= 6$

Since F_7 is a finite field we can find the zeros of any polynomial over F_7 by replacing the variable in turn by 0, 1, 2, 3, 4, 5, and 6.

Example 2: $P(x) = x^4 + 2x^2 + 1$ is a polynomial over F_7. Find the zeros, if any, of $P(x)$.

Solution: $P(0) = 0 + 2 \cdot 0 + 1 = 1$
$P(1) = 1^4 + 2 \cdot 1^2 + 1 = 4$
$P(2) = 2^4 + 2 \cdot 2^2 + 1 = 4$
$P(3) = 3^4 + 2 \cdot 3^2 + 1 = 2$
$P(4) = 4^4 + 2 \cdot 4^2 + 1 = 2$
$P(5) = 5^4 + 2 \cdot 5^2 + 1 = 4$
$P(6) = 6^4 + 2 \cdot 6^2 + 1 = 4$
$P(x)$ has no zeros.

Example 3: $P(x) = x^6 - 1$ is a polynomial over F_7. Find the zeros of $P(x)$.

Solution: $P(0) = 0^6 - 1 = 6$
$P(1) = 1^6 - 1 = 0$
$P(2) = 2^6 - 1 = 0$
$P(3) = 3^6 - 1 = 0$
$P(4) = 4^6 - 1 = 0$
$P(5) = 5^6 - 1 = 0$
$P(6) = 6^6 - 1 = 0$
$P(x)$ has six zeros; 1, 2, 3, 4, 5, and 6.

The equation that results from equating a polynomial over a field F to zero is called a **polynomial equation** over F. With each polynomial $P(x)$ over F there is associated a polynomial equation $P(x) = 0$.

We see that the values of x that are zeros of $P(x)$ are **solutions** of the polynomial equation $P(x) = 0$. The set of all the solutions of $P(x) = 0$ is called the **solution set** of the equation.

Example 4: Find the solution set over F_7 of $P(x) = x^6 - 1$.

Solution: Since the zeros of $P(x)$ over F_7 are 1, 2, 3, 4, 5, and 6, (see Example 3) the solution set over F_7 of $P(x) = 0$ is $\{1, 2, 3, 4, 5, 6\}$.

Exercise 5

1. Given the polynomial over F_7, $P(x) = x^6 + 3x + 2$, find
   (a) $P(1)$ (d) $P(5)$
   (b) $P(2)$ (e) $P(3)$
   (c) $P(0)$ (f) $P(4)$
2. Given the polynomial over F_7, $P(x) = x^5 + 3x^4 + 2x^3 + x + 1$, find
   (a) $P(2)$ (c) $P(5)$
   (b) $P(-3)$ (d) $P(-6)$
3. Find the zeros over F_7 of $P(x) = 4x^3 + 3x$.
4. Find the zeros over F_7 of $P(x) = 3x^6 - 2x^2 + 3x + 4$.

Find the solution sets over F_7:

5. $x^2 + 3 = 0$
6. $x^2 = 0$
7. $x^2 + 6 = 0$
8. $x^2 + 5x + 5 = 0$
9. $2x^2 + 5x + 5 = 0$
10. $3x^2 + 2x + 5 = 0$
11. $x^2 + 5x + 4 = 0$
12. $2x^2 + 5x + 4 = 0$
13. $3x^2 + 2x - 5 = 0$
14. $x^2 - 3x + 4 = 0$
15. $x^6 - 3 = 0$
16. $x^4 + 3x^3 + 2x^2 + x + 5 = 0$
17. $x^5 + x^4 + x^3 + x^2 + x + 1 = 0$
18. $x^3 + x^2 - x - 1 = 0$
19. $4x^3 + 2x^2 + 3 = 0$
20. $x^9 + 6 = 0$

7.
GRAPHS OF POLYNOMIAL FUNCTIONS OVER F_7

The coefficients of a polynomial over F_7 are elements of F_7. For every element $x \in F_7$, there is associated one and only one element $y \in F_7$ where $y = P(x)$. We see then that the set of ordered pairs (x, y) where $x \in F_7$ and $y = P(x)$ is a function over F_7. This function is called a **polynomial function over F_7**. The domain of a polynomial function over F_7 is the set F_7; the range of this function is a subset of F_7. The graph of this polynomial function over F_7 is called the **graph** of the polynomial $P(x)$.

We recall that the graph of a function is the graph of the set of all the ordered pairs belonging to the function. Let us consider the polynomial over F_7,

$$P(x) = 2x^2 + x + 3$$

Setting $P(x) = y$, we have

$$y = 2x^2 + x + 3$$

We would like to graph this polynomial over F_7. That is, we would like to graph the polynomial function over F_7

$$\{(x, y) \mid x \in F_7 \text{ and } y = 2x^2 + x + 3\}$$

The domain of this function is $F_7 = \{0, 1, 2, 3, 4, 5, 6\}$. The set of ordered pairs belonging to the function are

$$\{(0, 3), (1, 6), (2, 6), (3, 3), (4, 4), (5, 2), (6, 4)\}$$

We find this set of ordered pairs by replacing the variable x in $y = 2x^2 + x + 3$ by 0, 1, 2, 3, 4, 5, and 6 in turn and finding the values of y associated with each of these values of x. We now graph this set of ordered pairs on the Cartesian plane. The graph is shown in Figure 6.1.

Example 1: Graph the polynomial function over F_7 $\{(x, y) \mid x \in F_7$ and $y = 3x + 2\}$.

Solution: The graph is shown in Figure 6.2.

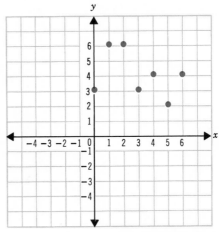

FIGURE 6.1

Example 2: Graph the polynomial over F_7, $P(x) = x^3 + 2x - 1$.

Solution: We are going to graph the polynomial function

$$\{(x, y) \mid x \in F_7 \text{ and } y = x^3 + 2x - 1\}$$

The graph is shown in Figure 6.3.

Observe that since F_7 has a finite number of elements, $F_7 \times F_7$ also has a finite number of elements. In fact, $F_7 \times F_7$ consists of $7^2 = 49$ ordered pairs. The graph of $F_7 \times F_7$ consists of 49 points of the Cartesian plane. Since every function on F_7 is a set of ordered pairs that is a subset

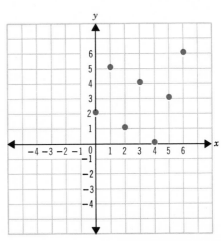

FIGURE 6.2

191

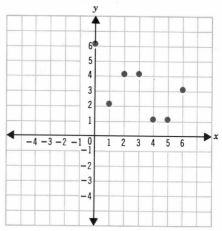

FIGURE 6.3

of $F_7 \times F_7$, we see that the graph of any function on F_7 consists of a finite number of points. This is clearly shown in the Examples above.

Exercise 6

Graph the following polynomials over F_7.
   1. $P(x) = x + 2$
   2. $P(x) = x^2 + 3x + 1$
   3. $P(x) = x^3$
   4. $P(x) = 2x^2 - 2x - 3$
   5. $P(x) = x^5 - x$
   6. $P(x) = x^3 + x^2 + x + 1$
   7. $P(x) = 5x^6 - 1$
   8. $P(x) = 3x^3 - 2x^2 + 5x + 3$
   9. $P(x) = 5x^4 - x^2 + 4$
   10. $P(x) = x^6 - x^5 + x^4 - x^3 + x^2 - x + 1$

8.
GROUPS

A less familiar abstract mathematical system than a field is a group. A **group,** G, is a set of elements, $\{a, b, c, \ldots\}$, with one binary operation denoted by the symbol "$*$" satisfying the following postulates:

G-1. **(Associative Property):**   *If a, b, and c are elements of G, then*

$$(a * b) * c = a * (b * c)$$

G-2. **(Identity Axiom):**   *There exists an element e of G, called the* **identity element** *such that for all elements a of G*

$$a * e = e * a = a$$

G-3. **(Inverse Axiom):**   *For each element a of G there is an element of G denoted by a^{-1} called the* **inverse** *of a such that*

$$a * a^{-1} = a^{-1} * a = e$$

where e is the identity element.

If, in addition to the above postulates, a group satisfies the following postulate, then we say that the group is a **commutative** or an **Abelian group.**

G-4. **(Commutative):**   *If a and b are elements of G, then*

$$a * b = b * a$$

Recall that the set of integers does not satisfy all the postulates of a field under the operations of addition and multiplication. This set, however, does satisfy all the postulates for a group with the operation $*$ interpreted as addition.

Postulate G-1 demands that the associative property of addition holds for the set of integers. From previous experience we know this is true.

G-2 is satisfied by using the number 0 for e.

Postulate G-3 is satisfied. Given an integer, its opposite is its inverse with respect to addition.

Since all the group postulates are satisfied with the operation $*$ interpreted as addition, we say that the set of integers is a group under addition or the set of integers is a group with respect to addition.

Since addition of integers is commutative we say that the set of integers with the operation of addition is a commutative or Abelian group.

Some other examples of groups are given below.

1. The set of all rational numbers different from zero and the operation of multiplication. Zero must be omitted from the set since it has no multiplicative inverse.
2. The set of real numbers and the operation of addition.

3. The set of even integers and the operation of addition.
4. The set $F_7 = \{0, 1, 2, 3, 4, 5, 6\}$ and addition (mod 7). Postulate G-2 is certainly satisfied. Postulate G-3 is satisfied taking $e = 0$. Postulate G-4 is satisfied since

$$0 + 0 = 0 \,(\text{mod } 7)$$
$$1 + 6 = 0 \,(\text{mod } 7)$$
$$3 + 4 = 0 \,(\text{mod } 7)$$

Since addition (mod 7) is commutative, this is a commutative group.

We comment here that if a set is a field then it is certainly a group with respect to addition. It is not a group with respect to multiplication because the additive identity has no multiplicative inverse.

Knowing the definition of a group, we can give an alternate definition of a field. A **field** is a set F together with two binary operations, addition and multiplication, satisfying the following conditions:

(1) F is a commutative group under addition.
(2) F, with the additive identity deleted, is a commutative group under multiplication.
(3) For all a, b, and c of F, $a(b + c) = ab + ac$.

Exercise 7

1. Is the set of positive integers a group with respect to addition? Why?
2. Is the set of integers a group with respect to multiplication? Why?
3. Which of the following sets are groups with respect to the given operation?
   (a) The set of nonnegative integers; addition.
   (b) The set of positive integers; multiplication.
   (c) The set of odd integers; addition.
   (d) The set of whole numbers; multiplication.
   (e) $\{0, 1, 2, 3\}$; addition (mod 4)
   (f) The set of rational numbers; addition
   (g) $\{1, 2, 3, 4, 5, 6\}$; multiplication (mod 7)
   (h) $\{0, 1, 2, 3\}$; multiplication (mod 4)

4. Given set $S = \{1, 2, 3, 4\}$ and the binary operation $*$ on the elements of S defined in the table below.

$*$	1	2	3	4
1	1	2	3	4
2	2	4	1	3
3	3	1	4	2
4	4	3	2	1

(a) Is this set a group with respect to the operation $*$?

(b) Does every element have an inverse? Name the inverse of each element if it exists.

5. Show that the set $\{1, 3, 5, 7\}$ is a group with respect to multiplication (mod 8).

6. If a set of elements is a field, is it a group under addition? Why?

7. Prove that the set of all integral multiples of a fixed integer m

$$\{\ldots, -2m, -m, 0, m, 2m, \ldots\}$$

is a commutative group with respect to the operation of addition.

8. If a set of elements is a field is it a group with respect to multiplication if the element 0 is deleted from the set? Why?

9. Given the set $\{t, -t, \frac{1}{t}, -\frac{1}{t}\}$ and the operation $*$ on the elements of the set where $*$ is defined as substituting the second factor for t in the first factor. For example:

$$t * \frac{1}{t} = \frac{1}{t}$$

$$\frac{1}{t} * -t = \frac{1}{-t}$$

Is this set a group under the operation $*$?

10. Name the inverse element of every element of the set in problem 9.

9.
THE GROUPS OF
SYMMETRIES OF A SQUARE

Everyone is familiar with the idea of symmetry. We shall now introduce an algebra of symmetry. Imagine a square cut from a piece of wood as shown in Figure 6.4. The corners of the square are numbered

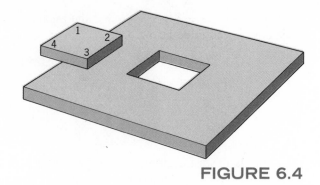

FIGURE 6.4

as shown in the figure. The numerals are marked on both sides of the square; the numeral 1 is marked on the back of the square directly under 1 on the front; 2 on the back is directly under 2 on the front, and so forth.

The square is lifted from the wood and then replaced in as many different ways as possible. We find that there are eight different ways of replacing the square in the wood. These eight different ways of replacing the square, called the **symmetries** of a square, are shown below. We name these symmetries as indicated. In the diagrams the figures on the left show the original position of the square; the ones on the right show the new position of the square when it is replaced.

We define R_0 to be the symmetry that rotates the square 360 degrees in a clockwise direction. Thus

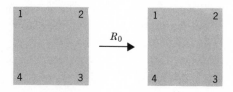

R_1 means to rotate the square 90 degrees in a clockwise direction. Thus

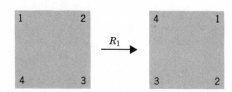

9. THE GROUPS OF SYMMETRIES OF A SQUARE

R_2 means to rotate the square 180 degrees in a clockwise direction. Thus

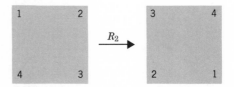

R_3 means to rotate the square 270 degrees in a clockwise direction. Thus

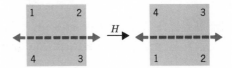

H means to rotate the square around its horizontal axis. Thus

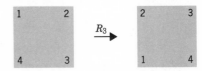

V means to rotate the square around its vertical axis. Thus

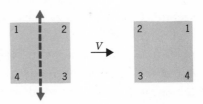

D means to rotate the square about the diagonal from the top right corner to the bottom left corner. Thus corners numbered 2 and 4 remain fixed and corners numbered 1 and 3 change places:

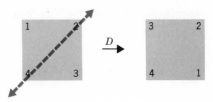

D' means to rotate the square about the diagonal from the top left corner to the bottom right corner. Thus

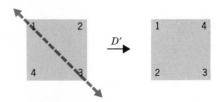

Let us consider the set of symmetries of the square:

$$\{R_0,\ R_1,\ R_2,\ R_3,\ H,\ V,\ D,\ D'\}$$

and define a binary operation denoted by $*$ on the elements of this set.

We define the binary operation $*$ as follows: $H * V$ means with the square in its original position perform H, and then on the result perform V. Thus

$$H * V = R_2$$

In $H * V = R_2$, we call H and V **factors** and R_2 the **product.**

$R_1 * R_2$ means to perform R_1, and then on the result perform R_2. Thus

$$R_1 * R_2 = R_3$$

We must be careful of the order here. $H * V$ does not mean the same as $V * H$; $H * V$ means do H first, V second. $V * H$ means do V first, H second. Care is needed because the end results may not be the same.

An operation table now may be made to exhibit all the products. This operation table is shown in Table 6.3. The elements in the left column of the table represent the first factor in the product; those in the top row, the second factor. (A good way to perform these operations is to use a cardboard square region and do the actual movements with it. Be sure that the numerals in the corners are on both sides.)

We now verify that this system is a group. Table 6.3 exhibits the presence of an element, R_0, which has no effect on the other elements. Hence R_0 is the identity. That the associative property holds can be verified by examining all the possible cases.† To illustrate this verification let us examine a few cases.

† There are 512 cases required to prove the associative property.

TABLE 6.3
Operation table for the symmetries of a square

		Second factor						
$*$	R_0	R_1	R_2	R_3	H	V	D	D'
R_0	R_0	R_1	R_2	R_3	H	V	D	D'
R_1	R_1	R_2	R_3	R_0	D	D'	V	H
R_2	R_2	R_3	R_0	R_1	V	H	D'	D
R_3	R_3	R_0	R_1	R_2	D'	D	H	V
H	H	D'	V	D	R_0	R_2	R_3	R_1
V	V	D	H	D'	R_2	R_0	R_1	R_3
D	D	H	D'	V	R_1	R_3	R_0	R_2
D'	D'	V	D	H	R_3	R_1	R_2	R_0

First factor (row labels)

(1) $(R_1 * R_2) * H = R_3 * H = D'$

$R_1 * (R_2 * H) = R_1 * V = D'$

$\therefore (R_1 * R_2) * H = R_1 * (R_2 + H)$

(2) $(V * D) * D' = R_1 * D' = H$

$V * (D * D') = V * R_2 = H$

$\therefore (V * D) * D' = V * (D * D')$

(3) $(R_3 * V) * H = D * H = R_1$

$R_3 * (V * H) = R_3 * R_2 = R_1$

$\therefore (R_3 * V) * H = R_3 * (V * H)$

We must now show that every element has an inverse. This also can be found from the table. The inverse of H is H because $H * H = R_0$ (the identity element); the inverse of V is V since $V * V = R_0$; the inverse of D is D because $D * D = R_0$; the inverse of D' is D' because $D' * D' = R_0$; the inverse of R_1 is R_3 because $R_1 * R_3 = R_0$; the inverse of R_2 is R_2 because $R_2 * R_2 = R_0$; the inverse of R_3 is R_1 because $R_3 * R_1 = R_0$.

This group of the symmetries of a square is not a commutative group as can be seen since, in particular,

$$H * D \neq D * H$$

Exercise 8

1. From Table 6.3 compute:
   (a) $R_1 * (H * R_3)$
   (b) $(V * R_1) * R_3$
   (c) $(D * D') * D$
   (d) $(H * D') * R_2$
   (e) $(H * V) * (D * R_3)$
   (f) $(H * V) * (D * D')$

2. Make a mathematical system for the symmetries of an isosceles triangle. There will be two symmetries: (1) a turn of 360 degrees in a clockwise direction (call it I), and (2) a turn around the vertical axis (call it V). Make an operation table of this system. Is it a group? If so, is it a commutative group?

3. Construct a mathematical system for the symmetries of an equilateral triangle. Make a triangle and label the vertices as shown in the diagram. There will be six symmetries. There will be three rotations in a clockwise direction: one of 120 degrees, one of 240 degrees, and one of 360 degrees. There will be three rotations about the axes (broken lines in the figure). Make an operation table for this system. Is it a group? If so, is it a commutative group?

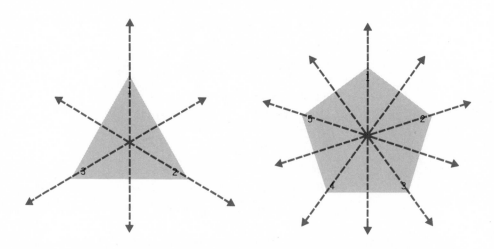

4. Construct a mathematical system for the symmetries of a regular pentagon. Make a pentagon and label it as shown in the diagram. There will be ten symmetries. There will be five rotations about the axes shown by broken lines in the figure, and five rotations in a clockwise direction of 72, 144, 216, 288, and 360 degrees. Make an operation table for this system. Is it a group?

5. How many symmetries are possessed by a regular
   (a) Octagon?
   (b) Decagon (10 sides)?
   (c) Dodecagon (12 sides)?
   (d) Polygon of n sides, $n \geq 3$ a natural number?

10.
SOME THEOREMS
ABOUT GROUPS

We now prove some theorems that are true for any group regardless of the elements or the operation.

THEOREM 6.6: *For all elements a, b, and c, of a group G, if b = c, then a * b = a * c.*

Proof:

$$a * c = a * c \qquad \text{Reflexive property of equality}$$
$$b = c \qquad \text{Given}$$
$$a * c = a * b \qquad \text{Substitution property of equality}$$

THEOREM 6.6a: *For all elements a, b, and c, of a group G, if b = c, then b * a = c * a.*

We prove this theorem in the same manner as we proved Theorem 6.6. The proof is left to the reader.

THEOREM 6.7: *For all elements a, b, and c of a group G, if a * b = a * c, then b = c.*

Proof:

$$a * b = a * c \qquad \text{Given}$$
$$a^{-1} \in G \qquad \text{Inverse axiom of a group}$$
$$a^{-1} * (a * b) = a^{-1} * (a * c) \qquad \text{Theorem 6.6}$$
$$(a^{-1} * a) * b = (a^{-1} * a) * c \qquad \text{Associative property of a group}$$
$$e * b = e * c \qquad \text{Inverse axiom of a group}$$
$$b = c \qquad \text{Identity axiom of a group}$$

THEOREM 6.7a: *For all elements a, b, and c of a group G, if b * a = c * a, then b = c.*

The proof of this theorem is left to the reader.

THEOREM 6.8: *The identity element of a group is unique.*

  Proof: By the identity axiom we know that there is an identity element in G. Let us call this identity element e_1. We wish to show that e_1 is the only identity element in G. Let us assume that there is another identity element, e_2, in G. Then

$$e_1 * e_2 = e_1 \qquad \text{Identity axiom of a group}$$
$$e_1 * e_2 = e_2 \qquad \text{Identity axiom of a group}$$
$$e_1 = e_2 \qquad \text{Substitution property of equality}$$

THEOREM 6.9: *The inverse element of any element of a group G is unique.*

  Proof: By the inverse axiom for every element a of G, there exists an inverse a^{-1} of G. Suppose for any element a of G, there are two inverse elements, a_1^{-1} and a_2^{-1}. Then

$$a * a_1^{-1} = e \qquad \text{Inverse axiom of a group}$$
$$a * a_2^{-1} = e \qquad \text{Inverse axiom of a group}$$
$$a * a_1^{-1} = a * a_2^{-1} \qquad \text{Substitution property of equality}$$
$$a_1^{-1} = a_2^{-1} \qquad \text{Theorem 6.7}$$

Exercise 9

Prove the following theorems.

1. For all elements a, b, and c of a group G, if $b * a = c * a$, then $b = c$.

2. For all elements a and b of a group G, $(a * b) * b^{-1} = a$.

3. For all elements a and b of a group G, if $a * x = a$, then $x = e$.

4. For all elements a and b of a group G, if $x * a = a$, then $x = e$.

5. In a group G, $e = e^{-1}$.

6. For all elements a and b of a group G, $(a * b)^{-1} = b^{-1} * a^{-1}$.

7. For all elements a and b of a group G, if $a = b$ then $a^{-1} = b^{-1}$.

8. In a group G, $a * x = b$ has a unique solution $x = a^{-1} * b$. (*Hint:* You must prove two things: (1) if $a * x = b$ then $x = a^{-1} * b$ is a solution and (2) if there are two solutions, x_1 and x_2, they are equal.)

9. In a group G, $y * a = b$ has a unique solution $y = b * a^{-1}$.

10. For all elements a in a group G, $(a^{-1})^{-1} = a$.

11.
PERMUTATION GROUPS

Let us consider a set of three elements, $\{A, B, C\}$. How many different arrangements can we make of the three elements of this set? We could, of course, write out all the possible arrangements of the three elements. If we did, we would find that there are six different ones.

Let us solve this problem using a method that will provide a rule for finding how many different arrangements we can make from a set of n different elements.

Let us suppose that we have three boxes into each of which we put one of the letters A, B, or C.

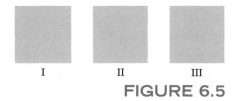

I   II   III

FIGURE 6.5

In Box I we can put any one of the three letters. Hence the choice of the first letter can be made in three ways. To show this we write "3" in Box I.

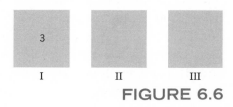

I   II   III

FIGURE 6.6

Having put one of the letters in Box I, we have two letters left. We can put any one of these two in Box II; that is, we can fill Box II in two ways. To show this we write "2" in Box II. We see, then, that there are $3 \cdot 2$

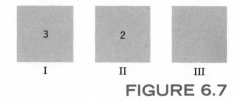

FIGURE 6.7

ways of filling Boxes I and II. Since only one element is left, there is only one way to fill Box III. To show this we write "1" in Box III. We now have $3 \cdot 2 \cdot 1 = 6$ arrangements that can be made from the three elements of the set.

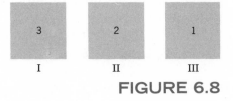

FIGURE 6.8

We see that we can form six different arrangements using three distinct objects; the objects, of course, are immaterial. We call each arrangement a **permutation** of three elements taken three at a time. There are six permutations that can be formed from three distinct objects.

Now let us generalize the above example and find how many permutations we can make using n distinct objects. Suppose we have n objects. Let us imagine that we have n boxes into each of which we can put one of the objects. In Box I we can put any one of n objects. Having put one of the objects in Box I, we have $(n - 1)$ objects left; so we can put any one of these in Box II. That is, we can fill Box II in $(n - 1)$ ways. This means that together there are $n(n - 1)$ ways to fill Boxes I and II. We now have $(n - 2)$ objects left; so we can fill Box III in $(n - 2)$ ways. This means that together there are $n(n - 1)(n - 2)$ ways of filling Boxes I, II, and III. Continuing in this fashion we find that the number of permutations of the n elements is

$$n(n - 1)(n - 2)(n - 3) \ldots (2)(1)$$

In general, then, **the number of permutations of n distinct objects, n a natural number, taken n at a time is**

$$n(n - 1)(n - 2)(n - 3) \ldots (2)(1)$$

We use the symbol $n!$ to denote the above product. We read this symbol "n factorial." Thus we see that

$$5! = 5 \cdot 4 \cdot 3 \cdot 2 \cdot 1 = 120$$
$$4! = 4 \cdot 3 \cdot 2 \cdot 1 = 24$$
$$3! = 3 \cdot 2 \cdot 1 = 6$$

We define $1!$ to equal 1. From the preceding we see that the number of permutations of n elements taken n at a time is $n!$. It should be emphasized that in a permutation we are concerned with the order of the elements; the permutation ABC is not the same as the permutation BCA.

Let us now consider the six permutations of three elements, A, B, C. These six permutations are

$$ABC \quad CAB \quad BCA \quad BAC \quad CAB \quad ACB$$

Let us call these permutations e, p, q, r, s, and t and denote them by the following symbols.

$$e = \begin{pmatrix} A & B & C \\ A & B & C \end{pmatrix} \quad p = \begin{pmatrix} A & B & C \\ C & A & B \end{pmatrix} \quad q = \begin{pmatrix} A & B & C \\ B & C & A \end{pmatrix}$$

$$r = \begin{pmatrix} A & B & C \\ B & A & C \end{pmatrix} \quad s = \begin{pmatrix} A & B & C \\ C & B & A \end{pmatrix} \quad t = \begin{pmatrix} A & B & C \\ A & C & B \end{pmatrix}$$

Consider the set of these six permutations:

$$G = \{e, p, q, r, s, t\}$$

We now define a binary operation denoted by $*$ on these elements. We define $p * q$ to mean that we first carry out p and then q. Thus

$$p * q = \begin{pmatrix} A & B & C \\ C & A & B \end{pmatrix} * \begin{pmatrix} A & B & C \\ B & C & A \end{pmatrix} = \begin{pmatrix} A & B & C \\ A & B & C \end{pmatrix} = e$$

Starting with the elements in the natural order, ABC, p replaces A by C and q replaces C by A, hence $p * q$ replaces A by itself; p replaces B by A and q replaces A by B, hence $p * q$ replaces B by B; p replaces C by B and q replaces B by C, hence $p * q$ replaces C by C. We see that $p * q$ results in the permutation ABC which is e. We call $p * q$ a "product" with the understanding that in this context "product" does not mean multiplication of two real numbers.

Using the above definition of the operation $*$ we have

$$r * p = \begin{pmatrix} A & B & C \\ B & A & C \end{pmatrix} * \begin{pmatrix} A & B & C \\ C & A & B \end{pmatrix} = \begin{pmatrix} A & B & C \\ A & C & B \end{pmatrix} = t$$

$$p * r = \begin{pmatrix} A & B & C \\ C & A & B \end{pmatrix} * \begin{pmatrix} A & B & C \\ B & A & C \end{pmatrix} = \begin{pmatrix} A & B & C \\ C & B & A \end{pmatrix} = s$$

With this understanding of the set G and the operation $*$ we find that G is a group under the operation $*$. It is called a **permutation group.** It is not a commutative group, for, in particular, $r * p \neq p * r$.

Example 1: How many permutations can be formed from the elements of the set $\{A, B, C, D\}$?

Solution: Since there are four distinct elements in the given set, there are

$$4! = (4)(3)(2)(1) = 24$$

permutations.

Example 2: We define a binary operation $*$ on the permutations of four distinct elements in the same way as we defined the binary operation $*$ on the permutations of three distinct elements. Find the product

$$\begin{pmatrix} A & B & C & D \\ B & A & C & D \end{pmatrix} * \begin{pmatrix} A & B & C & D \\ D & C & B & A \end{pmatrix}$$

Solution:

$$\begin{pmatrix} A & B & C & D \\ B & A & C & D \end{pmatrix} * \begin{pmatrix} A & B & C & D \\ D & C & B & A \end{pmatrix} = \begin{pmatrix} A & B & C & D \\ C & D & B & A \end{pmatrix}$$

The first permutation sends A into B; the second permutation sends B into C; hence the product sends A into C. The first permutation sends B into A; the second sends A into D; hence the product sends B into D. Similarly, we find that the product sends C into B and D into A.

Exercise 10

1. Complete the following operation table for the permutation group just discussed.

		Second factor				
*	e	p	q	r	s	t
e	e	p	q	r	s	t
p	p		e		s	
q	q					
r		t				q
s	s					
t	t					

First factor

2. Using the table constructed in problem 1, compute:
   (a) $(r * s) * t$
   (b) $(q * p) * e$
   (c) $(r * q) * q$
   (d) $(s * t) * (p * q)$
   (e) $(s * t) * (p * r)$
   (f) $(t * q) * (p * r)$

3. What is the identity element for the above permutation group?

4. Give the inverse element for every element of the permutation group discussed in Section 11.

5. How many elements would be in a permutation group when:
   (a) Two elements are permuted?
   (b) Five elements are permuted?
   (c) Four elements are permuted?

6. We define an operation * on permutations of n distinct elements in the same manner that we defined the operation * on permutations of three distinct elements. Find the following products.

(a) $\begin{pmatrix} A & B & C & D \\ D & C & B & A \end{pmatrix} * \begin{pmatrix} A & B & C & D \\ A & C & D & B \end{pmatrix}$

(b) $\begin{pmatrix} A & B & C & D \\ B & A & C & D \end{pmatrix} * \begin{pmatrix} A & B & C & D \\ C & D & A & B \end{pmatrix}$

(c) $\begin{pmatrix} A & B & C & D & E \\ B & C & A & E & D \end{pmatrix} * \begin{pmatrix} A & B & C & D & E \\ D & A & C & B & E \end{pmatrix}$

(d) $\begin{pmatrix} A & B & C & D & E \\ A & E & B & C & D \end{pmatrix} * \begin{pmatrix} A & B & C & D & E \\ E & A & B & C & D \end{pmatrix}$

7. Form all the permutations for a set of four elements A, B, C, and D.

8. What is the identity element for the set of permutations of four distinct elements A, B, C, D?

9. Determine x, y, z, and w such that

$$\begin{pmatrix} A & B & C & D \\ B & C & D & A \end{pmatrix} * \begin{pmatrix} A & B & C & D \\ x & y & z & w \end{pmatrix} = \begin{pmatrix} A & B & C & D \\ A & B & C & D \end{pmatrix}$$

207

10. Determine x, y, z, w, and t such that

$$\begin{pmatrix} A & B & C & D & E \\ B & C & E & A & D \end{pmatrix} * \begin{pmatrix} A & B & C & D & E \\ x & y & z & w & t \end{pmatrix} = \begin{pmatrix} A & B & C & D & E \\ A & B & C & D & E \end{pmatrix}$$

11. Let ABC be an equilateral triangle with vertices named clockwise, and let symmetries of this triangle be described by the following:

   (i) 120° clockwise rotation about the center;
  (ii) 240° clockwise rotation about the center;
 (iii) 360° clockwise rotation about the center;
  (iv) rotate about the line of symmetry passing through A;
   (v) rotate about the line of symmetry passing through B;
 (vi) rotate about the line of symmetry passing through C.

(a) Use the two-row notation of Section 11 to describe replacements of the vertices in each transformation. For example, (iv) is illustrated below:

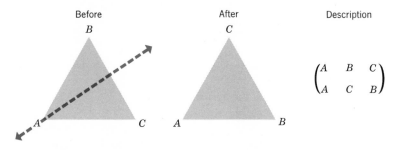

(b) Compare these descriptions with the six permutations in Exercise 8, problem 3 and label these transformations accordingly.

James A. Garfield (1831–1881), twentieth president of the United States, was born in Ohio the son of a farmer. He served in the Union Army during the Civil War and was promoted to Major General by Lincoln. In 1863 Garfield entered Congress. He was elected to the Senate in 1880, but before taking office he was elected to the presidency. His short administration was unhappy, terminating on July 2, 1881, when he was shot by Charles J. Guiteau. He died on September 19. In 1876, while a member of Congress, Garfield hit upon a solution to the Pythagorean Theorem during a mathematical discussion with other members of Congress. Since that time his proof, published in the *New England Journal of Education,* has been called Garfield's Demonstration.

THE
PYTHAGOREAN
THEOREM

7

1.
A BRIEF HISTORY

The theorem of Pythagoras is one of the most important theorems in elementary geometry. Many people, including the celebrated philosopher-mathematician Bertrand Russell, consider it to be the most important theorem in mathematics.

We now state this famous theorem.

THEOREM 7.1.   THE PYTHAGOREAN THEOREM: *In any right triangle, (the area of) the square on the hypotenuse equals the sum of (the areas of) the squares on the legs.*

This relationship between the hypotenuse and the legs of a right triangle with sides of lengths, 3, 4, and 5 was known long before the time of Pythagoras (540 B.C.). Figure 7.1 shows the truth of this theorem for a 3-4-5 triangle. A statement of this relationship can be found in Chinese literature written before 1100 B.C. From a papyrus dating about 2000 B.C., it is apparent that the Egyptians were aware that the 3-4-5 triangle was right angled and that they knew the numerical relation for special other cases, one being $1^2 + (\frac{3}{4})^2 = (1\frac{1}{4})^2$. The Hindus knew the Pythagorean

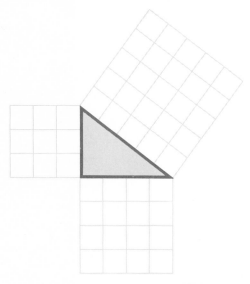

FIGURE 7.1

property long before the Christian era and the Babylonians applied it in 1600 B.C.

Although the proof of this important theorem is attributed to Pythagoras by many writers, including Plutarch and Cicero, there is no positive evidence that he was the first to prove the theorem.

Since the time of Pythagoras hundreds of different proofs of the theorem have been given and many descriptive titles have been applied to it. The theorem is sometimes called the pons asinorum† (bridge of fools) theorem. The Arabs called it the "Figure of the Bride." Other titles given to it are "The Bride's Chair," "The Carpenter's Theorem," "The 47th Proposition," and the "Hecatomb Proposition."

2.
PROOF OF THE PYTHAGOREAN THEOREM

To put the theorem of Pythagoras into modern algebraic form—which Pythagoras could not do—let us denote the measures of the legs by a and b, and the measure of the hypotenuse by c. Then the areas of the squares involved are a^2, b^2, and c^2, and the Pythagorean theorem states

$$a^2 + b^2 = c^2$$

We may ask, "How was Pythagoras led to the discovery of the theorem that bears his name?" History infers that Pythagoras began with the assumption that a triangle with sides 3, 4, and 5 was right angled. But knowing this, how did Pythagoras discover the general theorem? Knowing from the Egyptians that a triangle whose sides have measures of 3, 4, and 5 was a right triangle probably led him to consider whether a similar relation was true of a right triangle whose sides had different measures. The simplest case to investigate was that of a right triangle with legs of equal measure. Such a triangle is called an **isosceles right triangle.** The

---

† This name is usually given to the theorem, "In isosceles triangles the angles at the base are equal to one another," but it is not uncommonly applied to the Pythagorean Theorem by French writers. See D. E. Smith, *History of Mathematics,* Vol II. (Dover).

FIGURE 7.2

proof of the theorem in this particular case was from the construction of a figure. Figure 7.2 shows such a construction.

When it was discovered that an isosceles right triangle had the property in question, Pythagoras was led to establish the property for every right triangle.

To establish a proof of the Pythagorean theorem we start with a right triangle whose legs have measures a and b and whose hypotenuse has measure c (Figure 7.3). Now let us construct the square on the hypotenuse (Figure 7.4). Note that angle 1 and angle 2 are complementary angles (the sum of their measures is 90°), and that angle 3 is a right angle. Since the sum of the measures of angles 2, 3, and 4 is 180 degrees and the measure of angle 3 is 90 degrees, angles 2 and 4 are complementary. We now fit a right triangle congruent to the given right triangle along the base line as shown in Figure 7.5.

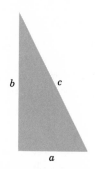

FIGURE 7.3

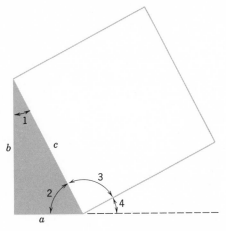

FIGURE 7.4

In a similar manner, two other triangles congruent to the given triangle may be fitted at strategic spots, completing a square (Figure 7.6) with sides of length $a + b$ and area $(a + b)^2$.

This square is made up of square I whose area is c^2 and four congruent right triangles each of whose areas is $\frac{1}{2}ab$. Hence

$$(a + b)^2 = c^2 + 4(\tfrac{1}{2}ab)$$

Now let us partition this square into appropriate areas as shown in Figure 7.7. This square is made up of square II whose area is b^2 and

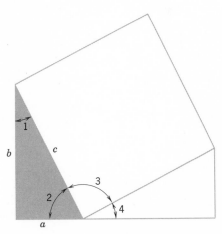

FIGURE 7.5

215

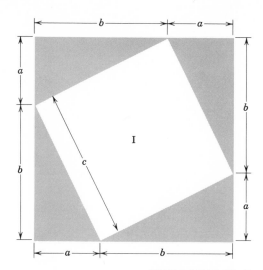

FIGURE 7.6

square III whose area is a^2 and four congruent right triangles each of whose areas is $\frac{1}{2}ab$. Then

$$(a + b)^2 = a^2 + b^2 + 4(\tfrac{1}{2}ab)$$

Since $(a + b)^2 = a^2 + b^2 + 4(\tfrac{1}{2}ab)$ and $(a + b)^2 = c^2 + 4(\tfrac{1}{2}ab)$, we have

$$a^2 + b^2 + 4(\tfrac{1}{2}ab) = c^2 + 4(\tfrac{1}{2}ab)$$

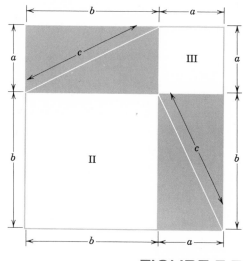

FIGURE 7.7

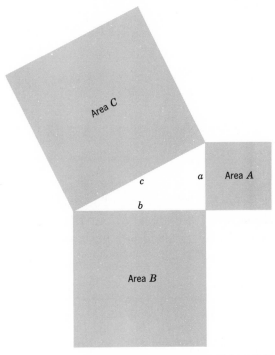

FIGURE 7.8

which reduces to

$$a^2 + b^2 = c^2$$

the algebraic statement of the Pythagorean theorem.

About 300 B.C., Euclid recorded a proof of the Pythagorean theorem. Euclid's proof of the theorem uses the diagram in which squares are constructed on each side of the right triangle (Figure 7.8). He then proceeded to prove that Area A + Area B = Area C. But since

$$\text{Area } A = a^2$$
$$\text{Area } B = b^2$$
$$\text{Area } C = c^2$$

it is true that

$$a^2 + b^2 = c^2$$

Another simple proof of the Pythagorean theorem is this: cut out four congruent right triangular regions as shown in Figure 7.9. Place these triangles in the pattern shown in Figure 7.10. Notice that a large square

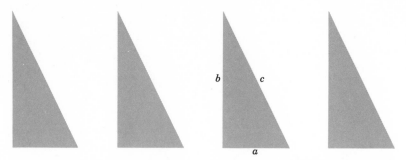

FIGURE 7.9

is formed. The measure of its side is c. Its area is c^2. But this area is made up of the areas of four right triangles plus the area of the center square. The center square has area $(b - a)^2$. Each right triangle has area $\frac{1}{2}ab$. Hence

$$
\begin{aligned}
c^2 &= (b - a)^2 + 4(\tfrac{1}{2}ab) \\
&= b^2 - 2ab + a^2 + 2ab \\
&= b^2 + a^2
\end{aligned}
$$

We shall demonstrate one more proof of the Pythagorean theorem. We start with the right triangle ABC (Figure 7.11). Now we drop a perpendicular, $\overline{CD}$, from the vertex of the right angle to the hypotenuse. Two pairs of similar triangles, $\triangle CDB$ and $\triangle ACB$, and $\triangle ADC$ and $\triangle ACB$, are formed. Triangle CDB and triangle ACB are similar (have the same shape) because both contain $\angle ABC$ and both contain right angles. Since corresponding sides of similar triangles are proportional, we have

$$
\frac{y}{b} = \frac{b}{c} \qquad \text{and} \qquad \frac{x}{a} = \frac{a}{c}
$$

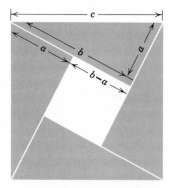

FIGURE 7.10

2. PROOF OF THE PYTHAGOREAN THEOREM

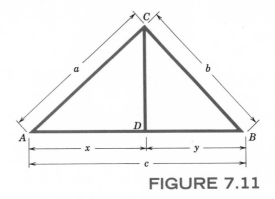

FIGURE 7.11

Solving $\frac{x}{a} = \frac{a}{c}$ and $\frac{y}{b} = \frac{b}{c}$ for x and y respectively, we have

$$x = \frac{a^2}{c}$$

$$y = \frac{b^2}{c}$$

Adding these two equations member by member we obtain

$$x + y = \frac{a^2}{c} + \frac{b^2}{c}$$

But $x + y = c$, hence

$$\frac{a^2}{c} + \frac{b^2}{c} = c$$

which reduces to

$$a^2 + b^2 = c^2$$

Exercise 1

1. Which of the following triples of numbers could be measures of the sides of right triangles?
    (a) (10, 24, 26)   (e) (1.5, 3.6, 3.9)
    (b) (8, 14, 17)   (f) $(1\frac{1}{2}, 2, 2\frac{1}{2})$
    (c) (7, 24, 25)   (g) $(4\frac{1}{4}, 2\frac{1}{2}, 5\frac{1}{4})$
    (d) (9, 40, 41)   (h) $(3\frac{1}{2}, 4\frac{1}{2}, 7\frac{1}{2})$

2. Show by the following demonstration that the Pythagorean theorem is unituitvely true.

THE PYTHAGOREAN THEOREM

(a) Prepare seven pieces of cardboard with the measurements shown in the figure below. Label each piece as shown.

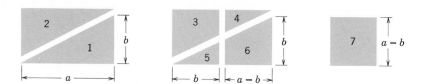

(b) Fit all the pieces together to make a square whose side is c.

(c) Fit pieces 1, 2, 4, 6, and 7 together to form a square whose side is a.

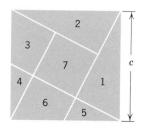

(d) Fit pieces 3 and 5 together to form a square whose side is b.

Explain how these steps give evidence of the truth of the Pythagorean theorem.

3. In the rectangular solid below, $AE = 1$, $AB = 2$, $AD = 2$. Find AH.

2. PROOF OF THE PYTHAGOREAN THEOREM

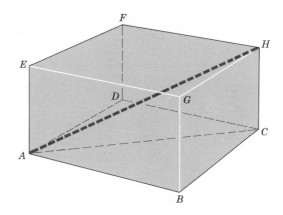

4. If the diagonal of a square is 15 inches, how long is each side of the square?

5. If the diagonal of a rectangle is 25 units and the length is twice the width, what are the dimensions of the rectangle?

6. A man walks 8 miles due north and then 2 miles due east. How far is he from his starting point?

7. A car travels 12 miles west and then 18 miles north. How far is it from its starting point?

8. In the rectangular solid indicated below, find the lengths of $\overline{AC}$ and $\overline{AD}$.

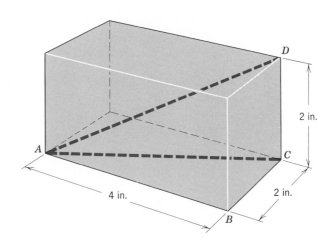

9. The lengths of the legs of a right triangle are 15 and 8. Find the length of the hypotenuse.

10. $\triangle ABC$ has an obtuse angle, $\angle B$. $AB = 6$, $BC = 14$, and $AC = 18$. Find the length of the altitude, h, to $\overline{AB}$.

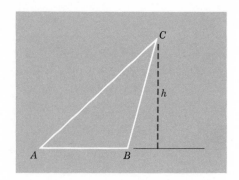

11. In right triangle ABC, $AB = 10$, $CB = 6$. $\overline{EF}$ is perpendicular to and bisects $\overline{AC}$. $AE = 5$. Find ED.

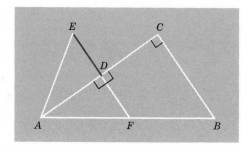

12. A proof of the Pythagorean theorem making use of the figure below was discovered by General James A. Garfield several years before he became President of the United States. It appeared about 1876 in the *New England Journal of Education.*

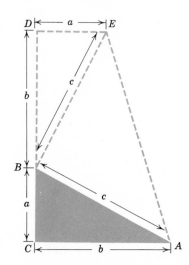

2. PROOF OF THE PYTHAGOREAN THEOREM

Prove that $a^2 + b^2 = c^2$ by stating algebraically that the area of the trapezoid equals the sum of the areas of the three triangles *ABC*, *BDE*, and *EBA*. The area of a trapezoid is found by taking one-half the product of the altitude and the sum of the bases. (How do you know $\angle EBA$ is a right angle?)

13. It is thought that the Babylonians arrived at the knowledge of the Pythagorean theorem by counting tiles in a pattern similar to the one shown below.

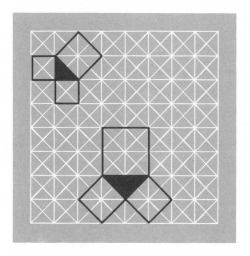

Check the truth of the theorem by counting tiles in the squares drawn on the sides of the shaded triangles.

14. A telephone pole is steadied by three guy wires. Each wire is to be fastened to the pole at a point 15 feet above the ground and anchored to the ground 8 feet from the base of the pole. How many feet of wire are needed for the three guy wires?

15. A gate is 4 feet wide and 6 feet high. How long is the brace that extends from *A* to *B*?

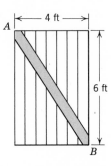

16. A softball diamond is square in shape. The bases are 60 feet apart. How far is it from home plate to second base.

3.
PYTHAGOREAN TRIPLES

The Pythagorean theorem may be expressed algebraically by the following equation:

$$x^2 + y^2 = z^2$$

where x and y are the measures of the legs and z is the measure of the hypotenuse.

It is always possible to find triples of real numbers x, y, and z that satisfy the Pythagorean relation $x^2 + y^2 = z^2$. Very interesting problems arise when we attempt to find only those solutions where x, y, and z are positive integers. This problem amounts to finding all triples of positive integers x, y, and z which satisfy the Pythagorean equation $x^2 + y^2 = z^2$.

It suffices to find only those triples that have no common divisor other than 1 (for example, 3, 4, 5 is such a solution; the only common divisor of these numbers is 1), since from each such solution we can find infinitely many other solutions by multiplying each of the three numbers by an arbitrary positive integer M. Thus since 3, 4, 5 is a solution

$$
\begin{array}{ccc}
6, & 8, & 10 \\
9, & 12, & 15 \\
12, & 16, & 20 \\
& \vdots & \\
3M, & 4M, & 5M
\end{array}
$$

are all solutions.

Solutions that have no common divisors other than 1 are called **primitive solutions.** In the following discussion we shall be looking for primitive solutions.

Since neither x nor y is zero, z must be greater than either x or y. Futhermore, for integral sides, x cannot be equal to y, for if $x = y = a$, then

$$
\begin{aligned}
z^2 &= a^2 + a^2 \\
&= 2a^2
\end{aligned}
$$

and

$$z = a\sqrt{2}$$

which is not an integer.

Moreover, for a primitive solution, any pair of the numbers x, y, and z must be relatively prime, that is, have greatest common divisor 1. If, for instance, x and y have a common factor $d \neq 1$, then for integers x_1 and y_1

$$x = dx_1$$
$$y = dy_1$$

and

$$
\begin{aligned}
x^2 + y^2 &= (dx_1)^2 + (dy_1)^2 \\
&= d^2 x_1^2 + d^2 y_1^2 \\
&= d^2(x_1^2 + y_1^2) \\
&= z^2
\end{aligned}
$$

Hence z^2 must be divisible by d^2 and z must be divisible by d, in which case the solution would not be primitive since x, y, and z have common divisor $d \neq 1$. In a similar manner we can prove that x and z cannot have a common factor other than 1, nor can y and z.

A consequence of this fact is that for a primitive solution, x and y cannot both be even.

We can also prove that both x and y cannot be odd if the solution is primitive. Suppose both x and y are odd. Since every odd number can be represented in the form $2n + 1$, for whole numbers h and k

$$x = 2h + 1 \quad \text{and} \quad y = 2k + 1$$

Then

$$
\begin{aligned}
x^2 + y^2 &= (2h + 1)^2 + (2k + 1)^2 \\
&= 4h^2 + 4h + 1 + 4k^2 + 4k + 1 \\
&= 4(h^2 + h + k^2 + k) + 2 \\
&= 4n + 2
\end{aligned}
$$

Since $x^2 + y^2 = z^2$, z^2 is of the form $4n + 2$, which is even. Since z^2 is even, z is even. But the square of an even number is divisible by 4, and $4n + 2$ is not divisible by 4. Hence both x and y cannot be odd.

From the above discussion, we see that if we are to find a primitive solution of $x^2 + y^2 = z^2$, x and y must be of **different parity,** that is, one is even and one is odd.

Let us assume that x is even and y is odd. Then z necessarily must be odd since an odd number squared is an odd number and an even number squared is an even number and the sum of an even number and an odd number is odd.

We now write $x^2 + y^2 = z^2$ as

$$z^2 - x^2 = (z - x)(z + x) = y^2$$

Since y is odd, y^2 is odd, and both $(z + x)$ and $(z - x)$ are odd. Furthermore, they are relatively prime, for every common divisor of $(z + x)$ and $(z - x)$ is a common divisor of their sum (Theorem 2.1)

$$(z + x) + (z - x) = 2z$$

and their difference (Theorem 2.2)

$$(z + x) - (z - x) = 2x$$

and hence is a divisor of the greatest common divisor of $2z$ and $2x$. But x and z are relatively prime, hence $2z$ and $2x$ have only the common divisor 2. Hence, if $(z + x)$ and $(z - x)$ are to have a common divisor other than 1, it could only be 2. But this is impossible since both x and z are odd.

When two numbers are relatively prime, their prime factors are different and their product cannot be a square unless each of them is a square. Since the product of $(z + x)$ and $(z - x)$ is a square, both are squares, so we can write

$$z + x = r^2$$
$$z - x = s^2$$

Since both $(z + x)$ and $(z - x)$ are odd, so are r and s. Since r and s are odd, both $(r + s)$ and $r - s$ are even, and hence divisible by 2. Since $(r + s)$ and $(r - s)$ are even $\frac{r+s}{2}$ and $\frac{r-s}{2}$ are positive integers. Let

(1) $$\frac{r + s}{2} = a \quad \text{and} \quad \frac{r - s}{2} = b$$

where a and b are positive integers. From this we find

$r = a + b$ (adding equations (1) member by member)
$s = a - b$ (subtracting equations (1) member by member)

Then

(2) $$z + x = r^2 = (a + b)^2 = a^2 + 2ab + b^2$$
(3) $$z - x = s^2 = (a - b)^2 = a^2 - 2ab + b^2$$

Solving these equations for z and x we obtain

$$z = a^2 + b^2$$
$$x = 2ab$$

Then

$$z^2 - x^2 = (a^2 + b^2)^2 - (2ab)^2$$
$$= (a^2 - b^2)^2$$
$$= y^2$$

and hence

$$y = a^2 - b^2$$

We now show that a and b are relatively prime and of different parity and that a is greater than b.

Since r and s are positive integers and

$$a = \frac{r + s}{2}$$

$$b = \frac{r - s}{2}$$

it is clear that $a > b$.

Let us assume that a and b are not of different parity. Then they are both odd or both even. If both a and b are odd

$$a^2 + b^2 = z$$

and

$$a^2 - b^2 = y$$

are even, which is impossible because y and z are both odd.

If a and b are both even they have a common factor, 2. This factor would divide

$$x = 2ab \qquad y = a^2 - b^2 \qquad z = a^2 + b^2$$

which is impossible if we are to have a primitive solution.

Hence for a primitive solution x, y, z we must have a and b relatively prime and of different parity. Summarizing the above discussion, we see that x, y, and z is a primitive solution of $x^2 + y^2 = z^2$ if and only if:

(1) $x = 2ab$, $y = a^2 - b^2$, and $z = a^2 + b^2$.
(2) a and b are relatively prime.
(3) a and b are of different parity.
(4) a is greater than b.

Some primitive solutions of $x^2 + y^2 = z^2$ are given in Table 7.1

TABLE 7.1
Pythagorean triples

a	b	x	y	z
2	1	4	3	5
3	2	12	5	13
4	1	8	15	17
4	3	24	7	25

Example 1: Find the primitive Pythagorean triple (x, y, z) when $a = 9$ and $b = 4$.

Solution:

$$x = 2ab = 2 \cdot 9 \cdot 4 = 72$$
$$y = a^2 - b^2 = 9^2 - 4^2 = 81 - 16 = 65$$
$$z = a^2 + b^2 = 9^2 + 4^2 = 81 + 16 = 97$$

Check: $65^2 + 72^2 = 4225 + 5184 = 9409 = 97^2$

Example 2: Is the triple of numbers (8, 15, 17) a Pythagorean triple?

Solution:

$$8^2 = 64$$
$$15^2 = 225$$
$$17^2 = 289$$

Since $225 + 64 = 289$, the given triple is a Pythagorean triple.

Exercise 2

1. Find all the primitive Pythagorean triples with $z \leq 100$.
2. Which of the following triples are primitive solutions of $x^2 + y^2 = z^2$?
   - (a) (1, 2, 3)
   - (b) (2, 3, 4)
   - (c) (3, 5, 8)
   - (d) (4, 5, 6)
   - (e) (6, 8, 10)
   - (f) (5, 12, 13)
   - (g) (7, 9, 12)
   - (h) (8, 15, 17)
   - (i) (9, 12, 15)
   - (j) (15, 20, 25)
3. Find the Pythagorean triples when:
   - (a) $a = 5, b = 4$
   - (b) $a = 7, b = 6$
   - (c) $a = 11, b = 8$
   - (d) $a = 25, b = 16$
   - (e) $a = 18, b = 7$
   - (f) $a = 13, b = 6$
   - (g) $a = 15, b = 4$
   - (h) $a = 11, n = 6$

4. Prove that if x, y, and z is a primitive solution of $x^2 + y^2 = z^2$, x and z cannot have a common factor $d \neq 1$.

5. Prove that if x, y, and z is a primitive solution of $x^2 + y^2 = z^2$, y and z cannot have a common factor $d \neq 1$.

6. How many primitive Pythagorean triples can you find when $a = 9$?

7. Prove that the only primitive Pythagorean triple in which x, y, and z are consecutive positive integers is 3, 4, 5.

8. How many primitive Pythagorean triples can you find when $a = 20$?

9. List all of the primitive Pythagorean triples with $z \leq 100$ for which two of the numbers in the triple are primes.

10. Is it possible to find a primitive Pythagorean triple in which all of the numbers are prime numbers? Why?

4.
PRIMITIVE
PYTHAGOREAN TRIANGLES

Any right triangle whose sides have measures that are Pythagorean triples is called a **Pythagorean triangle.** If the Pythagorean triple is a primitive solution of the Pythagorean equation $x^2 + y^2 = z^2$, the triangle is called a **primitive Pythagorean triangle.**

Many problems suggest themselves about these triangles. We shall discuss a few of them. In a Pythagorean triangle, the longest side is obviously the **hypotenuse.** The other two sides are called **legs.** If the legs of a Pythagorean triangle are x and y (their lengths) and the hypotenuse is z (its length), then by the theorem of Pythagoras, $x^2 + y^2 = z^2$.

It is easy to verify that the measure of one leg of a primitive Pythagorean triangle is always divisible by 3.

THEOREM 7.2: *If x and y are (the lengths of) the legs of a primitive Pythagorean triangle, then either x or y is divisible by 3.*

Proof: From Theorem 2.6, we know that every square is either a multiple of 3 or of the form $3k + 1$. If either x^2 or y^2 is a multiple of 3, then x or y is a multiple of 3 and the theorem is verified.

229

Suppose neither x^2 nor y^2 is a multiple of 3. Then both of them are of the form $3k + 1$. Let

$$x^2 = 3k + 1 \qquad \text{and} \qquad y^2 = 3h + 1$$

Then

$$
\begin{aligned}
x^2 + y^2 &= (3k + 1) + (3h + 1) \\
&= 3k + 3h + 2 \\
&= 3(k + h) + 2
\end{aligned}
$$

Since $x^2 + y^2 = z^2$, z^2 is of the form $3n + 2$, from the above. This is impossible since every square is either a multiple of 3 or of the form $3k + 1$. Hence our assumption is false and either x or y is a multiple of 3.

THEOREM 7.3: *The length of one side of a primitive Pythagorean triangle is divisible by 5.*

Proof: First we shall show that every square is either a multiple of 5 or is of the form $5k + 1$ or $5k + 4$. All whole numbers may be written in one of the forms $5k$, $5k + 1$, $5k + 2$, $5k + 3$, or $5k + 4$. Squaring each of these in turn we have

$$
\begin{aligned}
(5k)^2 &= 25k^2 = 5M \\
(5k + 1)^2 &= 25k^2 + 10k + 1 = 5M + 1 \\
(5k + 2)^2 &= 25k^2 + 20k + 4 = 5M + 4 \\
(5k + 3)^2 &= 25k^2 + 30k + 9 = 5M + 4 \\
(5k + 4)^2 &= 25k^2 + 40k + 16 = 5M + 1
\end{aligned}
$$

If the measure of one of the legs of a primitive Pythagorean triangle is a multiple of 5, the theorem is verified. Suppose neither leg is a multiple of 5. We shall show that this implies that the hypotenuse must be a multiple of 5.

If both x^2 and y^2 are of the form $5M + 1$, we have

$$
\begin{aligned}
x^2 + y^2 &= (5n + 1) + (5p + 1) \\
&= 5M + 2
\end{aligned}
$$

Hence z^2 is of the form $5M + 2$, which is impossible. From this we know that x^2 and y^2 cannot both be of the form $5M + 1$.

Now, suppose both x^2 and y^2 are of the form $5M + 4$. We have

$$\begin{aligned}
x^2 + y^2 &= (5n + 4) + (5p + 4) \\
&= 5n + 5p + 8 \\
&= 5n + 5p + 5 + 3 \\
&= 5M + 3
\end{aligned}$$

Hence z^2 is of the form $5M + 3$, which is impossible.

Now let x be of the form $5M + 1$, and y be of the form $5M + 4$. Then

$$\begin{aligned}
x^2 + y^2 &= (5n + 1) + (5p + 4) \\
&= 5n + 5p + 5 \\
&= 5M
\end{aligned}$$

and z^2 is divisible by 5; hence z is divisible by 5.

In examining Table 7.1 we see that there are cases in which the measure of the hypotenuse of a primitive Pythagorean triangle exceeds the measure of one of the legs by 1. In looking at these cases we see that in each case listed when $z = x + 1$, $a = b + 1$. This is true in general.

THEOREM 7.4: *If $x = 2ab$, $y = a^2 - b^2$ and $z = a^2 + b^2$ are the measures of the legs and the hypotenuse, respectively, of a primitive Pythagorean triangle and $z = x + 1$, then $a = b + 1$.*

Proof: Since $z = x + 1$, $z - x = 1$. But $z = a^2 + b^2$ and $x = 2ab$, hence

$$z - x = a^2 + b^2 - 2ab = 1$$

Since $a^2 + b^2 - 2ab = (a - b)^2$, we have

$$(a - b)^2 = 1$$

and $a - b = 1$ or $a - b = -1$. Since $a > b$ and both are positive integers, $a - b$ cannot equal -1, hence $a - b = 1$, and $a = b + 1$.

5.
FINDING SIDES
OF PYTHAGOREAN
TRIANGLES GIVEN
THE LENGTH OF ONE SIDE

Now we ask: "if we are given the measure of the odd leg of a primitive Pythagorean triangle, can we find the measures of the other two sides?"

Since

$$y = a^2 - b^2$$
$$= (a - b)(a + b)$$

given y, we factor it into the product of unique, relatively prime factors in all possible ways, and in each case we take the larger factor as $a + b$, and the smaller factor as $a - b$.

For example, given $y = 15$, then

$$15 = 3 \cdot 5$$
$$15 = 1 \cdot 15$$

Then $a + b = 5$ and $a - b = 3$, or $a + b = 15$ and $a - b = 1$.

When $a + b = 5$ and $a - b = 3$, we have $a = 4$ and $b = 1$. Hence $x = 8$, $y = 15$, and $z = 17$.

In the case $a + b = 15$ and $a - b = 1$, $a = 8$ and $b = 7$. Then $x = 112$, $y = 15$, and $z = 113$.

If we are given the measure of the even leg of a primitive Pythagorean triangle, can we find the measures of the other two sides?

Since $x = 2ab$ and a and b are of opposite parity, at least one of them is even. Hence x must have a factor 4.

We now factor ab into the product of two relatively prime factors of different parity in all possible ways. We then take the larger factor as a and the smaller factor as b and determine y and z.

For example, suppose $x = 24$. Since x is a multiple of 4, it may be the measure of one leg of a primitive Pythagorean triangle. Then

$$x = 24 = 2 \cdot 12$$

5. FINDING SIDES OF PYTHAGOREAN TRIANGLES

We now factor 12 into two factors of different parity and whose g.c.d. is 1. The possible factors are 1 and 12, and 3 and 4 (we do not consider 2 and 6 because they are both even and hence are not of different parity and are not relatively prime).

Using $a = 4$ and $b = 3$, we find $y = 7$ and $z = 25$.

Using $a = 12$ and $b = 1$, we find $y = 143$ and $z = 145$.

It is more difficult to answer the question: Given the measure of the hypotenuse of a primitive Pythagorean triangle, can we find the measures of the legs? It can be proved, although we shall not do it here, that given a positive integer n, there exists a primitive Pythagorean triangle with hypotenuse of measure n if n has at least one prime factor of the form $4k + 1$.

Looking at Table 7.1 we see that each value of z given in the table has a prime factor of the form $4k + 1$. For example $25 = 5^2$ and $5 = 4 \cdot 1 + 1$.

Example 1: Given $z = 65$. Is there a primitive Pythagorean triangle with hypotenuse of measure z?

Solution: Since $65 = 5 \cdot 13$ and 5 is of the form $4k + 1$ (so is 13), it is possible to find positive integers x and y such that $x^2 + y^2 = 65^2$.

Example 2: Can the hypotenuse of a primitive Pythagorean triangle have measure 27?

Solution: Since $27 = 3^3$ and 3 is of the form $4k + 3$, it is impossible to find positive integers x and y such that $x^2 + y^2 = 27^2$.

Exercise 3

1. Find all the primitive Pythagorean triangles when x is equal to:
   (a) 28
   (b) 8
   (c) 12
   (d) 16
   (e) 20
   (f) 32
   (g) 36
   (h) 40
2. Find all the primitive Pythagorean triangles when y is equal to:
   (a) 7
   (b) 9
   (c) 27
   (d) 33
   (e) 25
   (f) 35
   (g) 63
   (h) 75

3. Which of the following are possible measures of the hypotenuse of a primitive Pythagorean triangle?

(a) 65   (d) 207

(b) 79   (e) 429

(c) 87   (f) 851

4. Find all the Pythagorean triangles (not necessarily primitive) with the measures of one leg equal to a natural number less than 10 and $z \leq 100$.

5. If m is an arbitrary positive integer, it can be proved that there exist m Pythagorean triangles such that the hypotenuses are successive positive integers $n, n + 1, n + 2, \ldots, n + m - 1$. For example, for $m = 3$, we have the three Pythagorean triangles (15, 36, 39), (24, 32, 40) and (9, 40, 41). Find all the Pythagorean triangles when $m = 4$.

6. Find two primitive Pythagorean triangles whose legs have measures that are successive positive integers. For example, (3, 4, 5) is a Pythagorean triangle whose legs have measures that are consecutive positive integers, 3 and $3 + 1 = 4$.

7. Find all the primitive Pythagorean triangles with one leg of measure 56.

8. Find all the primitive Pythagorean triangles with one leg of measure 105.

9. Find all the primitive Pythagorean triangles with one leg of measure 68.

10. Find all the primitive Pythagorean triangles whose legs have measures that are n and $n + 1$, n a positive integer, and $z \leq 200$.

6.
SQUARE ROOTS FROM PYTHAGOREAN TRIANGLES

If the measure of each of the legs of an isosceles right triangle is 1, the measure of the hypotenuse is not a whole number. In fact, it is not even a **rational number** (a rational number is one that may be expressed as $\frac{a}{b}$, where a and b are integers and $b \neq 0$). We see that if $x = y = 1$, then

$$1^2 + 1^2 = z^2$$
$$1 + 1 = z^2$$
$$2 = z^2$$

and $z = \sqrt{2}$.

There is no whole number that can be multiplied by itself to give 2. Thus the length of the hypotenuse of an isosceles right triangle is **incommensurable**. By commensurable lengths x and z we mean that positive integers m and n can be found such that $mx = nz$. Thus the hypotenuse of an isosceles right triangle is incommensurable with either leg. Another way of saying that is that $\sqrt{2}$ cannot be expressed as the ratio of two integers. Such numbers as $\sqrt{2}$ are called **irrational numbers.**†

It is not true that the square root of every whole number is irrational. Thus $\sqrt{4} = 2$, $\sqrt{9} = 3$. However, if the square root of a whole number is not a whole number, it is an irrational number.

A geometric method for finding the square root of a whole number makes use of the Pythagorean theorem. If an isosceles right triangle is constructed with 1 for the measure of each of the legs, the measure of the hypotenuse will be $\sqrt{2}$. We now construct another right triangle as shown in Figure 7.12. The hypotenuse of this second triangle has a measure $\sqrt{3}$ since by the Pythagorean theorem it is the square root of $(\sqrt{2})^2 + 1^2$.

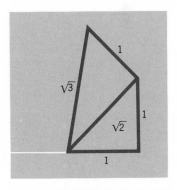

FIGURE 7.12

Continuing in this fashion as shown in Figure 7.13, we can construct right triangles the measures of whose hypotenuses are

$$\sqrt{4}, \ \sqrt{5}, \ \sqrt{6}, \ldots$$

Continuing this construction indefinitely, we construct a figure called the **square root spiral,** with ever-increasing lines extending out from 0 and forming ever-decreasing angles.

It is impractical to find the square root of large numbers by the construction of the square root spiral. If we construct the square root spiral so that we have the square roots of the natural numbers less than or equal

---

† See Chapter 5, Section 2.

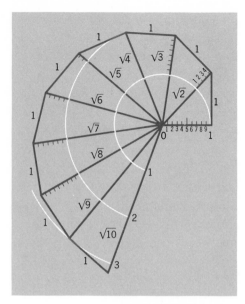

FIGURE 7.13

to 10, it is possible to use the spiral to find the square roots of other numbers. For example, suppose we wish to find $\sqrt{15}$. The square root of 15 can be found from $\sqrt{8}$ and $\sqrt{7}$ using Figure 7.13. This construction is shown in Figure 7.14.

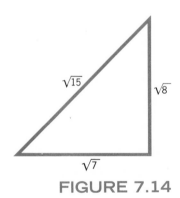

FIGURE 7.14

Exercise 4

1. If the measures of the legs of a right triangle are 1 and n, find the measure of the hypotenuse by use of the Pythagorean theorem.

2. If the measures of the legs of a right triangle are $\sqrt{p}$ and $\sqrt{q}$, what is the measure of the hypotenuse?

3. Using the square root spiral construct:

(a) $\sqrt{13}$ (e) $\sqrt{19}$

(b) $\sqrt{11}$ (f) $\sqrt{12}$

(c) $\sqrt{20}$ (g) $\sqrt{16}$

(d) $\sqrt{17}$ (h) $\sqrt{18}$

4. Triangle ABC is a right triangle. The measures of the legs are a and b; the measure of the hypotenuse is c. Find the measure of the hypotenuse if:

(a) $a = 6$ and $b = 8$

(b) $a = 4$ and $b = 6$

5. In an equilateral triangle each side is 15 inches. How long is the altitude?

6. The sides of a triangle are 6 inches, 9 inches, and 11 inches. Is it a right triangle? If it is a right triangle, which side is the hypotenuse?

7. If r and s are lengths of the legs of a right triangle and t is the length of the hypotenuse, show that for any positive integer n, the numbers nr, ns, and nt are also lengths of the sides of a right triangle.

8. How long is the diagonal of a square if its side has measure:

(a) 6?   (d) 13?

(b) 9?   (e) $\sqrt{6}$?

(c) $\sqrt{2}$?   (f) 17?

7.
THE PYTHAGOREAN THEOREM AND THE DISTANCE BETWEEN TWO POINTS

Let us consider two points P and Q on the Cartesian plane, with coordinates (x_1, y_1) and (x_2, y_2) respectively, as shown in Figure 7.15. We draw line $\overleftrightarrow{PR}$ parallel to the x-axis and line $\overleftrightarrow{QR}$ parallel to the y-axis. These two lines intersect in point R whose coordinates are (x_2, y_1). We see that triangle PQR is a right triangle.

Using the Pythagorean theorem we can find the length of the line segment with P and Q as end points. That is, we can find the distance between the points P and Q. We see that the length of $\overline{QR}$ is $(y_2 - y_1)$ and the length of $\overline{PR}$ is $(x_2 - x_1)$. Since triangle PQR is a right triangle,

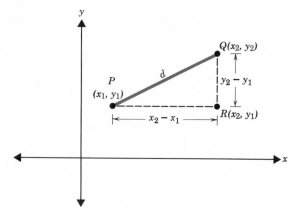

FIGURE 7.15

the distance $\overline{PQ}$, which we shall denote by d, can be found using the Pythagorean relation:

$$d^2 = (x_2 - x_1)^2 + (y_2 - y_1)^2$$

and

$$d = \sqrt{(x_2 - x_1)^2 + (y_2 - y_1)^2}$$

This equation is called the **distance formula.**

We use this distance formula to find the lengths of line segments in the Cartesian plane knowing the coordinates of the end points of the segment.

Example 1: Find the length of the line segment whose end points are $(4, 5)$ and $(-1, -7)$.

Solution: Using the distance formula we have

$$
\begin{aligned}
d &= \sqrt{(x_2 - x_1)^2 + (y_2 - y_1)^2} \\
&= \sqrt{(-1 - 4)^2 + (-7 - 5)^2} \\
&= \sqrt{(-5)^2 + (-12)^2} \\
&= \sqrt{25 + 144} \\
&= \sqrt{169} \\
&= 13
\end{aligned}
$$

Example 2: Show that $(1, 0)$, $(3, 3)$, $(9, -1)$ and $(7, -4)$ are the coordinates of the vertices of a rectangle.

Solution: In Figure 7.16 we see that geometric figure $ABCD$ is a rectangle if $AB = DC$, $AD = BC$, and all the angles of the

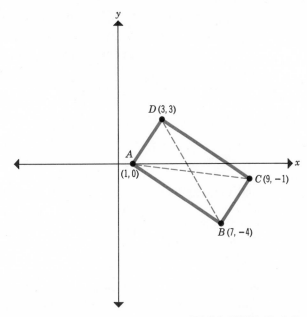

$D(3, 3)$

A

$(1, 0)$

$C(9, -1)$

$B(7, -4)$

FIGURE 7.16

figure are right angles. Using the distance formula we can find the lengths of the sides of the figure.

$$AB = \sqrt{(7 - 1)^2 + (-4 - 0)^2} = \sqrt{52}$$
$$DC = \sqrt{(9 - 3)^2 + (-1 - 3)^2} = \sqrt{52}$$
$$AD = \sqrt{(3 - 1)^2 + (3 - 0)^2} = \sqrt{13}$$
$$BC = \sqrt{(9 - 7)^2 + (-1 - (-4))^2} = \sqrt{13}$$

We see that the opposite sides of the figure are equal in length. We now use the Pythagorean theorem to show that all of the angles of $ABCD$ are right angles. If triangle ABC is a right angle,

$$AC^2 = AB^2 + BC^2$$

Using the distance formula we find

$$AC^2 = (9 - 1)^2 + (-1 - 0)^2 = 65$$

and

$$AB^2 + BC^2 = (\sqrt{52})^2 + (\sqrt{13})^2 = 52 + 13 = 65$$

Since triangle ABC is a right triangle, triangle ACD is also

239

a right triangle. Why? In a similar manner we can show that triangles *DCB* and *DAB* are right triangles.

We see from the above that all four angles of *ABCD* are right angles and the lengths of its opposite sides are equal, hence it is a rectangle.

Exercise 5

1. Find the lengths of the line segments whose end points have the given coordinates.

(a) (6, 2), (3, 6)

(b) (0, 0), (4, 3)

(c) (−1, 2), (2, 6)

(d) (−2, −3), (6, 3)

(e) (−1, 9), (3, 3)

(f) (−4, 2), (4, −3)

2. Show that (−1, 4), (2, −1) and (12, 5) are the vertices of a right triangle.

3. Show that (2, −1), (3, 4) and (−7, 6) are the vertices of a right triangle.

4. Find the lengths of the sides of a triangle whose vertices are (−3, −1), (−10, 13) and (5, 5).

5. Find the lengths of the sides of a triangle whose vertices are (2, 4), (6, 1) and (11, 13).

6. Show that (3, 7), (5, −5) and (−2, 0) are the vertices of an isosceles triangle.

7. Find the perimeter of a triangle whose vertices are (6, 2), (4, 3) and (0, 1).

8. Show that (−2, −2), (2, 2) and $(2\sqrt{3}, -2\sqrt{3})$ are the vertices of an equilateral triangle.

9. Show that (2, −2), (4, 0), (2, 2), and (0, 0) are the vertices of a square.

10. Find the *x*-coordinate of point *B* if $\overline{AB}$ is 13 units long and *A* has coordinates (2, 5) and the *y*-coordinate of *B* is −7. (There are two possible values for *x*.)

11. If the distance from the point (2, *y*) to the point (−1, 3) is 5, find *y*. (There are two possible values for *y*.)

12. Find a point on the *y*-axis equidistant from the points (1, 2) and (5, 6). (*Hint:* Let the point on the *y*-axis have coordinates (0, *y*).)

13. The points (3, −1), (−1, 3) and (7, −5) are collinear. Show that the point (3, −1) is the midpoint of the line segment with (−1, 3) and (7, −5) as endpoints.

14. The center of a circle is at the point (0, 0). A point on the circle has coordinates (3, 4). Find the length of the radius of the circle.

15. Show that the points (4, 4) and (5, 3) lie on a circle whose center is (2, 1). (*Hint:* Two points on the same circle are equidistant from the center of the circle.)

16. Show that *A*(−3, 1), *B*(1, 2), and *C*(5, 3) are collinear, that is, lie on the same straight line. (*Hint:* Show that the distance from *A* to *C* is equal to the sum of the distances from *A* to *B* and from *B* to *C*.)

17. Determine whether $(4, 4)$, $(4, -2)$, and $(-4, -2)$ are the vertices of a right triangle. Find the area of this triangle.

18. Is the triangle with vertices $(1, 7)$, $(-7, -5)$, and $(9, -6)$ isosceles?

19. Are the four points $(0, 0)$, $(4, 25)$, $(-1, 26)$, and $(-5, 1)$ the vertices of a rectangle?

20. Show that $(7, 1)$, $(1, 3)$, $(-1, -3)$, and $(5, -5)$ are the vertices of a square.

Jerome Cardan (1501–1576) or Girolamo Cardano, the ille-
gitimate son of an Italian jurist, was a pioneer in the field
of probability. At one time or another he was a professor
at the University of Bologna, the rector of the College of
Physicians of Milan, and an inmate of an almshouse. He
was indeed a man of contrasts. He was an astrologer as
well as a student of philosophy, a gambler as well as an
algebraist, a physician as well as the father of a murderer,
and a man of genius as well as a man devoid of principles.
Cardan's book, *Liber de Ludo Aleae* (*Book on Games of Chance*),
is really a gambler's manual and may be considered the first
book on probability. In this book Cardan discussed how
many successes in how many trials, mathematical expec-
tation, and the additive properties of probability. He correctly showed the number of ways that two dice may be
thrown and also listed the ways in which three dice may
be thrown. As a dramatic ending to this unprincipled genius,
Cardan predicted the day of his own death. When the day
arrived and he was still alive, he committed suicide to make
his prediction come true.

PROBABILITY AND STATISTICS

8

THE FUNDAMENTAL COUNTING PRINCIPLE

Let us consider the following problem. At the Rolling Hills Racetrack the entries in the first two races are:

First Race	**Second Race**
1. Miss J	1. George's Choice
2. Butter Up	2. Funny Face
3. Fade Out	3. L.S.D.
4. June Bride	4. Cope Out
5. Little Maggie	5. Be Fast
	6. Faker

In how many ways can the winners of the two races be picked? From Figure 8.1 we see that there are five possible choices for the winner of the first race and each of these five choices can be paired with any one of the six horses in the second race. Hence there are $5 \cdot 6 = 30$ possible ways to pick the winners of the two races.

The above problem illustrates a very important principle in the study of probability called the fundamental counting principle. The **fundamental counting principle** states that if a first choice can be made in any one of n_1 ways, and following this a second choice can be made in any one of n_2 ways, and so on until the kth choice can be made in n_k ways, then the number of choices in the order indicated can be made in $n_1 n_2 \ldots n_k$ ways.

The diagram in Figure 8.1 used to determine the 30 ways of picking the winners of the first two races is called a **tree diagram.** Notice that the winner of the first race can be chosen in any one of 5 ways and the winner of the second race can be chosen in any one of 6 ways; hence, applying the fundamental counting principle the winners of the two races can be picked in any one of $5 \cdot 6 = 30$ ways.

Example 1: A baseball team has 8 pitchers and 4 catchers. In how many ways can a pitcher and a catcher be chosen for a game?

 Solution: Since the pitcher can be chosen in 8 ways and the catchers

1. THE FUNDAMENTAL COUNTING PRINCIPLE

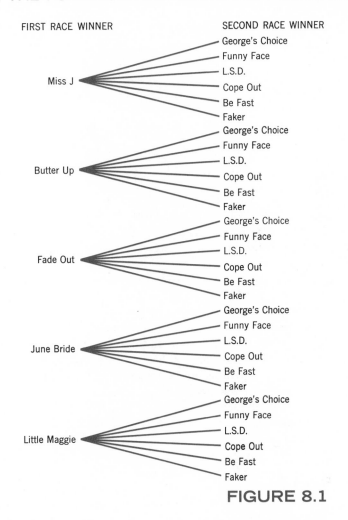

FIRST RACE WINNER

SECOND RACE WINNER

Miss J
- George's Choice
- Funny Face
- L.S.D.
- Cope Out
- Be Fast
- Faker

Butter Up
- George's Choice
- Funny Face
- L.S.D.
- Cope Out
- Be Fast
- Faker

Fade Out
- George's Choice
- Funny Face
- L.S.D.
- Cope Out
- Be Fast
- Faker

June Bride
- George's Choice
- Funny Face
- L.S.D.
- Cope Out
- Be Fast
- Faker

Little Maggie
- George's Choice
- Funny Face
- L.S.D.
- Cope Out
- Be Fast
- Faker

FIGURE 8.1

can be chosen in 4 ways, by the fundamental counting principle there are $8 \cdot 4 = 32$ ways of picking a pitcher-catcher combination.

Example 2: A penny and a nickel are tossed. How many different ways can they fall?

Solution: The penny can fall showing heads or tails. The nickel can fall showing heads or tails. Hence the penny and the nickel can fall $2 \cdot 2 = 4$ ways. The tree diagram in Figure 8.2 shows how we determined these four ways.

245

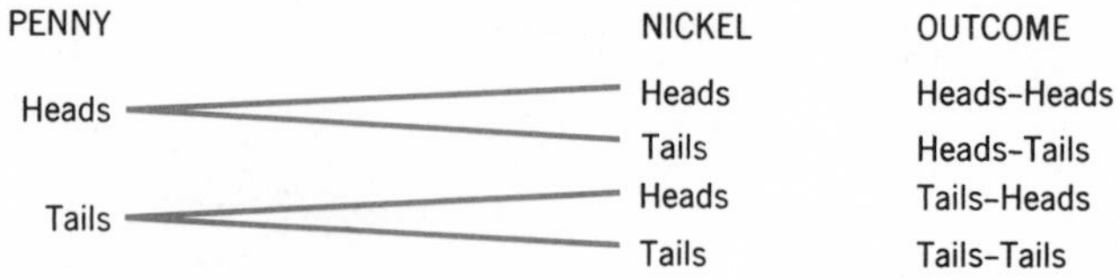

PENNY	NICKEL	OUTCOME
Heads	Heads	Heads–Heads
	Tails	Heads–Tails
Tails	Heads	Tails–Heads
	Tails	Tails–Tails

FIGURE 8.2

Exercise 1

1. From a set of three boys and six girls a committee of one boy and one girl is chosen. How many possible committees are there?

2. In how many ways can the offices of president, vice-president, and secretary be filled from a club with twelve members assuming any one of the members is eligible for any office?

3. There are 16 members, 10 men and 6 women, in the Outriders Motorcycle Club. In how many ways can the president, vice-president, secretary, and treasurer of the club be chosen if the by-laws state that a man must be president and a woman must be secretary?

4. A girl has 6 sweaters and 4 skirts. Each sweater may be worn with any of the skirts. How many outfits can she select?

5. A signal consisting of 4 flags hung in a vertical line is to be made from 7 signal flags. How many are possible?

6. A license number consists of one letter from the English alphabet followed by three digits chosen from 0, 1, 2, 3, 4, 5, 6, 7, 8, and 9. How many license numbers can be made?

7. A hot dog stand sells hot dogs and hamburgers with a choice of mustard, relish, horse-radish or catsup, on it for 35¢. How many different sandwiches are possible?

8. A cafeteria has a blue-plate special that consists of a choice of meat, potato, and one vegetable. There are two choices of meat, a choice of baked, mashed, or french fried potatoes, and as a choice of vegetable, beets, corn, peas, or eggplant. How many different choices of the blue-plate special are there?

9. A penny, a nickel, and a dime are tossed. How many possible ways can the three coins fall?

10. How many different three-digit numerals can be formed from the set of digits $\{1, 2, 3, 4, 5, 6\}$ if:

    (a) Repetitions of digits are allowed?

    (b) No repetition of digits is allowed?

11. The extension numbers of the phones of the offices of the mathematics department of a large university all consist of four digits. All the extensions are arrangements of the digits 0, 1, 4, and 6. How many extension numbers are possible if zero is not used as the first digit?

12. An ice cream parlor features 52 different flavors of ice cream and 16 different toppings. How many different sundaes (ice cream covered with topping) can be made?

2. PERMUTATIONS†

Suppose we have a set

$$A = \{a, b, c\}$$

How many arrangements of the three elements of this set can we make? Since there are three elements, any one of the three may be selected for first place in the arrangement. After the choice of first place has been made, there are two possible choices for second place. Hence the first and second places can be made in $3 \cdot 2 = 6$ ways. Having made our choices for first and second places, there is only one way to make the third choice. There are then $3 \cdot 2 \cdot 1 = 6$ ways to arrange the three elements of set A. These six arrangements, called **permutations,** of the three elements are: *abc, acb, bac, bca, cab, cba.* Observe that the elements themselves have no bearing on the number of permutations we can make. We see then that there are six permutations possible from three distinct elements taken three at a time.

Now let us generalize the above example and determine how many permutations we can make from n elements taken n at a time. The first choice can be made in n ways. Having made the first choice, we can make the second choice in $(n - 1)$ ways. We see that the first two choices can be made in $n(n - 1)$ ways. After the first two choices have been made, we can make the third choice in $(n - 2)$ ways. There are $n(n - 1)(n - 2)$ ways of making the first three choices. Continuing in this fashion we find that the number of permutations of n elements taken n at a time, denoted by the symbol $P(n; n)$ is

$$P(n; n) = n(n - 1)(n - 2) \cdot \ldots \cdot 2 \cdot 1$$

We use the symbol $n!$, read: n factorial, to denote the product

---

† This topic was introduced in Chapter 6, Section 12.

$n(n-1) \cdot \ldots \cdot 2 \cdot 1$. Then,

$$2! = 2 \cdot 1 = 2$$
$$3! = 3 \cdot 2 \cdot 1 = 6$$
$$4! = 4 \cdot 3 \cdot 2 \cdot 1 = 24$$
$$5! = 5 \cdot 4 \cdot 3 \cdot 2 \cdot 1 = 120, \text{ and so forth}$$

We define 1! to be 1. Then

$$P(n; n) = n!$$

Example 1: In how many ways can the five horses in the first race (See Section 1) finish the race?

Solution: We are looking for the number of permutations of 5 things taken 5 at a time. Then

$$P(5; 5) = 5! = 5 \cdot 4 \cdot 3 \cdot 2 \cdot 1 = 120$$

There are 120 ways in which the horses in the first race can finish.

In a horse race there are three winning positions, first, second, and third, called win, place, and show. Considering the five horses in the first race in Section 1, in how many different ways can the three winners be chosen? We are looking for the number of permutations of 5 things taken 3 at a time. There are 5 possible choices for the winner. After the winner is decided, there are 4 possible choices for the second place, making $5 \cdot 4 = 20$ ways for the first and second places to be selected. There are now 3 possible horses for the third place. Hence there are $5 \cdot 4 \cdot 3 = 60$ ways that the winners of the race may be chosen.

Now let us derive a general formula for the number of permutations of n distinct things taken k at a time, $k \leq n$. We use the symbol $P(n; k)$ to denote the number of permutations of n objects taken k at a time. Since there are n objects, the first choice can be made in any one of n ways. After the first choice has been made, the second choice can be made in $(n-1)$ ways. There are $n(n-1)$ ways to make the first and second choices. The third choice can be made in $(n-2)$ ways. Hence there are $n(n-1)(n-2)$ ways to make the first three choices. Continuing in this manner we see that the kth choice can be made in $(n-k+1)$ ways. Then

(1) $$P(n; k) = n(n-1)(n-2) \ldots (n-k+1)$$

Multiplying the right member of equation (1) by $\frac{(n-k)!}{(n-k)!}$ we obtain

$$P(n; k) = n(n-1)\ldots(n-k+1)\cdot\frac{(n-k)!}{(n-k)!}$$

$$= \frac{n(n-1)\ldots(n-k+1)(n-k)(n-k-1)\ldots(2)(1)}{(n-k)!}$$

and

$$P(n; k) = \frac{n!}{(n-k)!}$$

The formula $P(n; k) = \frac{n!}{(n-k)!}$ becomes, when $k = n$

$$P(n; n) = \frac{n!}{(n-n)!} = \frac{n!}{0!}$$

But $P(n; n) = n!$, hence

$$n! = \frac{n!}{0!}$$

and we see that $0! = 1$.

Example 2: In how many ways can the win, place, and show positions be made for the six horses in the second race in Section 1?

Solution: We are looking for the number of permutations of 6 things taken 3 at a time. Then

$$P(6; 3) = 6\cdot5\cdot4 = 120$$

There are 120 ways of picking the three winners in the race.

Example 3: All of the faculty and administration offices of a certain college have 7-digit phone numbers, but in each case the first three digits in the phone numbers are 543. How many different extension numbers can be formed from the digits 0, 1, 2, . . . , 9?

Solution: Since the first three digits of each phone number are fixed, we are seeking the number of permutations of ten things (the ten digits) taken four at a time (the last four digits in each phone number). Then

$$P(10; 4) = \frac{10!}{(10-4)!} = 5040$$

1. Evaluate.

(a) $\dfrac{4!}{2!}$

(c) $\dfrac{(12-7)!}{6!}$

(b) $\dfrac{6!3!}{4!}$

(d) $\dfrac{30!7!}{28!9!}$

2. Evaluate

(a) $P(6; 6)$ (c) $P(8; 2)$ (e) $P(18; 16)$

(b) $P(5; 3)$ (d) $P(12; 4)$ (f) $P(100; 98)$

3. Find the number of different arrangements of the letters in the word "police" if they are taken

(a) Six at a time (c) Five at a time

(b) Two at a time (d) Three at a time

4. In how many different ways can 8 posters be hung on a wall in a straight line?

5. The president and the eight directors of the executive board of an organization are lined up in a row to have their picture taken. In how many ways can they be lined up if the president is to stand in the center of the group?

6. In how many ways can a disc jockey choose six records from a stack of ten records?

7. In how many ways can the manager of a baseball team arrange his batting order if the pitcher must bat last and the catcher bat first?

8. How many odd integers each consisting of three digits can be made from the digits 1, 2, 3, 4, 5, and 6 if repetitions of a digit are not allowed?

9. There are twelve men available for the four backfield positions of a football team. If we assume that any of the 12 men can fill any position, in how many ways can these positions be filled?

10. How many different "four letter words" can be formed from the letters of the word "routines"?

3.
COMBINATIONS

In a permutation the order of the objects is important. For example, *abc* and *acb* are different permutations of the letters *a*, *b*, and *c*. Now let us consider the elements in an arrangement rather than the order in which the elements appear. Suppose we consider the set

$$S = \{a, b, c, d\}$$

The permutations of these four elements when they are taken three at a time are:

$$
\begin{array}{cccc}
abc & abd & acd & bcd \\
acb & adb & adc & bdc \\
bca & bda & cda & cbd \\
bac & bad & cad & cdb \\
cab & dab & dac & dbc \\
cba & dba & dca & dcb
\end{array}
$$

Of these twenty-four permutations, only four

$$
\begin{array}{cccc}
abc & abd & acd & bcd
\end{array}
$$

contain different elements. Each of these is called a **combination** of the four elements a, b, c, d, taken three at a time. *In a permutation the order of the elements is important; in a combination the elements and not their order is important.*

We shall now prove that the number of combinations of n objects taken k at a time, denoted by $\binom{n}{k}$ is given by the formula

$$\binom{n}{k} = \frac{n!}{k!(n-k)!}$$

We know that the number of permutations of n objects taken k at a time is $\frac{n!}{(n-k)!}$. Each of the combinations that can be formed from the k elements furnished $k!$ of these. In the example above, the combination abc furnished $3! = 6$ of the twenty-four permutations. Then

$$k! \cdot \binom{n}{k} = \frac{n!}{(n-k)!}$$

and

$$\binom{n}{k} = \frac{n!}{k!(n-k)!}$$

Example 1: Evaluate $\binom{7}{3}$.

Solution:

$$\binom{7}{3} = \frac{7!}{3!(7-3)!}$$

$$= \frac{7!}{3!4!}$$

$$= 35$$

Example 2: Find the number of four-men committees that can be chosen from a group of eight men.

Solution: Since a committee consisting of *A*, *B*, *C*, and *D* is the same as one consisting of *B*, *C*, *A*, and *D*, we are looking for the number of combinations of eight persons taken 4 at a time.

$$\binom{8}{4} = \frac{8!}{4!(8-4)!}$$

$$= \frac{8!}{4!4!}$$

$$= 70$$

Consider **Pascal's Triangle** shown in Table 8.1. The entries in the table are the values of $\binom{n}{r}$. Table 8.1 shows part of Pascal's Triangle for the values of *n* from 0 through 8. The rows of the table correspond to the

TABLE 8.1
Pascal's triangle

n \ r	0	1	2	3	4	5	6	7	8
0	1								
1	1	1							
2	1	2	1						
3	1	3	3	1					
4	1	4	6	4	1				
5	1	5	10	10	5	1			
6	1	6	15	20	15	6	1		
7	1	7	21	35	35	21	7	1	
8	1	8	28	56	70	56	28	8	1

values of n; the columns, to the values of r. The first and last entries in each row are 1 because $\binom{n}{0} = \binom{n}{n} = 1$. The entry other than the first or last in each row is the sum of the entry immediately above it and the entry to the left of that one. Thus, for example, the entry (6) for $n = 4$, $r = 2$, appearing in the fifth row and the third column of the body of the table is the sum of the entry in the fourth row and the third column (3) and that in the fourth row and the second column (3).

Exercise 3

1. Evaluate.

(a) $\binom{4}{2}$ (c) $\binom{8}{2}$ (e) $\binom{12}{9}$

(b) $\binom{7}{3}$ (d) $\binom{10}{5}$ (f) $\binom{20}{8}$

2. In how many ways can a committee of eight men be chosen from a group of fifteen men?

3. A television dealer has twelve new TV sets to display. If he has room for only three at a time in his show window, in how many different ways can he select three of the TV sets to display?

4. How many three-men teams can be selected from eighteen men who are eligible for the team?

5. There are six points, no three of which are on the same straight line. How many lines are determined by these points?

6. How many five-card poker hands can be dealt from a deck of 52 playing cards?

7. An employer wishes to fill nine positions with five men and four women. In how many different ways can these positions be filled if 12 men and 10 women who are equally qualified apply?

8. A box contains 12 size D flashlight batteries, 10 good ones and 2 bad ones. In how many different ways can three batteries be chosen such that 2 are good and 1 is bad?

9. Find the number of ways seven bulbs, three green and four red, can be selected from a box containing ten green bulbs and eight red ones.

10. A committee of three men is to be chosen from a group of six men. If a certain pair of men must serve on the same committee, but do not have to serve on all committees, how many committees can be formed.

11. Write out the entries that would be in the rows for $n = 9$ and $n = 10$ in Pascal's Triangle, Table 8.1.

4.
SAMPLE SPACES
AND EVENTS

Suppose we toss a coin. When a coin is tossed there are two possible ways that it can fall; showing heads or showing tails. Each of these two ways is called an **outcome** of the **experiment** of tossing a coin.

Now let us perform the experiment of tossing a die. A die is a small cube symmetrical in shape and uniform in material, whose faces are marked with 1, 2, 3, 4, 5, and 6 dots as shown in Figure 8.3. When a die is tossed the possible outcomes are 1, 2, 3, 4, 5, or 6 dots showing on the top face.

When an experiment is performed a set containing all possible outcomes of the experiment is called a **sample space** of the experiment. A sample space of the experiment of tossing a coin is

$$S_1 = \{H, T\}$$

where H stands for heads showing and T stands for tails showing. A sample space for the experiment of tossing a die is

$$S_2 = \{1, 2, 3, 4, 5, 6\}$$

Any subset of a sample space is called an **event** of the sample space. For example, the event, E, "an even number showing" of the sample space S_2 is

$$E = \{2, 4, 6\}$$

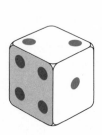

FIGURE 8.3

Example 1: Suppose two coins are tossed. What is a sample space of the experiment?

Solution: We can find all possible outcomes of this experiment by constructing a tree diagram as shown in Figure 8.4.

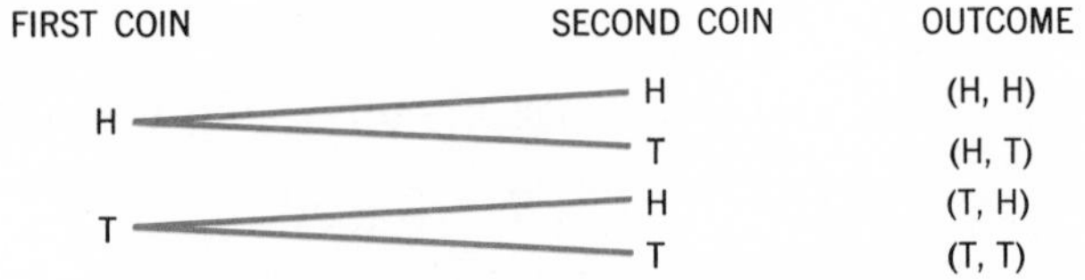

FIRST COIN		SECOND COIN	OUTCOME
H		H	(H, H)
		T	(H, T)
T		H	(T, H)
		T	(T, T)

FIGURE 8.4

The possible outcomes are (H, H)† (meaning heads on the first coin and heads on the second coin), (H, T) (meaning heads on the first coin and tails on the second coin), (T, H), and (T, T). A sample space is

$$\{(H, H), (H, T), (T, H), (T, T)\}$$

Example 2: Two dice are tossed. How many possible outcomes of this experiment are there?

Solution: The possible outcomes on the first die are 1, 2, 3, 4, 5, or 6 dots showing. The possible outcomes on the second die are 1, 2, 3, 4, 5, or 6 dots showing. By the fundamental counting principle there are $6 \cdot 6 = 36$ outcomes.

Example 3: Two cards are drawn from a deck of six cards numbered 1, 2, 3, 4, 5, and 6. How many outcomes are there? List the elements of a sample space.

Solution: There are 15 outcomes:

$$\binom{6}{2} = \frac{6!}{2!4!} = 15$$

A sample space is

$$\{(1, 2), (1, 3), (1, 4), (1, 5), (1, 6), (2, 3), (2, 4), (2, 5),$$
$$(2, 6), (3, 4), (3, 5), (3, 6), (4, 5), (4, 6), (5, 6)\}$$

where $(1, 2)$ means 1 on one card and 2 on the other card.

† (H, H) is often denoted by *HH*.

5.
PROBABILITY
OF AN EVENT

A sample space of an experiment is the set of all possible outcomes, denoted by $o_1, o_2, \ldots, o_n$:

$$S = \{o_1, o_2, o_3, \ldots, o_n\}$$

An event, E, of S, containing exactly one outcome is called a **simple event** of S. Any event that is not a simple event is called a **compound event.** Every compound event can be expressed as the union of simple events. For example, the compound event

$$F = \{o_1, o_2, o_3, o_4\} = \{o_1\} \cup \{o_2\} \cup \{o_3\} \cup \{o_4\}$$

To each simple event $E = \{o_i\}$ of S, we assign a number called the **probability** of E. We denote the probability of an event E by the symbol $P(E)$, read: the probability of event E.

If each of n outcomes of an experiment is as likely to happen as any other we say that each outcome is **equally likely** to occur and we agree to assign probability $\frac{1}{n}$ to each simple event. In this book all of the experiments shall be such that all of the outcomes are equally likely to occur. With this agreement, it can be proved, although we shall not do it here, that the probability of any event E of S is

$$P(E) = \frac{\text{number of elements in } E}{\text{number of elements in } S}$$

Denoting the number of outcomes in E by $n(E)$ and the number of outcomes in S by $n(S)$, we have

$$P(E) = \frac{n(E)}{n(S)}$$

Note that

$$P(S) = \frac{n(S)}{n(S)} = 1$$

and

$$P(\emptyset) = \frac{n(\emptyset)}{n(S)} = 0$$

We see from the above that *the probability of an event that is certain to occur is 1* and *the probability of an event that cannot occur is 0*. The probability of an event, E, of S, then is a number between 0 and 1 inclusive. If $E = S$, then $P(E) = 1$, if $E = \emptyset$, then $P(E) = 0$.

Example 1: Consider the experiment of tossing a single die with sample space

$$S = \{1, 2, 3, 4, 5, 6\}$$

What is the probability of the event, E, "a number less than 5 showing"?

Solution: The event, E, "a number less than 5 showing" is

$$E = \{1, 2, 3, 4\}$$

and $n(E) = 4$. Then

$$P(E) = \frac{n(E)}{n(S)} = \frac{4}{6} = \frac{2}{3}$$

Example 2: A box contains 4 red beads and 3 blue beads. The beads are exactly alike except for color. Two beads are selected at random from the box. An item is selected **at random** from a group of items if the selection procedure is such that each item in the group is equally likely to be selected. What is the probability that both beads are red?

Solution: Since 2 beads are selected from 7 beads, the number of possible outcomes is the number of combinations of 7 things taken 2 at a time. Then

$$n(S) = \binom{7}{2} = \frac{7!}{2!5!} = 21.$$

The number of outcomes in the event E, "both of the beads selected are red" is the number of combinations of 4 things taken 2 at a time. Then

$$n(E) = \binom{4}{2} = 6$$

and

$$P(E) = \frac{6}{21} = \frac{2}{7}$$

Since an event E of S is a subset of S, its complement, denoted by E' and called the **complementary event** of E, is the set of all outcomes in S that are not elements of E. If $n(E) = k$, and $n(S) = n$, $k \leq n$, then $n(E') = n - k$. We see that

$$P(E') = \frac{n(E')}{n(S)} = \frac{n - k}{n} = \frac{n}{n} - \frac{k}{n} = 1 - P(E)$$

hence

$$P(E') + P(E) = 1$$

Observe that if E occurs, then E' cannot occur. We may use the notation E' to state that an event E cannot occur.

Example 3: Two coins are tossed. What is the probability that the event, E, "two heads showing" does not occur?

Solution: The event, E, "two heads showing" is

$$E = \{HH\}$$

and the sample space S is

$$S = \{HH,\ HT,\ TH,\ TT\}$$

Then

$$P(E) = \frac{n(E)}{n(S)} = \frac{1}{4}$$

The probability that the event "two heads showing" does not occur is

$$P(E') = 1 - P(E) = 1 - \frac{1}{4} = \frac{3}{4}$$

Exercise 4

1. A coin is tossed. What is a sample space?
2. Two coins are tossed.
   (a) What is a sample space?
   (b) List the elements of the event "heads showing on both coins."
   (c) List the elements of the event "heads showing on the first coin."
3. Two dice are tossed.
   (a) List the elements in a sample space. (Use the notation $(1, 2)$ to denote 1 showing on the first die and 2 showing on the second.)
   (b) How many elements are in the sample space?

(c) List the elements of the event "the sum of the numbers showing on the two dice is 7."

(d) List the elements of the event "a 6 showing on one of the dice."

4. A card is drawn from an ordinary deck of 52 playing cards.

(a) How many possible outcomes are there?

(b) How many elements are in the event "drawing a queen"?

(c) How many elements are in the event "drawing a face card"? (A face card is a jack, queen, or king.)

5. A penny, a nickel, and a dime are tossed.

(a) Use a tree diagram to list the possible outcomes of the experiment.

(b) List the elements of the event "exactly two heads showing."

(c) List the elements of the event "one head and two tails showing."

(d) List the elements of the event "at least one tail showing."

6. A box contains 20 screws. A carpenter selects 5 at random. How many elements are in a sample space?

7. Four coins are tossed. What is the subset of the sample space defining each of the following events?

(a) All heads showing.

(b) An equal number of heads and tails showing.

(c) An even number of tails showing. (*Hint:* zero is an even number.)

8. Four burned-out Christmas tree bulbs are put in a box containing 21 good bulbs. The burned out bulbs cannot be distinguished from the good bulbs. Two bulbs are selected at random from the box. What is the probability that both bulbs will be good?

9. Two dice are tossed. What is the probability of each of the following events?

(a) The sum of the numbers showing is 7.

(b) A double is showing (a double means both dice show the same number).

(c) A 6 showing on one die.

(d) The sum of the number showing is divisible by 4.

(e) The sum of the number showing is less than 9.

10. A bag contains 25 jelly beans, 6 red, 5 green, 12 white, and 2 black. Two jelly beans are picked at random from the bag. What is the probability of each of the following events?

(a) Both are red.   (d) One is red and one is white.

(b) Both are green.   (e) One is green and one is red.

(c) Both are white.   (f) Both are black.

11. Three letters are chosen at random from the word "gamble." Find the probability of each of the following events.

(a) One of the letters is g.

(b) One of the letters is a vowel.

(c) The letter g is not selected.

12. A box contains ten slips of paper numbered 1, 2, 3, 4, 5, 6, 7, 8, 9, and 10. A slip is drawn at random from the box. Find the probability of each of the following events.

(a) Drawing a slip numbered 10.

(b) Drawing a slip with a number less than 8.

(c) Drawing a slip with a prime number on it.

13. An employer wishes to fill two positions from a group of 15 equally qualified employees, 8 men and 7 women. If he selects the two without knowing the identity of the employees, say by assigning a number to each and picking two numbers from a hat, what is the probability of the following events?

(a) One of the employees is a woman.

(b) One of the employees is a man.

(c) Both of the employees are women.

(d) Both of the employees are men.

14. The probability that a student will receive a grade of A in his mathematics class is 0.125. What is the probability that he will not receive a grade of A?

15. Lee's mother and Lee have very dissimilar taste in clothes. On her birthday Lee's mother sends Lee a dress for a present. The probability that Lee will like the dress is 0.01. What is the probability that Lee will not like the dress?

16. Two regular tetrahedra (a regular tetrahedron is a regular solid having four congruent faces) have faces marked with the numerals 1, 2, 3, and 4. They are tossed in the air and allowed to fall on a table. An outcome is described by the numeral on the bottom face.

(a) How many possible outcomes are there?

(b) What is the probability of the event "the sum of the numbers is 8"?

(c) What is the probability of the event "the sum of the numbers is 5"?

17. Two regular octrahedra (a regular octahedron is a regular solid having eight congruent faces) have faces numbered serially 1 through 8. We toss them and allow them to fall on a hard surface. We describe an outcome by the numerals on the bottom faces. What are the probabilities of the following events?

(a) The sum of the numbers is 8.

(b) The sum of the numbers is 16.

(c) The sum of the numbers is 7.

(d) The sum of the numbers is 11.

18. Two pairs of shoes have been thrown on the floor of a dark closet. You pick up the first two shoes you feel. What is the probability that you will have a matched pair?

19. There are twenty slips of paper in a hat. Each slip has a different name written on it. A slip of paper is drawn at random from the hat, the name on it read, and the slip returned to the hat. Let the names on the slips be represented by the first twenty letters of the alphabet: A, B, C, . . . , T.

(a) Is $P(A)$ equal to $P(T)$?

(b) What is $P(S) + P(J)$?

(c) What is $P(H) + P(Q) + P(B)$?

(d) What is $P(A) + P(B) + P(C) + \cdots + P(T)$?

20. An urn contains 100 ping pong balls. Fifty of the balls are numbers 1 through 50, inclusive. The rest are blank. Two balls are drawn at random from the urn. What is the probability of each of the following events?

(a) Both balls are blank?
(b) The balls numbered 6 and 8 are selected?
(c) Both balls have numerals on them.
(d) One ball is blank.

6.
MUTUALLY
EXCLUSIVE EVENTS

In computing probabilities the relation of two events of the sample space must be taken into consideration. If S is a sample space and E_1 and E_2 are two events of S, E_1 and E_2 are related in one of the following ways:

(1) E_1 and E_2 have no elements in common.
(2) E_1 and E_2 have some elements in common.

If E_1 and E_2 have no elements in common, that is $E_1 \cap E_2 = \emptyset$, they are called **mutually exclusive** events. If E_1 and E_2 are mutually exclusive events, then the probability of both of them occurring at the same time, that is the probability of the event E_1 and E_2, denoted by $E_1 \cap E_2$, is 0, since

$$P(E_1 \cap E_2) = \frac{n(E_1 \cap E_2)}{n(S)} = \frac{n(\emptyset)}{n(S)} = 0$$

If E_1 and E_2 are not mutually exclusive then $n(E_1 \cap E_2) \neq 0$ and

$$P(E_1 \cap E_2) = \frac{n(E_1 \cap E_2)}{n(S)}$$

Example 1: Two dice are tossed. What is the probability of the event "a sum of 7 and a 4 showing on one of the dice"?

Solution: There are 36 outcomes for this experiment, hence $n(S) = 36$. Let E_1 be the event "a sum of 7." Then

$$E_1 = \{(1, 6), (2, 5), (3, 4), (4, 3), (5, 2), (6, 1)\}$$

Let E_2 be the event "a 4 showing on one of the dice." Then

$$E_2 = \{(4, 1), (4, 2), (4, 3), (4, 4), (4, 5), (4, 6), (1, 4), (2, 4), (3, 4), (5, 4), (6, 4)\}$$

We are interested in the probability of

$$E_1 \cap E_2 = \{(3, 4), (4, 3)\}$$

We find

$$P(E_1 \cap E_2) = \frac{n(E_1 \cap E_2)}{n(S)} = \frac{2}{36} = \frac{1}{18}$$

Example 2: Two dice are tossed. What is the probability of the event "a sum of 10 and a 1 showing on one of the dice"?

Solution: Again $n(S) = 36$. The event, E_1, "a sum of 10" is

$$E_1 = \{(4, 6), (5, 5), (6, 4)\}$$

and the event, E_2, "a 1 showing on one of the dice" is

$$E_2 = \{(1, 1), (1, 2), (1, 3), (1, 4), (1, 5), (1, 6), (2, 1), (3, 1), (4, 1), (5, 1), (6, 1)\}$$

Since E_1 and E_2 are mutually exclusive events, $E_1 \cap E_2 = \emptyset$ and

$$n(E_1 \cap E_2) = 0, \text{ hence } P(E_1 \cap E_2) = 0$$

Suppose we are interested in event E_1 or event E_2 occurring. In this case, we are satisfied if either E_1 or E_2 or both events occur. We denote the event E_1 or E_2 by $E_1 \cup E_2$. It can be proved, although we shall not do it here, that

$$P(E_1 \cup E_2) = P(E_1) + P(E_2) - P(E_1 \cap E_2)$$

If E_1 and E_2 are mutually exclusive events, then $P(E_1 \cap E_2) = 0$, and the above equation reduces to

$$P(E_1 \cup E_2) = P(E_1) + P(E_2)$$

We shall illustrate the truth of the above formulas by examples.

Suppose we roll a pair of dice. What is the probability of the event "a sum of 7 or a sum of 11"? Here E_1 is the event "a sum of 7" and E_2 is the event "a sum of 11." Then

$$E_1 = \{(1, 6), (2, 5), (3, 4), (4, 3), (5, 2), (6, 1)\}$$
$$E_2 = \{(5, 6), (6, 5)\}$$

Notice that E_1 and E_2 are mutually exclusive events. Now

$$E_1 \cup E_2 = \{(1, 6), (2, 5), (3, 4), (4, 3), (5, 2), (6, 1), (5, 6), (6, 5)\}$$

The sample space is the familiar sample space with thirty-six elements. Hence

$$P(E_1) = \frac{6}{36}, \qquad P(E_2) = \frac{2}{36}$$

and

$$P(E_1 \cup E_2) = \frac{8}{36}$$

Notice

$$P(E_1 \cup E_2) = \frac{8}{36} = \frac{6}{36} + \frac{2}{36} = P(E_1) + P(E_2)$$

Now let us find the probability of the event "a sum of 7 or a 4 showing on one die." Then

$$E_1 = \{(1, 6), (2, 5), (3, 4), (4, 3), (5, 2), (6, 1)\}$$
$$E_2 = \left\{ \begin{array}{l} (1, 4), (2, 4), (3, 4), (4, 4), (5, 4), (6, 4), \\ (4, 1), (4, 2), (4, 3), (4, 5), (4, 6) \end{array} \right\}$$
$$E_1 \cap E_2 = \{(3, 4), (4, 3)\}$$
$$E_1 \cup E_2 = \left\{ \begin{array}{l} (1, 6), (2, 5), (3, 4), (4, 3), (5, 2), (6, 1), (1, 4), (2, 4), \\ (4, 4), (5, 4), (6, 4), (4, 1), (4, 2), (4, 5), (4, 6) \end{array} \right\}$$

Then

$$P(E_1 \cup E_2) = \frac{15}{36}$$

But

$$P(E_1) = \frac{6}{36} = \frac{1}{6}$$

$$P(E_2) = \frac{11}{36}$$

$$P(E_1 \cap E_2) = \frac{2}{36}$$

and

$$P(E_1 \cup E_2) = P(E_1) + P(E_2) - P(E_1 \cap E_2) = \frac{6}{36} + \frac{11}{36} - \frac{2}{36} = \frac{15}{36}$$

Exercise 5

1. Two dice are tossed. What is the probability of each of the following events?
(a) A sum of 3 or a sum of 2.
(b) A sum of 7 or a sum of 9.
(c) A sum of 8 or a 4 showing on one of the dice.
(d) A sum of 3 and a sum of 2.
(e) A sum of 8 and a 4 showing on one of the dice.

2. There are 4 green, 5 yellow, and 2 red beads in a box. All of the beads are the same size and shape. One bead is picked at random from the box.
(a) What is the probability of picking a yellow bead?
(b) What is the probability of picking a red bead?
(c) What is the probability of picking a green or a red bead?
(d) What is the probability of picking a yellow or a green bead?
(e) What is the probability of picking a black bead?

3. The names of 29 cats are placed in a box at a kennel; 4 of the cats are Persian; 12 are Siamese; 5 are Manx; and 8 are domestic short hairs (alley cats). Each cat is owned by a different person. A name is drawn at random from the box. A free week-end of boarding is awarded to the owner of the cat whose name is drawn.
(a) What is the probability that the winner will own a Persian cat?
(b) What is the probability that the winner will own a Siamese cat?
(c) What is the probability that the winner will own a Manx or an alley cat?
(d) What is the probability that the winner will own a Persian or a Siamese cat?

4. A bag has been filled with 100 jelly beans of assorted colors: there are 30 red, 25 yellow, 20 green, 20 pink, and 5 black. The jelly beans are thoroughly mixed so that the chance of getting any one jelly bean is as likely as another. If you pick one bean from the bag, what is the probability you get:
(a) Black?   (c) Pink or red?
(b) Red or green?   (d) Yellow or green?

5. Two coins are tossed. If E_1 is the event "tails showing on the first coin" and E_2 is the event "the coins match," what is the probability of tails on the first coin or the coins match?

6. Two dice, one red and one green are rolled. What is the probability of a sum of 11 or a number different from 6 on the green die?

7. King Dumkoff has five daughters, Ann, Bertha, Carrie, Doris, and Edna. He has chosen three of his favorite knights as husbands for three of his daughters. He selects the three to be married at random by writing their names on cards, placing the cards in the Cardinal's hat, and selecting three at random. What is the probability that both Ann and Bertha or both Carrie and Doris will be chosen to marry the knights?

8. Two cards are drawn from an ordinary deck of 52 playing cards. What is the probability that both are red or that both are queens?

9. A jacket factory manufactures jackets in three sizes: small, medium, and large. In a lot of 1000 small jackets, 50 have been made too small and 20 have been made too large. The remainder are acceptable according to factory standards. An inspector selects a jacket at random from the lot.

(a) What is the probability that it will be too small?

(b) What is the probability that it will be too large?

(c) What is the probability that it will be too small or too large?

10. Two cards are drawn at random from an ordinary deck of 52 playing cards. What is the probability of each of the following events?

(a) Both cards are black.

(b) Both cards are aces.

(c) Both cards are black aces.

(d) Both cards are black or aces.

7.
CONDITIONAL
PROBABILITY

Often we need to determine the probability of an event when we know that another event has already occurred. For example, suppose we toss a single die. We may be asked the probability of a 6 showing knowing that an even number is showing.

A sample space of this experiment is

$$S = \{1, 2, 3, 4, 5, 6\}$$

To find the probability of a 6 showing when we know that an even number is showing, we don't need all of the outcomes in S. We can work with the event E_2 whose elements are the outcomes that are even numbers. Thus

$$E_2 = \{2, 4, 6\}$$

The number of outcomes in E_2 is three. Of these outcomes only one is 6, the outcome we desire. Then the event, E_1, "a 6 showing" is

$$E_1 = \{6\}$$

The probability of E_1 is

$$P(E_1) = \frac{n(E_1)}{n(E_2)} = \frac{1}{3}$$

Example 1: Suppose two dice are tossed. What is the probability of a double knowing that the sum showing is 8?

Solution: A sample space contains the familiar 36 outcomes. But to find the probability of a double knowing that a sum of 8 is showing, we do not need all of these 36 outcomes; we can work with the outcomes of the event E_2 "a sum of 8 showing" since we know that event E_2 has occurred. Now

$$E_2 = \{(2, 6), (3, 5), (4, 4), (5, 3), (6, 2)\}$$

Of the outcomes of E_2 those that are doubles are the elements of E_1

$$E_1 = \{(4, 4)\}$$

The probability of E_1 is

$$P(E_1) = \frac{n(E_1)}{n(E_2)} = \frac{1}{5}$$

We can derive a general formula for finding the probability in problems like those above. Let S be a sample space of an experiment. Let E_1 and E_2 be two events of S, $E_2 \neq \emptyset$. We are interested in the probability of E_1 knowing that event E_2 has occurred. We call this the **conditional probability** of E_1 given that E_2 has already occurred and write $P(E_1 | E_2)$. We read this symbol "The probability of event E_1 when E_2 has occurred."

Since we know that event E_2 has occurred, not all of the outcomes of S can occur. Only those outcomes in S that are also in E_2 can occur. We see also that not all of the outcomes of E_1 can now occur. Only those outcomes that are also in E_2 can possibly occur. This set of outcomes are those elements of $E_1 \cap E_2$. Then

(1) $$P(E_1 | E_2) = \frac{n(E_1 \cap E_2)}{n(E_2)}$$

Dividing numerator and denominator of the right member of equation (1) by $n(S)$ we obtain

$$P(E_1 | E_2) = \frac{\dfrac{n(E_1 \cap E_2)}{n(S)}}{\dfrac{n(E_2)}{n(S)}}$$

but

$$\frac{n(E_1 \cap E_2)}{n(S)} = P(E_1 \cap E_2) \qquad \text{(Relative to } S\text{)}$$

and

$$\frac{n(E_2)}{n(S)} = P(E_2) \qquad \text{(Relative to } S\text{)}$$

Hence equation (1) may be written

$$\boldsymbol{P(E_1 | E_2) = \frac{P(E_1 \cap E_2)}{P(E_2)}}$$

Example 2: A box contains four cards numbered 1, 2, 3, and 4. When two cards are drawn at random from the box, what is the probability that the card numbered 3 is one of them knowing that one of the cards drawn is numbered 1?

Solution: Here the number of possible outcomes is the number of combinations of four things taken two at a time. Hence $n(S) = 6$. But

$$E_2 = \{(1, 2), (1, 3), (1, 4)\}$$
$$E_1 = \{(1, 3), (2, 3), (3, 4)\}$$
$$E_1 \cap E_2 = \{(1, 3)\}$$

Then

$$P(E_2) = \frac{1}{2} \qquad \text{and} \qquad P(E_1 \cap E_2) = \frac{1}{6}$$

hence

$$P(E_1 | E_2) = \frac{\frac{1}{6}}{\frac{1}{2}} = \frac{1}{3}$$

We defined conditional probability by the formula

$$P(E_1 | E_2) = \frac{P(E_1 \cap E_2)}{P(E_2)}$$

Multiplying each member of the above equation by $P(E_2)$ we obtain

(2) $$P(E_2) P(E_1 | E_2) = P(E_1 \cap E_2)$$

By replacing E_1 by E_2 and E_2 by E_1 in equation (2) we have

$$P(E_1)P(E_2 \mid E_1) = P(E_2 \cap E_1)$$

Since $E_1 \cap E_2 = E_2 \cap E_1$ we have

$$\boldsymbol{P(E_1 \cap E_2) = P(E_1)P(E_2 \mid E_1)}$$

This formula is called the **theorem of compound probabilities.**

Example 3: A bag contains three balls identical except for color. Two are red and one is green. Two balls are taken from the bag, one after the other. What is the probability that both are red?

Solution: Let E_1 be the event "first ball red" and E_2 be the event "second ball red". Then we want $P(E_1 \cap E_2)$. We use the formula

$$P(E_1 \cap E_2) = P(E_1)P(E_1 \mid E_2)$$

The probability that the first ball drawn is red is

$$P(E_1) = \frac{2}{3}$$

If the first ball drawn is red, there remains for the second draw two balls, one red and one green, Then

$$P(E_1 \mid E_2) = \frac{1}{2}$$

Hence

$$P(E_1 \cap E_2) = \frac{2}{3} \cdot \frac{1}{2} = \frac{1}{3}$$

Example 3: Bowl I contains 4 red and 6 green balls. Bowl II contains 4 red, 3 white, and 2 blue balls. A bowl is selected at random and one ball is drawn from the bowl selected. What is the probability that a red ball is obtained?

Solution: The data in the problem is summarized in Figure 8.5. Since we select a bowl at random, the probability that either bowl is selected is $\frac{1}{2}$ (these probabilities are shown on the branches from the starting point to the two bowls). Given the bowl selected, we are now given the conditional probabilities of

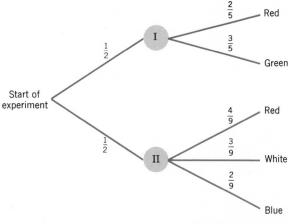

FIGURE 8.5

drawing a ball of a specified color. These probabilities are shown on each branch leading from a bowl to an outcome of the second trial (color of the ball).

The event, E, obtaining a red ball, can occur in two mutually exclusive ways: (E_1) selecting Bowl I and drawing a red ball; (E_2) selecting Bowl II and drawing a red ball. We see that the event E "obtaining a red ball" is the union of two mutually exclusive events, E_1 and E_2. But since E_1 and E_2 are mutually exclusive events

$$P(E_1 \cup E_2) = P(E_1) + P(E_2)$$

and

$$P(E) = P(\text{Bowl I and red}) + P(\text{Bowl II and red})$$

To find P(Bowl I and red) and P(Bowl II and red) we use the formula for the probability of the intersection of two events.

$$P(\text{Bowl I and red}) = \frac{1}{2} \cdot \frac{2}{5}$$

$$P(\text{Bowl II and red}) = \frac{1}{2} \cdot \frac{4}{9}$$

Then

$$P(E) = \frac{1}{2} \cdot \frac{2}{5} + \frac{1}{2} \cdot \frac{4}{9} = \frac{19}{45}$$

269

Exercise 6

1. Two dice are thrown. Given that a 4 is showing on one die, what is the probability that the sum is 10?

2. A bag contains 25 jelly beans, of which 5 are red, 10 are pink, 8 are white, and 2 are black. Two jelly beans are picked at random from the bag. If one is known to be red, what is the probability that the other is pink?

3. A box contains four cards marked A, B, C, and D. Three cards are drawn at random from the box. We know that C is on one card. What is the probability that D is on one of the cards?

4. Three coins are tossed and we are told that not all three show heads. What is the probability that there is an odd number of heads showing?

5. Jar I contains 2 black balls and 1 white ball; Jar II has 1 white and 3 red balls. A jar is selected at random and a ball is drawn. What is the probability that the ball selected is white?

6. There are 15 boys and 5 girls in one section of Mathematics 101 and 12 boys and 18 girls in another section. One student is selected at random from the two classes. What is the probability that the student selected will be a boy?

7. A store receives a lot of 50 batteries. There are 5 defective batteries in the shipment. A clerk picks two batteries at random. What is the probability that the first battery he selects is good and the second defective?

8. In a pantry there are two shelves containing foodstuffs. On one shelf there are 3 boxes of cherry gelatine and 5 boxes of lime gelatine. On the second shelf there are 2 boxes of lime gelatine and 3 boxes of strawberry. Mrs. Wise goes to the pantry and selects a box of gelatine at random. What is the probability that the box she selects is lime?

9. An instructor has 25 students in his class. Unknown to him, 5 are repeating the course. He selects two students at random to do a problem on the board. What is the probability that the first student he selects is a repeater and the second will be taking the course for the first time?

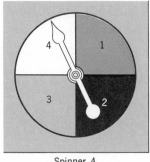
Spinner *A*

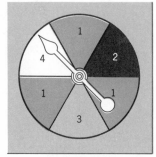
Spinner *B*

FIGURE 8.6

10. In a game there are two spinners as shown in Figure 8.6. A player may select either spinner when he plays. He puts the pointer in motion and moves on the game board the appropriate number of spaces. What is the probability that the spinner the first player picks will stop on the region marked by a 1?

8.
MEASURES OF
CENTRAL TENDENCY

Jerome Cardan's book *Liber de Ludo Aleae* (Book on Games of Chance) was probably the first book written about probability. Cardan was a very colorful personality. He was a lawyer and a physician by profession, but he spent so much time gambling and pursuing interests outside of his profession that his medical practice suffered. His writings include such diverse topics as mathematics, games of chance, physics, astronomy, and death (he predicted the date of his death, and when the day arrived and he was still alive, he committed suicide to make his prediction come true).

Although Cardan was one of the earliest pioneers of the theory of probability, Blaise Pascal and Pierre Fermat are usually credited with putting the study of probability on a scientific basis. Probability, in turn, laid the foundation on which modern statistics is based.

One of the fundamental concepts of statistics is the idea of **average** which is familiar to everyone. No matter how little we know of mathematics, we are at home with batting averages, average salaries, average families, and the like. We realize that the purpose of an average is to represent a set of individual values or measures by one measure that is most representative of the members of the set. The idea of average is so handy that it is not surprising that several kinds of averages have been invented.

Averages are called **measures of central tendency.** The three measures of central tendency in common use are the **arithmetic mean,** the **median,** and the **mode.** All three measures are averages. Each has its advantages and limitations. The nature of the particular distribution of the data and the purpose for which the average is chosen determine which measure of central tendency is used.

9.
THE ARITHMETIC MEAN

The **arithmetic mean,** usually referred to as the **mean,** of a set of numbers is obtained by adding the numbers and dividing the sum by the number of numbers in the set. If the symbols X_1, X_2, X_3, $\ldots$, X_n denote the numbers in a set of N numbers, the arithmetic mean, designated by $\overline{X}$, is given by

$$\overline{X} = \frac{X_1 + X_2 + X_3 + \cdots + X_n}{N}$$

A highly useful symbol in mathematics is the uppercase Greek letter sigma, Σ, used in **summation notation.** The sum $X_1 + X_2 + X_3 + \cdots + X_n$ is compactly denoted by the notation

$$\sum_{i=1}^{n} X_i$$

The symbol $\displaystyle\sum_{i=1}^{n} X_i$ is read "the sum for $i = 1$ to $i = n$ of X_i." We see, then, that the formula for the arithmetic mean may be denoted by

$$\overline{X} = \frac{1}{N} \sum_{i=1}^{n} X_i$$

Suppose the weights of five wrestlers to the nearest pound are: 204, 190, 240, 211, and 240. Then the mean weight of the wrestlers is 217 pounds:

$$\frac{204 + 190 + 240 + 211 + 240}{5} = \frac{1085}{5} = 217$$

If the numbers to be added in finding the mean are large, a considerable amount of work may be saved in the computation by averaging deviations from an **assumed mean.** We illustrate this method in computing the mean weight of the wrestlers. First let us assume a value of the mean, say 200 pounds. Next we find how much each weight differs from the assumed mean. In the example of finding the mean weight of the wrestlers, these deviations are:

$$204 - 200 = +4$$
$$190 - 200 = -10$$
$$240 - 200 = +40$$
$$211 - 200 = +11$$
$$240 - 200 = +40$$

We now obtain the mean of these deviations.

$$\frac{4 + (-10) + 40 + 11 + 40}{5} = \frac{85}{5} = 17$$

Adding this correction to the assumed mean we obtain the true mean

$$200 + 17 = 217$$

If we designate the assumed mean as X_0 and the mean of the deviations as $\overline{Y}$, the above example shows that the mean, $\overline{X}$, is equal to the assumed mean plus the mean of the deviations, $\overline{Y}$. Then

$$\overline{X} = X_0 + \overline{Y}$$

Example 1: The attendance figures for four games of a recent world series were 55,942, 55,947, 54,445, and 54,458. What was the daily average attendance?

Solution: Here "average" means the mean attendance. Looking at the given figures, we see that the average daily attendance was around 55,000. Using this figure as the assumed mean we have the following deviations:

$$55,942 - 55,000 = +942$$
$$55,947 - 55,000 = +947$$
$$54,445 - 55,000 = -555$$
$$54,458 - 55,000 = -542$$

Then the mean of the deviations, $\overline{Y}$, is

$$\overline{Y} = \frac{942 + 947 - 555 - 542}{4} = \frac{792}{4} = 198$$

The mean daily attendance, $\overline{X}$, was

$$\overline{X} = 55,000 + 198 = 55,198$$

Example 2: To obtain a grade of C in a mathematics course, Jake needs a mean score of 72 on four examinations. On the first three examinations he has scores of 64, 88, and 70. What is the lowest score Jake can make on the fourth examination and receive a grade of C in the course?

Solution: Since the mean score needed for a C grade is 72, the total number of points needed on the four examinations is $4 \cdot 72 = 288$. The sum of the scores on Jake's three examinations is

$$64 + 88 + 70 = 222$$

The difference between 288 and 222 is 66. This is the lowest Jake can get on the fourth examination and get a grade of C.

Check: $$\frac{64 + 88 + 70 + 66}{4} = 72$$

Example 3: The Dog House Pet Motel and Grooming Shop has six employees, whose annual salaries are given below.

Position	Number	Salary
Manager-owner	1	$10,000
Groomer	2	6,000
Bookkeeper	1	4,100
Handymen	2	3,000

What is the mean annual salary of the employees of the shop?

Solution: We find the sum of all the salaries of the employees. This is done as follows.

$$
\begin{array}{l}
1 \cdot 10,000 = 10,000 \\
2 \cdot 6,000 = 12,000 \\
1 \cdot 4,100 = 4,100 \\
2 \cdot 3,000 = \underline{6,000} \\
 32,100
\end{array}
$$

Then the mean annual salary is $5,350:

$$\frac{32,100}{6} = 5,350$$

10.
THE MEDIAN

The **median** of a set of N numbers is the middle number when they are arranged in sequential order. The median of the numbers

165, 185, 199, 242, 300, 307, 310

is 242.

If N is even, it is customary to take the median as the mean of the two middle numbers. The median of

126, 134, 157, 199, 199, 204

is the mean of 157 and 199.

$$\frac{157 + 199}{2} = \frac{356}{2}$$
$$= 178$$

11.
THE MODE

The **mode** of a set of N numbers is the number that occurs most frequently. It is possible for a set of numbers to have more than one mode. If two or more numbers occur the maximum number of times, each number is called a mode. The mode of

18, 19, 24, 24, 27, 29, 31, 31, 31, 37, 39

is 31.

The set

26, 27, 29, 29, 29, 34, 37, 37, 37, 39, 42

has two modes, 29 and 37.

12.
COMPARING
THE THREE AVERAGES

The most frequently used measure of central tendency is the arithmetic mean. It is easy to compute and define, takes all the measures into consideration, and is well defined for algebraic manipulation. The mean is a magnitude average. It is the value each measure would have if all were equal. One of the chief advantages of the mean is its reliability in sampling. The mean is not typical if the data include a few extreme values in one direction.

The median is a positional average. The number of measures greater than the median is the same as the number that is less. The median, although easy to calculate and define, is not influenced by extreme measures. If measures are concentrated in distinct and widely separated groups, the median may be of little value as a measure of central tendency.

The mode is thought to be the most typical measure of all (since it is the most frequently occurring), but it does not take account of the other values in the data. It is less important than the median because of its ambiguity.

Exercise 7

1. Find the mean, median, and mode of each of the following sets.
(a) 2, 3, 4, 7, 5, 3, 3, 10, 4, 9.
(b) 24, 21, 20, 25, 21, 27.
(c) 3, 3, 1, 2, 2, 3, 6, 4, 2, 3, 4, 4, 2, 3, 3.

2. Find the mean of the numbers 1 through 10.

3. Find the mean of 1461, 1464, 1460, 1467, 1470, and 1462. Use the assumed mean 1465.

4. Find the mean of 14.22, 14.25, 14.56, 14.77, 14.11, and 14.32. Use the assumed mean 14.25.

5. The barrels of crude oil produced daily in four of the United States is given below. What is the mean number of barrels of crude oil produced in these four states?

Texas	54,400
Oklahoma	23,300
Louisiana	12,400
New Mexico	7,800

6. An elevator is designed to carry a maximum load of $2\frac{1}{2}$ tons. If it is loaded with 30 passengers having an average (mean) weight of 150 pounds is there any danger that it might fail?

7. Listed below are the annual salaries in a small retail store.

Manager	$14,000
Assistant Manager	12,000
Buyer	10,000
Bookkeeper	6,000
Sales Clerk	5,000
Stock Clerk	3,000
Janitor	2,500

(a) Find the mean annual salary of the store employees.

(b) Find the median annual salary of the store employees.

8. The daily output of fifteen factory workers on a piece-work job was: 111, 119, 116, 128, 115, 140, 121, 128, 116, 118, 128, 130, 131, 133, and 131.

(a) What was the mean number of pieces produced?

(b) What is the median number of peices produced?

(c) What was the mode number of pieces produced?

9. A drugstore owner sold the following five brands of shampoo. In reordering would the druggist be interested in mean, median, or mode? Why?

Brand	Number of bottles sold
A	4
B	6
C	30
D	12
E	15

10. If $\overline{X}$ is the mean of $X_1, X_2, \ldots, X_n$, what is the mean of:

(a) $cX_1, cX_2, \ldots, cX_n$?

(b) $X_1 + k, X_2 + k, \ldots, X_n + k$?

11. The following wages were paid by a factory:

Annual wage	Number receiving this wage
$50,000	1
25,000	1
15,000	2
10,000	5
8,000	12
5,000	40
3,000	4

(a) What is the mean wage?

(b) What is median wage?

(c) What is the mode wage?

(d) Which average would a union leader use in presenting arguments for salary increases?

(e) Which average would the factory owner use in advertising for help?

(f) Which average would the Internal Revenue Service use in speaking of the taxable income?

12. Suppose you are a buyer for a department store. Which type of average would probably be of the greatest value to you?

13. What average would be used in the following?

(a) In meteorology, for obtaining the average rainfall.

(b) In the garment industry, for determining the number of shirts of each size to manufacture?

(c) In business, for computing average wages.

(d) In education, for determining the average age of sixth grade children.

14. The contributions by alumni of a small college to a building fund are given below.

Amount in dollars	Number of contributions
1,000	1
500	2
100	8
50	12
25	20
15	5
10	7

(a) What is the mean contribution?

(b) What is the median contributions?

(c) What is the mode contribution?

(d) What is the "usual" contribution?

(e) What average best represents this data?

15. To obtain a grade of B in a mathematics course, Joyce needs a mean score of 84 on five examinations. On the first four examinations she has scores of 93, 80, 79, and 98. What is the lowest score that Joyce can make on the fifth examination and get a grade of B in the course?

16. Six pennies are tossed 100 times. The number of pennies (N) falling with heads showing and the corresponding frequencies (f) are as follows.

$$N \quad 0 \quad 1 \quad 2 \quad 3 \quad 4 \quad 5 \quad 6$$
$$f \quad 2 \quad 10 \quad 24 \quad 35 \quad 22 \quad 6 \quad 1$$

Find the mean number of heads for this distribution.

17. The mean of a certain set of numbers is 82 and the median is 76. If each number in the set is increased by 4, what is the

(a) Mean of the new set?

(b) Median of the new set?

18. Hope needs a grade of A in her English class if she expects to graduate in June. To get a grade of A she must have a mean score of 92 on six examinations.

Her scores on the first five examinations are 94, 87, 91, 88, and 90. Is it possible for Hope to get an A in English if the highest score possible on the sixth examination is 100 points? Why?

19. A customer asked Maisie, a waitress at the Old Salt Bar, what her average tip was. She told him $1.00. The customer gave her a tip of $1.00. After he left, Maisie told the bartender she hoped that she served that same customer again as he was a generous tipper. Did Maisie lie about her average tip or could the customer and Maisie have been thinking about different types of averages? Suppose Maisie had received tips as follows: 10 of 25¢, 9 of 50¢, 4 of 75¢, 5 of $1.00, 1 of $5.00, and 2 of $10.00. (The last three were for special services given to the customer.) What average was Maisie thinking of when she said the customer was generous? What average was the customer thinking of?

20. The Hayes Moving Company has eight employees. The median weekly salary is $115; the mean of the weekly salaries is $123; and the mode of the weekly salaries is $110. What is the total amount of the weekly payroll?

13.
MEASURES
OF VARIABILITY

The grades for two tests given to a mathematics class were the following:

Test 1	Test 2
100	80
90	75
80	75
75	70
75	70
70	70
70	70
70	70
65	65
65	65
65	65
55	60
40	60
40	55
40	55

The mean grade in Test 1 is 66.7 which we round to 67. On Test 2 the mean is again 67. If we knew only the mean we probably would assume that the class did equally well on each test. Looking at the grades on the two tests, however, we see that this is not the case. There are both higher and lower grades on the first test. There is less variability on the second test. If each student had received a grade of 67 on the second test, the instructor would certainly not consider this performance of the class comparable to the performance of the class on the first test. We need, in addition to the measures of central tendency, a statistical measure to indicate the extent to which the variates tend to spread out. Such measures are called **measures of variability.**

The simplest measure of variability is the **range.** It tells the difference between the highest and the lowest measures. The range for Test 1 is $100 - 40 = 60$. The range for Test 2 is $80 - 55 = 25$. This indicates more variability on Test 1.

The range is a poor measure of variability because it depends only on two measures, telling us nothing about the remaining measures. Because the range has shortcomings as a measure of variability, others have been devised. One of them is the standard deviation denoted by the Greek letter σ.

To see intuitively how the standard deviation was derived, let us consider deviation from the mean by examining three sets of data:

$$\text{Set 1: } 15, \ 15, \ 15, \ 15, \ 15, \ 15$$
$$\text{Set 2: } 12, \ 12, \ 12, \ 18, \ 18, \ 18$$
$$\text{Set 3: } 10, \ 12, \ 14, \ 16, \ 18, \ 20$$

In each case, the mean is 15. What does this average mean? We might predict any number chosen at random from the set will not deviate much from 15. In fact, the errors we incur in selecting numbers greater than 15 are compensated by the errors in selecting numbers less than 15. For example, in Set 3,

$$\text{The difference between 10 and 15 is } 10 - 15 = -5$$
$$\text{The difference between 12 and 15 is } 12 - 15 = -3$$
$$\text{The difference between 14 and 15 is } 14 - 15 = -1$$
$$\text{The difference between 16 and 15 is } 16 - 15 = 1$$
$$\text{The difference between 18 and 15 is } 18 - 15 = 3$$
$$\text{The difference between 20 and 15 is } 20 - 15 = 5$$

Each of these difference is called a **deviation** from the mean. The total of these deviations is

$$-5 + (-3) + (-1) + 1 + 3 + 5 = 0$$

Similarly, the sum of the deviations in each of the other two sets of measures is 0. We see then for a set of measures, the sum of the deviations from the mean is always zero, and hence is of no value in measuring variability. It does, however, furnish a useful means for checking the accuracy of a calculated mean. The sum of the deviations from the mean tells us nothing about the dispersion. That is, it doesn't tell us whether all the scores are close to the mean or whether there is a great deal of variability.

In describing the variability it is immaterial whether a measure is a certain distance above or below the mean. Hence we need a method of converting positive and negative quantities into quantities that are all positive. Such a method is the process of squaring.

After we have found each deviation for a set of measures we square each of these quantities. We then find the average of the sum of these squares, and take the square root of this average to find the **standard deviation.** The formula for finding the standard deviation is

$$\sigma = \sqrt{\frac{1}{N} \sum_{i=1}^{n} (X_i - \overline{X})^2}$$

To illustrate the use of the above formula let us find the standard deviation for Test 1 (Table 8.2) and for Test 2 (Table 8.3).

Substituting the values from Table 8.2 in the standard deviation formula, we have

$$\sigma = \sqrt{\frac{4283.5}{15}} = 16.9$$

The values from Table 8.3 give

$$\sigma = \sqrt{\frac{740}{15}} = 7.0$$

These two results indicate that Test 1, with the higher standard deviation, gives more variable results than Test 2. That is, the scores on Test 1 have a greater tendency to diverge from the mean score.

TABLE 8.2

Score	Deviation from mean (66.7)	Deviation squared
100	33.3 (100 − 66.7)	1108.9
90	23.3 (90 − 66.7)	542.9
80	13.3	176.9
75	8.3	68.9
75	8.3	68.9
70	3.3	10.9
70	3.3	10.9
70	3.3	10.9
65	−1.7	2.9
65	−1.7	2.9
65	−1.7	2.9
55	−11.7	136.9
40	−26.7	712.9
40	−26.7	712.9
40	−26.7	712.9
$N = 15$		4283.5

TABLE 8.3

Score	Deviation from mean (67)	Deviation squared
80	13	169
75	8	64
75	8	64
70	3	9
70	3	9
70	3	9
70	3	9
70	3	9
65	−2	4
65	−2	4
65	−2	4
60	−7	49
60	−7	49
55	−12	144
55	−12	144
$N = 15$		740

The standard deviation has the very remarkable property that in many sets of measures approximately two-thirds of the measures fall between $\overline{X} - \sigma$ and $\overline{X} + \sigma$. In general, *a relatively small standard deviation indicates that the measures tend to cluster close to the mean, and a relatively high one shows that the measures are widely scattered from the mean.*

Now that we know how to calculate the standard deviation, let us see what use we can make of it. Suppose the mean score on a mathematics test is 42 and the standard deviation is 6 points. If you have a score of 48, you are 6 points above the mean, or as many points as one standard deviation above the mean. Suppose that the mean on a second test is 50 and the standard deviation is 10. If you score 60 on the test, you would be 10 points, or one standard deviation above the mean. Since you are one standard deviation above the mean on each test, you know that you did equally well on both tests.

Example 1: Two mathematics tests were given to a class. On the first test the mean was 65 and the standard deviation was 6. On the second the mean was 45 and the standard deviation was 4. Joe received a score of 87 on the first test and a score of 53 on the second. On which test did he do better?

Solution: On the first test Joe's score was two standard deviations [65 + (2)(6) = 87] above the mean. On the second test his score was also two standard deviations above the mean [47 + (2)(4) = 53]. Hence he did equally well on both tests.

Example 2: Tom took two history tests and received scores of 50 on both tests. The mean on the first test was 80 with a standard deviation of 5. The mean on the second test was 74 with a standard deviation of 6. Did Tom improve on the second test?

Solution: On the first test Tom's score was six standard deviations below the mean. On the second test his score was four standard deviations below the mean. Hence he improved on the second test.

Exercise 8

1. Compute the standard deviation for the following scores.
   (a) 16, 24, 21, 17, 10, 19, 23, 20, 15, 12
   (b) 17, 16, 15, 14, 13, 12, 11

2. On a mathematics test the following scores were made in a class of fifteen students. Find the mean, mode, median, range, and standard deviation of these scores: 98, 65, 80, 75, 88, 95, 90, 40, 82, 60, 78, 90, 80, 70, 50

3. A machine turns out three-inch bolts. Samples taken during the day were measured with a micrometer with the following results: 3.03, 3.05, 2.99, 3.02, 3.06, 2.95, 2.98, 3.08, 3.00, 3.02, 2.94, 2.99. Find the mean and standard deviation of this sample.

4. Two tests were given. On the first test the mean was 80 and the standard deviation was 2. On the second test the mean was 75 and the standard deviation was 4. On which of the two tests did the following students do better?

Student	First Test Score	Second Test Score
(a) Mark	80	80
(b) Jane	40	50
(c) Kay	50	18
(d) Leo	95	90
(e) Deane	76	82

5. Suppose you are 10 pounds heavier than the average person of your age. Suppose the average weight of persons of your age is 130 pounds and the standard deviation is 5 pounds. How many standard deviations is your weight above the mean?

6. Two chemistry tests were given. On the first test the mean was 54 and the standard deviation was 8. On the second test the mean was 28 and the standard deviation 10. The scores of six students who took the test are given below.

Student	First Score	Second Score
Jan	62	38
Lela	66	23
Fred	46	28
Arthur	54	33
Gayle	70	43
Walter	50	18

(a) Which students improved in regard to the second test?
(b) Which student improved most on the second test?
(c) Which students did worse on the second test than on the first?
(d) Which students did equally well on both tests?

Arthur Cayley (1821–1895), left, and James Joseph Sylvester (1814–1897), right, were close friends who inspired each other in their work on the theory of invariants and matrices. The two men were of very different character. Cayley was gentle, generous, and led a serene life. Sylvester, however, was hot tempered and spent much of his life "fighting the world." With two such different temperaments, the friendship was not always a happy one. Sylvester was often on the point of exploding but Cayley was the safety valve. Sylvester paid grateful tribute to Cayley in a lecture at Oxford in 1885 by saying, "Cayley, who, though younger than myself, is my spiritual progenitor—who first opened my eyes and purged them of dross so that they could see and accept the higher mysteries of our common Mathematical Faith." Cayley mentioned Sylvester frequently in his memoirs but never in such glowing terms as those used by Sylvester. In 1858 Cayley invented matrices and their algebra. Some sixty years later, in 1925, Heisenberg recognized in the algebra of matrices the tool he needed in his work on quantum mechanics.

MATRICES

9

1.
HISTORY

We are all familiar with sets on which binary operations† are defined. For example, we are familiar with the set, W, of whole numbers

$$W = \{0, 1, 2, 3, \ldots\}$$

and the binary operations of addition and multiplication on these numbers. We are also familiar with the set of rational numbers and their operations. In Chapter 4 we became acquainted with the set

$$\{0, 1, 2, \ldots, (m - 1)\}$$

and addition and multiplication modulo m.

In this chapter we shall study a mathematical system created in the nineteenth century by Arthur Cayley (1821–1895) and James Joseph Sylvester (1814–1897). The results of the discoveries of these men furnished a broad foundation upon which the modern theory of matrices was built.

In the early 1940's an upsurge of interest occurred in the study of matrix theory. This branch of mathematics is extremely useful not only to mathematicians but also to persons interested in biology, sociology, economics, engineering, physics, psychology, and statistics. We shall study this system as a matter of interest and also to introduce a new and powerful symbolism.

2.
THE FORM OF A MATRIX

A (real) **matrix** (plural: matrices) is a rectangular array of real numbers, denoted by the symbol

$$\begin{pmatrix} a_{11} & a_{12} & \cdots & a_{1n} \\ a_{21} & a_{22} & \cdots & a_{2n} \\ \vdots & & & \\ a_{m1} & a_{m2} & \cdots & a_{mn} \end{pmatrix}$$

† See Chapter 6, Section 1.

In the above symbol the letters $a_{11}, a_{12}, \ldots, a_{mn}$ represent real numbers and m and n are positive integers. Observe that when we represent such an array of real numbers we put parentheses around it. The matrix represented above has m rows and n columns. We call a matrix with m rows and n columns a **m by n** matrix. We write this $m \times n$ (read: m by n), and call m and n the **dimensions** of the matrix. If $m = n$ the matrix has the same number of rows as columns and is called a **square** matrix.

Each real number in a matrix is called an **element** of the matrix. The double subscript used in denoting the elements of a matrix is a very convenient notation. The symbol a_{12} denotes the real number which appears in the first row and the second column of the matrix. In general, a_{ij} denotes the element in the ith row and the jth column of the matrix. We call i the **row subscript** and j the **column subscript.** When we use the double subscript notation we call i and j the **indices.** In the $m \times n$ matrix above we see that $1 \le i \le m$ and $1 \le j \le n$. Although we used i and j for the indices we could just as well have used other symbols for these subscripts. Thus we could have the elements of the matrix represented by a_{pq} where a_{pq} is the element in the pth row and the qth column. The thing to remember is that the first subscript refers to the number of the row and the second subscript refers to the number of the column in which the element appears.

Some examples of matrices are shown below. Matrix (a) has dimensions 1×4 and is called a 1 by 4 matrix; matrix (b) is a 2×2 matrix; matrix (c) is a 3×3 matrix; matrix (d) is a 3×2 matrix; and matrix (e) is a 4×1 matrix.

$$
(1 \ \ 2 \ \ 3 \ \ 4) \qquad
\begin{pmatrix} 1 & 2 \\ 3 & 4 \end{pmatrix} \qquad
\begin{pmatrix} 1 & 7 & -3 \\ -2 & 0 & 6 \\ -7 & -8 & -5 \end{pmatrix} \qquad
\begin{pmatrix} 1 & 2 \\ 3 & 8 \\ 1 & 7 \end{pmatrix} \qquad
\begin{pmatrix} 2 \\ 4 \\ 5 \\ 6 \end{pmatrix}
$$

$$\text{(a)} \qquad\qquad \text{(b)} \qquad\qquad\qquad \text{(c)} \qquad\qquad\qquad \text{(d)} \qquad \text{(e)}$$

We use capital letters such as A, B, C, etc., to denote matrices. Let A be a $m \times n$ matrix. Then

$$
A = \begin{pmatrix}
a_{11} & a_{12} & \cdots & a_{1n} \\
a_{21} & a_{22} & \cdots & a_{2n} \\
\vdots & & & \\
a_{m1} & a_{m2} & \cdots & a_{mn}
\end{pmatrix}
$$

MATRICES

We commonly abbreviate the above notation and simply write

$$A = (a_{ij})_{m \times n}$$

We say two matrices A and B, are **equal,** denoted by $A = B$, if and only if they have the same dimensions and the elements in the corresponding positions are equal. This means that

$$A = (a_{ij})_{m \times n} \quad \text{and} \quad B = (b_{ij})_{m \times n}$$

are equal if and only if $a_{ij} = b_{ij}$. Thus

$$\begin{pmatrix} 1 & 2 \\ 3 & 4 \end{pmatrix} = \begin{pmatrix} 1+0 & 1+1 \\ 2+1 & 3+1 \end{pmatrix}$$

because $1 = 1 + 0$, $2 = 1 + 1$, $3 = 2 + 1$ and $4 = 3 + 1$.

Example 1: Write a 4×3 matrix $(a_{ij})_{4 \times 3}$ where $a_{ij} = i + j + 1$.

Solution:
$$\begin{pmatrix} 1+1+1 & 1+2+1 & 1+3+1 \\ 2+1+1 & 2+2+1 & 2+3+1 \\ 3+1+1 & 3+2+1 & 3+3+1 \\ 4+1+1 & 4+2+1 & 4+3+1 \end{pmatrix} = \begin{pmatrix} 3 & 4 & 5 \\ 4 & 5 & 6 \\ 5 & 6 & 7 \\ 6 & 7 & 8 \end{pmatrix}$$

Example 2: Determine the values of x and y that make the following a true statement.

$$\begin{pmatrix} 3 & 4 \\ -3 & x+3 \\ 2 & 1 \end{pmatrix} = \begin{pmatrix} y-1 & 4 \\ -3 & 7 \\ 2 & 1 \end{pmatrix}$$

Solution: Since two matrices are equal if and only if the elements in the corresponding positions are equal,

$$x + 3 = 7 \text{ and } y - 1 = 3$$

Hence $x = 4$ and $y = 4$.

Exercise 1

1. Consider the 4×3 matrix below and name the following entries.

$$\begin{pmatrix} 3 & -2 & 4 \\ 5 & 3 & -6 \\ -7 & 0 & 2 \\ -1 & -2 & 4 \end{pmatrix}$$

(a) a_{23} (c) a_{43}

(b) a_{31} (d) a_{12}

2. Which of the following pairs of matrices are equal?

(a) $\begin{pmatrix} 0 & 3 \\ 5 & 4 \end{pmatrix}$; $\begin{pmatrix} 0 & -3 \\ -5 & -4 \end{pmatrix}$

(b) $\begin{pmatrix} 0.75 & 0.5 \\ 0.2 & 0.1 \end{pmatrix}$; $\begin{pmatrix} \frac{3}{4} & \frac{1}{2} \\ \frac{1}{5} & \frac{1}{10} \end{pmatrix}$

(c) $\begin{pmatrix} \sqrt{9} & \sqrt{4} \\ \sqrt{16} & \sqrt{25} \\ \sqrt{1} & \sqrt{100} \end{pmatrix}$; $\begin{pmatrix} 3 & 2 \\ 4 & 5 \\ 1 & 10 \end{pmatrix}$

(d) $\begin{pmatrix} 2 & 3 & 5 \\ 1 & 0 & -1 \\ 22 & 6 & 7 \end{pmatrix}$; $\begin{pmatrix} 2 & 1 & 22 \\ 3 & 0 & 6 \\ 5 & -1 & 7 \end{pmatrix}$

3. Give the values of x, y, and z that make the following pairs of matrices equal.

(a) $\begin{pmatrix} 1 & 3 \\ 5 & y \end{pmatrix} = \begin{pmatrix} x & 3 \\ 5 & -2 \end{pmatrix}$

(b) $\begin{pmatrix} 7 & 2 \\ x+3 & 7 \end{pmatrix} = \begin{pmatrix} 7 & y+6 \\ 9 & 7 \end{pmatrix}$

(c) $\begin{pmatrix} -1 & 2 & x-1 \\ 3 & y+6 & -4 \\ -5 & -2 & 1 \end{pmatrix} = \begin{pmatrix} -1 & 2 & -7 \\ 3 & -6 & z+2 \\ -5 & -2 & 1 \end{pmatrix}$

(d) $\begin{pmatrix} 5 & 2 \\ x+2\sqrt{3} & y+3\sqrt{6} \\ \sqrt{7} & 2z\sqrt{5} \end{pmatrix} = \begin{pmatrix} 5 & 2 \\ 5\sqrt{3} & -2\sqrt{6} \\ \sqrt{7} & -8\sqrt{5} \end{pmatrix}$

4. Give the dimensions of the following matrices.

(a) $\begin{pmatrix} 1 & 2 \\ 3 & 4 \end{pmatrix}$ (c) $(1 \quad -2 \quad 3 \quad 4)$

(b) $\begin{pmatrix} -2 & 3 \\ -1 & -2 \\ 3 & -4 \\ 1 & 0 \end{pmatrix}$ (d) $\begin{pmatrix} 1 & 2 & 4 & 9 \\ 0 & 1 & 0 & 1 \end{pmatrix}$

5. Write out the 3×3 matrix $(a_{ij})_{3\times3}$ where

(a) $a_{ij} = i + j$ (d) $a_{ij} = ij$

(b) $a_{ij} = i^2 - j^2$ (e) $a_{ij} = i^3 + j^3$

(c) $a_{ij} = \sqrt{ij}$ (f) $a_{ij} = 2i - 3j$

6. Write out the 4×5 matrix $(a_{ij})_{4 \times 5}$ where $a_{ij} = (-1)^{i+j}$.

7. A certain automobile agency has its salesmen send in 3×5 matrices as sales reports where the rows stand, in order, for the number of hardtops, sedans, and station wagons sold, and the columns, in order, for the colors black, blue, red, white, and green. The company received the following reports from two of its salesmen:

$$A: \begin{pmatrix} 2 & 3 & 4 & 1 & 0 \\ 3 & 3 & 1 & 1 & 4 \\ 1 & 1 & 5 & 6 & 2 \end{pmatrix} \qquad B: \begin{pmatrix} 4 & 1 & 0 & 0 & 0 \\ 3 & 2 & 3 & 5 & 1 \\ 7 & 4 & 1 & 0 & 4 \end{pmatrix}$$

(a) How many hardtops did A sell?
(b) How many green hardtops did B sell?
(c) How many sedans did B sell?
(d) Which man sold the most station wagons?
(e) How many cars did B sell?
(f) How many white cars did B sell?
(g) How many cars did A sell?
(h) How many red cars where sold by A and B?

8. Albert, Burton, and Cox go to the college bookstore and purchase the following articles:

> Albert: 2 notebooks, 4 textbooks, 2 football tickets
> Burton: 5 notebooks, 1 textbook, 1 football ticket
> Cox: 6 notebooks, 6 textbooks, 4 football tickets

Write a 3×3 matrix whose rows represent in order notebooks, textbooks, and football tickets and whose columns represent in order Albert, Burton, and Cox.

9. The following matrix gives the vitamin content of three foods in chosen units. The rows represent, in order, foods I, II, and III. The columns represent, in order, vitamins A, B, C, and B_1.

$$\begin{pmatrix} 0.3 & 0.3 & 0 & 0 \\ 0.4 & 0 & 0.2 & 0.1 \\ 0.5 & 0.5 & 0.1 & 0.6 \end{pmatrix}$$

(a) How much vitamin B is in food I?
(b) How much vitamin B_1 is in food III?
(c) How much vitamin C would you get if you ate one unit of each of the three foods?

10. An instructor used the following matrix to indicate the grades received by five graduate students during a semester. The rows represent, in order, students I, II, III, IV, and V. The columns represent, in order, grades A, B, C, D, and F.

$$\begin{pmatrix} 2 & 3 & 1 & 0 & 0 \\ 4 & 1 & 1 & 0 & 0 \\ 3 & 1 & 1 & 1 & 1 \\ 5 & 0 & 0 & 0 & 1 \\ 1 & 0 & 3 & 1 & 1 \end{pmatrix}$$

(a) How many A's did the instructor give during the semester?
(b) How many A's and B's did student II receive?
(c) How many students received F's during the semester?
(d) How many C's did the instructor give during the semester?
(e) How many D's and F's did the instructor give during the semester?

3.
ADDITION OF MATRICES

Matrices of the same dimensions are added elementwise. For example

$$\begin{pmatrix} 2 & 3 \\ 4 & 5 \end{pmatrix} + \begin{pmatrix} 4 & 1 \\ 3 & 1 \end{pmatrix} = \begin{pmatrix} 2+4 & 3+1 \\ 4+3 & 5+1 \end{pmatrix} = \begin{pmatrix} 6 & 4 \\ 7 & 6 \end{pmatrix}$$

In general

$$\begin{pmatrix} a_{11} & a_{12} & \cdots & a_{1n} \\ a_{21} & a_{22} & \cdots & a_{2n} \\ \vdots & & & \\ a_{m1} & a_{m2} & \cdots & a_{mn} \end{pmatrix} + \begin{pmatrix} b_{11} & b_{12} & \cdots & b_{1n} \\ b_{21} & b_{22} & \cdots & b_{2n} \\ \vdots & & & \\ b_{m1} & b_{m2} & \cdots & b_{mn} \end{pmatrix}$$

$$= \begin{pmatrix} a_{11}+b_{11} & a_{12}+b_{12} & \cdots & a_{1n}+b_{1n} \\ a_{21}+b_{21} & a_{22}+b_{22} & \cdots & a_{2n}+b_{2n} \\ \vdots & & & \\ a_{m1}+b_{m1} & a_{m2}+b_{m2} & \cdots & a_{mn}+b_{mn} \end{pmatrix}$$

Matrices of different dimensions cannot be added. If we wish to define addition of $m \times n$ matrices in words, we say that the sum of two $m \times n$ matrices is a $m \times n$ matrix whose elements are the sums of the elements in the corresponding places in the matrices to be added. Thus

$$(a_{ij})_{m \times n} + (b_{ij})_{m \times n} = (a_{ij} + b_{ij})_{m \times n}$$

MATRICES

Let us consider the set $M_{m \times n}$ of all $m \times n$ matrices. Matrix addition as we defined it is a binary operation (See Chapter 6, Section 1) on $M_{m \times n}$.

Since the elements of two matrices are real numbers, the order in which we add them does not matter; the result is the same. Hence the order in which we add two matrices does not matter. For example,

$$\begin{pmatrix} 1 & 2 \\ 3 & 1 \\ 0 & -1 \end{pmatrix} + \begin{pmatrix} -2 & 3 \\ 1 & 0 \\ 4 & -2 \end{pmatrix} = \begin{pmatrix} 1+(-2) & 2+3 \\ 3+1 & 1+0 \\ 0+4 & -1+(-2) \end{pmatrix} = \begin{pmatrix} -1 & 5 \\ 4 & 1 \\ 4 & -3 \end{pmatrix}$$

$$\begin{pmatrix} -2 & 3 \\ 1 & 0 \\ 4 & -2 \end{pmatrix} + \begin{pmatrix} 1 & 2 \\ 3 & 1 \\ 0 & -1 \end{pmatrix} = \begin{pmatrix} -2+1 & 3+2 \\ 1+3 & 0+1 \\ 4+0 & -2+(-1) \end{pmatrix} = \begin{pmatrix} -1 & 5 \\ 4 & 1 \\ 4 & -3 \end{pmatrix}$$

Because the order in which we add matrices does not affect the sum, we say that matrix addition is **commutative.** In general,

$$(a_{ij})_{m \times n} + (b_{ij})_{m \times n} = (b_{ij})_{m \times n} + (a_{ij})_{m \times n}$$

Let us now consider a special $m \times n$ matrix all of whose elements are 0:

$$\begin{pmatrix} 0 & 0 & \cdots & 0 \\ 0 & 0 & \cdots & 0 \\ \vdots & & & \\ 0 & 0 & \cdots & 0 \end{pmatrix}_{m \times n}$$

A matrix of this form is called a **zero matrix** for $M_{m \times n}$. Notice the result when a $m \times n$ zero matrix is added to a $m \times n$ matrix:

$$\begin{pmatrix} a_{11} & a_{12} & \cdots & a_{1n} \\ a_{21} & a_{22} & \cdots & a_{2n} \\ \vdots & & & \\ a_{m1} & a_{m2} & \cdots & a_{mn} \end{pmatrix} + \begin{pmatrix} 0 & 0 & \cdots & 0 \\ 0 & 0 & \cdots & 0 \\ \vdots & & & \\ 0 & 0 & \cdots & 0 \end{pmatrix} = \begin{pmatrix} a_{11}+0 & a_{12}+0 & \cdots & a_{1n}+0 \\ a_{21}+0 & a_{22}+0 & \cdots & a_{2n}+0 \\ \vdots & & & \\ a_{m1}+0 & a_{m2}+0 & \cdots & a_{mn}+0 \end{pmatrix}$$

$$= \begin{pmatrix} a_{11} & a_{12} & \cdots & a_{1n} \\ a_{21} & a_{22} & \cdots & a_{2n} \\ \vdots & & & \\ a_{m1} & a_{m2} & \cdots & a_{mn} \end{pmatrix}$$

Since the sum of $(a_{ij})_{m \times n}$ and the $m \times n$ zero matrix is $(a_{ij})_{m \times n}$, the zero matrix is called the **additive identity** for $M_{m \times n}$.

We recall that every real number a has an additive inverse, $-a$, such that

$$a + (-a) = 0$$

Thus

$$2 + (-2) = 0$$
$$(-3) + [-(-3)] = 0$$
$$\tfrac{1}{2} + (-\tfrac{1}{2}) = 0$$
$$\sqrt{2} + (-\sqrt{2}) = 0$$

We now ask ourselves: Does every $m \times n$ matrix have an additive inverse? Since every real number has an additive inverse and since the elements of matrices are real numbers, we are assured that every $m \times n$ matrix has an additive inverse. Observe that

$$\begin{pmatrix} 3 & 2 \\ 4 & 1 \\ 5 & 6 \end{pmatrix} + \begin{pmatrix} -3 & -2 \\ -4 & -1 \\ -5 & -6 \end{pmatrix} = \begin{pmatrix} 3 + (-3) & 2 + (-2) \\ 4 + (-4) & 1 + (-1) \\ 5 + (-5) & 6 + (-6) \end{pmatrix} = \begin{pmatrix} 0 & 0 \\ 0 & 0 \\ 0 & 0 \end{pmatrix}$$

In general

$$\begin{pmatrix} a_{11} & a_{12} & \cdots & a_{1n} \\ a_{21} & a_{22} & \cdots & a_{2n} \\ \vdots & & & \\ a_{m1} & a_{m2} & \cdots & a_{mn} \end{pmatrix} + \begin{pmatrix} -a_{11} & -a_{12} & \cdots & -a_{1n} \\ -a_{21} & -a_{22} & \cdots & -a_{2n} \\ \vdots & & & \\ -a_{m1} & -a_{m2} & \cdots & -a_{mn} \end{pmatrix}$$

$$= \begin{pmatrix} a_{11} + (-a_{11}) & a_{12} + (-a_{12}) & \cdots & a_{1n} + (-a_{1n}) \\ a_{21} + (-a_{21}) & a_{22} + (-a_{22}) & \cdots & a_{2n} + (-a_{2n}) \\ \vdots & & & \\ a_{m1} + (-a_{m1}) & a_{m2} + (-a_{m2}) & \cdots & a_{mn} + (-a_{mn}) \end{pmatrix}$$

$$= \begin{pmatrix} 0 & 0 & \cdots & 0 \\ 0 & 0 & \cdots & 0 \\ \vdots & & & \\ 0 & 0 & \cdots & 0 \end{pmatrix}_{m \times n}$$

We see that the additive inverse of $A = (a_{ij})_{m \times n}$ is $-A = (-a_{ij})_{m \times n}$.

MATRICES

Notice the following additions

(a)
$$\left[\begin{pmatrix} 1 & 2 \\ 3 & 4 \\ -1 & 2 \end{pmatrix} + \begin{pmatrix} -2 & 1 \\ 1 & 0 \\ 0 & 1 \end{pmatrix}\right] + \begin{pmatrix} -3 & 4 \\ 0 & 5 \\ -1 & -1 \end{pmatrix} = \begin{pmatrix} -1 & 3 \\ 4 & 4 \\ -1 & 3 \end{pmatrix} + \begin{pmatrix} -3 & 4 \\ 0 & 5 \\ -1 & -1 \end{pmatrix}$$

$$= \begin{pmatrix} -4 & 7 \\ 4 & 9 \\ -2 & 2 \end{pmatrix}$$

(b)
$$\begin{pmatrix} 1 & 2 \\ 3 & 4 \\ -1 & 2 \end{pmatrix} + \left[\begin{pmatrix} -2 & 1 \\ 1 & 0 \\ 0 & 1 \end{pmatrix} + \begin{pmatrix} -3 & 4 \\ 0 & 5 \\ -1 & -1 \end{pmatrix}\right] = \begin{pmatrix} 1 & 2 \\ 3 & 4 \\ -1 & 2 \end{pmatrix} + \begin{pmatrix} -5 & 5 \\ 1 & 5 \\ -1 & 0 \end{pmatrix}$$

$$= \begin{pmatrix} -4 & 7 \\ 4 & 9 \\ -2 & 2 \end{pmatrix}$$

Hence

$$\left[\begin{pmatrix} 1 & 2 \\ 3 & 4 \\ -1 & 2 \end{pmatrix} + \begin{pmatrix} -2 & 1 \\ 1 & 0 \\ 0 & 1 \end{pmatrix}\right] + \begin{pmatrix} -3 & 4 \\ 0 & 5 \\ -1 & -1 \end{pmatrix}$$

$$= \begin{pmatrix} 1 & 2 \\ 3 & 4 \\ -1 & 2 \end{pmatrix} + \left[\begin{pmatrix} -2 & 1 \\ 1 & 0 \\ 0 & 1 \end{pmatrix} + \begin{pmatrix} -3 & 4 \\ 0 & 5 \\ -1 & -1 \end{pmatrix}\right]$$

In general

(1) $[(a_{ij})_{m \times n} + (b_{ij})_{m \times n}] + (c_{ij})_{m \times n} = (a_{ij})_{m \times n} + [(b_{ij})_{m \times n} + (c_{ij})_{m \times n}]$

The truth of this statement is assured because of the associative property of addition of real numbers. The elements of $[(a_{ij})_{m \times n} + (b_{ij})_{m \times n}] + (c_{ij})_{m \times n}$ are

$$(a_{ij} + b_{ij}) + c_{ij} \qquad 1 \leq i \leq m, \; 1 \leq j \leq n$$

The elements of $(a_{ij})_{m \times n} + [(b_{ij})_{m \times n} + (c_{ij})_{m \times n}]$ are

$$a_{ij} + (b_{ij} + c_{ij}) \qquad 1 \leq i \leq m, \; 1 \leq j \leq n$$

Since a_{ij}, b_{ij}, and c_{ij} are real numbers,

$$(a_{ij} + b_{ij}) + c_{ij} = a_{ij} + (b_{ij} + c_{ij}) \qquad 1 \le i \le m, 1 \le j \le n$$

because of the associative property of addition of real numbers. We see then that statement 1 above is true.

Since the order in which we group matrices in matrix addition does not affect the sum we say that matrix addition is **associative.** Since matrix addition is associative, we may omit the grouping symbols. Thus $[(a_{ij})_{m \times n} + (b_{ij})_{m \times n}] + (c_{ij})_{m \times n}$ is usually written $(a_{ij})_{m \times n} + (b_{ij})_{m \times n} + (c_{ij})_{m \times n}$.

Exercise 2

1. Find the sums.

(a) $\begin{pmatrix} -2 & 3 \\ 4 & -4 \end{pmatrix} + \begin{pmatrix} 3 & -4 \\ 2 & 0 \end{pmatrix}$

(b) $\begin{pmatrix} -9 & 21 \\ 71 & -82 \end{pmatrix} + \begin{pmatrix} 49 & 101 \\ -89 & -26 \end{pmatrix}$

2. Find the sums.

(a) $\begin{pmatrix} -3 & 4 \\ 2 & 0 \\ -4 & 6 \end{pmatrix} + \begin{pmatrix} 7 & 2 \\ -6 & -2 \\ -3 & 4 \end{pmatrix}$

(b) $\begin{pmatrix} 3 & 4 & 2 \\ 1 & -4 & 6 \end{pmatrix} + \begin{pmatrix} -2 & -3 & -4 \\ 0 & 1 & 2 \end{pmatrix}$

3. Find the sums.

(a) $\begin{pmatrix} \frac{5}{6} & \frac{4}{5} & \frac{1}{3} \\ \frac{3}{8} & \frac{5}{6} & \frac{1}{2} \\ \frac{3}{4} & \frac{1}{3} & \frac{2}{5} \end{pmatrix} + \begin{pmatrix} \frac{1}{2} & \frac{3}{10} & \frac{5}{6} \\ \frac{3}{4} & \frac{2}{3} & \frac{1}{2} \\ \frac{5}{8} & \frac{7}{12} & \frac{9}{10} \end{pmatrix}$

(b) $\begin{pmatrix} 1.9 & 2.6 \\ 3.1 & 4.8 \\ 7.1 & -3.4 \end{pmatrix} + \begin{pmatrix} -1.7 & -3.4 \\ -2.6 & 7.8 \\ 1.3 & -8.7 \end{pmatrix}$

4. Find the sums.

(a) $\begin{pmatrix} \frac{3}{4} & \frac{5}{8} & -\frac{1}{2} \\ -\frac{2}{3} & -\frac{7}{12} & \frac{9}{16} \\ \frac{7}{8} & \frac{1}{3} & -\frac{5}{9} \\ \frac{3}{4} & \frac{2}{3} & \frac{2}{5} \end{pmatrix} + \begin{pmatrix} -\frac{3}{4} & -\frac{5}{8} & \frac{1}{2} \\ \frac{2}{3} & \frac{7}{12} & -\frac{9}{16} \\ -\frac{7}{8} & -\frac{1}{3} & \frac{5}{9} \\ -\frac{3}{4} & -\frac{2}{3} & -\frac{2}{5} \end{pmatrix}$

(b) $\begin{pmatrix} -7.3 & -8.6 & -1.7 & 5.9 \\ 3.8 & 6.1 & 7.5 & 3.9 \end{pmatrix} + \begin{pmatrix} 1.1 & 2.2 & -4.7 & 3.8 \\ 6.9 & -4.8 & -1.1 & -8.7 \end{pmatrix}$

5. Find the sums.

(a) $\begin{pmatrix} -\sqrt{5} & -\sqrt{2} \\ -3\sqrt{5} & 2\sqrt{2} \end{pmatrix} + \begin{pmatrix} 3\sqrt{5} & 4\sqrt{2} \\ -4\sqrt{5} & 6\sqrt{2} \end{pmatrix}$

(b) $\begin{pmatrix} \sqrt{3} & \sqrt{2} & \sqrt{5} \\ 3\sqrt{2} & -2\sqrt{5} & -\sqrt{3} \\ \sqrt{5} & -5\sqrt{5} & 0 \end{pmatrix} + \begin{pmatrix} 4\sqrt{3} & -2\sqrt{2} & -4\sqrt{5} \\ 6\sqrt{2} & 7\sqrt{5} & 3\sqrt{3} \\ \sqrt{5} & 5\sqrt{5} & -2 \end{pmatrix}$

6. Give the replacements for x, y, z, and w that will result in true statements.

(a) $\begin{pmatrix} x + 3 & 1 \\ 3 & 9 \end{pmatrix} + \begin{pmatrix} 4 & y - 7 \\ z + 5 & w - 2 \end{pmatrix} = \begin{pmatrix} 0 & 0 \\ 0 & 0 \end{pmatrix}$

(b) $\begin{pmatrix} x - 2 & 4 & z - 3 \\ -2 & 6 & y + 8 \end{pmatrix} + \begin{pmatrix} 7 & w + 2 & -8 \\ 2 & -6 & -5 \end{pmatrix} = \begin{pmatrix} 0 & 0 & 0 \\ 0 & 0 & 0 \end{pmatrix}$

7. Give the additive inverse of each of the following matrices.

(a) $\begin{pmatrix} 3 & -2 \\ -4 & 1 \end{pmatrix}$

(d) $\begin{pmatrix} \frac{1}{2} & \frac{2}{3} & -\frac{3}{4} \\ -2 & 0 & 1 \end{pmatrix}$

(b) $\begin{pmatrix} -1.8 & -3.2 \\ 6.7 & 1.5 \end{pmatrix}$

(e) $\begin{pmatrix} -1 \\ 3 \\ 2 \end{pmatrix}$

(c) $\begin{pmatrix} 0 & 0 & 3 \\ -2 & 7 & -9 \end{pmatrix}$

(f) $(-1 \quad \frac{1}{2} \quad 5 \quad -2)$

8. Find the values of x, y, z, and w that make the following true statements.

(a) $\begin{pmatrix} 2 & 3 \\ 7 & 6 \end{pmatrix} + \begin{pmatrix} x & y \\ z & w \end{pmatrix} = \begin{pmatrix} 6 & -2 \\ -4 & 9 \end{pmatrix}$

(b) $\begin{pmatrix} 3 & -2 \\ 7 & 6 \\ -4 & 5 \end{pmatrix} + \begin{pmatrix} x & y \\ z & -2 \\ 3 & w \end{pmatrix} = \begin{pmatrix} 8 & 4 \\ -2 & 4 \\ 1 & 6 \end{pmatrix}$

(c) $\begin{pmatrix} -1.11 & x & y \\ -2.23 & 1.42 & 7.11 \end{pmatrix} + \begin{pmatrix} 2.66 & 3.77 & -8.42 \\ z & w & -1.04 \end{pmatrix} = \begin{pmatrix} 1.55 & -2.86 & -3.12 \\ 7.21 & -3.04 & 6.07 \end{pmatrix}$

9. Find the following sums.

(a) $\begin{pmatrix} 3 & 2 \\ 1 & 2 \end{pmatrix} + \begin{pmatrix} -1 & 3 \\ -2 & 4 \end{pmatrix} + \begin{pmatrix} 6 & -3 \\ -2 & -1 \end{pmatrix}$

(b) $\begin{pmatrix} -3 & 2 & -1 \\ 4 & -2 & 3 \\ -6 & 7 & 0 \end{pmatrix} + \begin{pmatrix} 4 & -2 & 3 \\ 8 & -2 & -1 \\ 0 & 0 & 1 \end{pmatrix} + \begin{pmatrix} 5 & 4 & 2 \\ -1 & -1 & -1 \\ 0 & 0 & 0 \end{pmatrix}$

(c) $\begin{pmatrix} \frac{1}{2} & \frac{2}{3} \\ \frac{3}{4} & -\frac{1}{2} \\ \frac{5}{6} & \frac{7}{8} \end{pmatrix} + \begin{pmatrix} -\frac{1}{2} & \frac{1}{3} \\ \frac{1}{4} & -\frac{1}{2} \\ \frac{1}{6} & -\frac{3}{8} \end{pmatrix} + \begin{pmatrix} \frac{1}{4} & \frac{5}{6} \\ \frac{3}{8} & \frac{3}{4} \\ -\frac{5}{6} & -\frac{3}{4} \end{pmatrix}$

10. Let $A = (a_{ij})_{3\times4}$ and $B = (b_{ij})_{3\times4}$ where $a_{ij} = i + j$ and $b_{ij} = i^2 + j^2$. Find $A + B$.

11. A manufacturer has three factories, one in the east, one in the midwest, and one in the west. Each factory produces dresses in junior sizes and misses sizes in two price ranges labeled 1 and 2. The quantities in each factory are given in the following matrices:

	East (E)			West (W)			Midwest (M)	
	1	2		1	2		1	2
Junior	200	150	Junior	150	100	Junior	500	300
Misses	300	200	Misses	300	150	Misses	450	400

(a) Express the total production in a single matrix T.
(b) How many price 1 dresses are manufactured by the company?
(c) How many junior dresses are manufactured by the company?
(d) How many price 2 dresses are manufactured by the company?
(e) How many misses dresses are manufactured by the company?

12. A city has two junior colleges, one in the northern section of the city and one in the southern section. Each college has two types of programs, a liberal arts program and a vocational program. The students enrolled in each program are indicated in the following matrices.

	North (N)			South (S)	
	L.A.	Voc.		L.A.	Voc.
Freshmen	300	400	Freshmen	550	675
Sophomores	250	300	Sophomores	480	520

Express the total enrollment in a single matrix E.

13. Let

$$A = \begin{pmatrix} 1 & 3 & 4 & -2 \\ -1 & 2 & -3 & 4 \end{pmatrix} \qquad B = \begin{pmatrix} 3 & 0 & -2 & -3 \\ 4 & -2 & 1 & 5 \end{pmatrix}$$

$$C = \begin{pmatrix} -2 & -5 & -7 & 1 \\ -1 & 0 & -1 & 0 \end{pmatrix} \qquad D = \begin{pmatrix} 0 & 0 & 0 & 0 \\ 0 & 0 & 0 & 0 \end{pmatrix}.$$

Show that

(a) $A + B = B + A$ (c) $(A + B) + C = A + (B + C)$
(b) $C + (-C) = D$ (d) $B + D = B$

14. An aircraft company has two plants in a certain city. It has salaried and hourly employees. The number of men and women employed at the two plants is shown in the following matrices:

	Plant I			Plant II	
	Men	Women		Men	Women
Salaried	$\begin{pmatrix} 1200 \\ \text{Hourly} \quad 2400 \end{pmatrix}$				

Plant I
          Men   Women
Salaried ⎛1200   875 ⎞
Hourly   ⎝2400   1250 ⎠

Plant II
          Men   Women
Salaried ⎛ 650   300 ⎞
Hourly   ⎝1525   850 ⎠

Express the total number of employees in a single matrix S.

4.
MATRIX MULTIPLICATION

Before we define matrix multiplication let us consider the following problem. (All units in this problem are purely fictitious.)

A landscape architect specializes in the building of swimming pools and patios. In each swimming pool he estimates that he uses 8 units of steel, 240 units of concrete, 8 units of wood, and 0 units of glass. In each patio he uses 3 units of steel, 12 units of concrete, 25 units of wood, and 6 units of glass. The architect uses the matrix, M, below to conveniently summarize the above date. The rows represent, in order, swimming pools and patios. The columns represent, in order, units of steel, concrete, wood and glass.

$$M = \begin{pmatrix} 8 & 240 & 8 & 0 \\ 3 & 12 & 25 & 6 \end{pmatrix}$$

He finds that steel cost $20 per unit, concrete $12 per unit, wood $8 per unit, and glass $6 per unit. These costs are summarized in the 4×1 matrix P below.

$$P = \begin{pmatrix} 20 \\ 12 \\ 8 \\ 6 \end{pmatrix}$$

To find the cost of materials needed for a swimming pool we have

$$(8)(20) + (240)(12) + (8)(8) + (0)(6) = 3104$$

The total cost for material for a patio is

$$(3)(20) + (12)(12) + (25)(8) + (6)(6) = 440$$

We can represent this total cost of material by the 2×1 matrix T:

$$T = \begin{pmatrix} 3104 \\ 440 \end{pmatrix}$$

We are going to define matrix multiplication so that the product of matrices M and P, in that order, equals T, that is $M \cdot P = T$ or

$$\begin{pmatrix} 8 & 240 & 8 & 0 \\ 3 & 12 & 25 & 6 \end{pmatrix} \cdot \begin{pmatrix} 20 \\ 12 \\ 8 \\ 6 \end{pmatrix} = \begin{pmatrix} 3104 \\ 440 \end{pmatrix}$$

Let us now leave our model and formally define the product $A \cdot B$ of two matrices A and B. In order to multiply two matrices A and B, in that order, that is to find the product $A \cdot B$, the number of columns of A and the number of rows of B must be the same. Thus, if A is an $m \times k$ matrix, in order to find $A \cdot B$, B must have k rows. Suppose

$$A = (a_{ij})_{m \times k} \quad \text{and} \quad B = (b_{ij})_{k \times n}$$

We are going to define the product $A \cdot B$ by defining the element in the ith row and the jth column of $A \cdot B$. Let us denote the product $A \cdot B$ by the matrix $C = (c_{ij})$. We define the element c_{ij} in the ith row and the jth column of $A \cdot B = C$ by

$$a_{i1}b_{1j} + a_{i2}b_{2j} + a_{i3}b_{3j} + \cdots + a_{ik}b_{kj}$$

Notice that since A is an $m \times k$ matrix and B is a $k \times n$ matrix, all possible choices of i and j ($1 \leq i \leq m$ and $1 \leq j \leq n$) gives us a matrix with m rows and n columns. Thus the product $A \cdot B$ is an $m \times n$ matrix.

To clarify this definition let us look at an example. Let us consider

$$A = \begin{pmatrix} a_{11} & a_{12} \\ a_{21} & a_{22} \\ a_{31} & a_{32} \end{pmatrix} \quad \text{and} \quad B = \begin{pmatrix} b_{11} & b_{12} & b_{13} \\ b_{21} & b_{22} & b_{23} \end{pmatrix}$$

Notice that A has dimensions 3×2 and B has dimensions 2×3. Since the number of columns of A is the same as the number of rows of B, we can find the product $A \cdot B = C = (c_{ij})_{3\times3}$. Also notice that according to our definition we cannot find the product $B \cdot A$.

To find c_{11} we use the first row of A and the first column of B. Thus

$$
\begin{pmatrix} a_{11} & a_{12} \\ * & * \\ * & * \end{pmatrix} \cdot \begin{pmatrix} b_{11} & * & * \\ b_{21} & * & * \end{pmatrix} = \begin{pmatrix} a_{11}b_{11} + a_{12}b_{21} & * & * \\ * & & * & * \\ * & & * & * \end{pmatrix}
$$

To find c_{12} we use the first row of A and the second column of B:

$$
\begin{pmatrix} a_{11} & a_{12} \\ * & * \\ * & * \end{pmatrix} \cdot \begin{pmatrix} * & b_{12} & * \\ * & b_{22} & * \end{pmatrix} = \begin{pmatrix} * & a_{11}b_{12} + a_{12}b_{22} & * \\ * & * & * \\ * & * & * \end{pmatrix}
$$

To find c_{23} we use the second row of A and the third column of B:

$$
\begin{pmatrix} * & * \\ a_{21} & a_{22} \\ * & * \end{pmatrix} \cdot \begin{pmatrix} * & * & b_{13} \\ * & * & b_{23} \end{pmatrix} = \begin{pmatrix} * & * & * \\ * & * & a_{21}b_{13} + a_{22}b_{23} \\ * & * & * \end{pmatrix}
$$

Continuing in this fashion we find

$$
A \cdot B = \begin{pmatrix} a_{11} & a_{12} \\ a_{21} & a_{22} \\ a_{31} & a_{32} \end{pmatrix} \cdot \begin{pmatrix} b_{11} & b_{12} & b_{13} \\ b_{21} & b_{22} & b_{23} \end{pmatrix}
$$

$$
= \begin{pmatrix} a_{11}b_{11} + a_{12}b_{21} & a_{11}b_{12} + a_{12}b_{22} & a_{11}b_{13} + a_{12}b_{23} \\ a_{21}b_{11} + a_{22}b_{21} & a_{21}b_{12} + a_{22}b_{22} & a_{21}b_{13} + a_{22}b_{23} \\ a_{31}b_{11} + a_{32}b_{21} & a_{31}b_{12} + a_{32}b_{22} & a_{31}b_{13} + a_{32}b_{23} \end{pmatrix}
$$

Example 1: Given $A = \begin{pmatrix} 1 & 2 & 3 \\ 2 & 3 & 4 \end{pmatrix}$ and $B = \begin{pmatrix} 3 & 1 & 2 \\ 2 & 1 & 0 \\ 1 & 0 & 1 \end{pmatrix}$, find $A \cdot B$.

Solution:

$$A \cdot B = \begin{pmatrix} 1 & 2 & 3 \\ 2 & 3 & 4 \end{pmatrix} \cdot \begin{pmatrix} 3 & 1 & 2 \\ 2 & 1 & 0 \\ 1 & 0 & 1 \end{pmatrix}$$

$$= \begin{pmatrix} 1 \cdot 3 + 2 \cdot 2 + 3 \cdot 1 & 1 \cdot 1 + 2 \cdot 1 + 3 \cdot 0 & 1 \cdot 2 + 2 \cdot 0 + 3 \cdot 1 \\ 2 \cdot 3 + 3 \cdot 2 + 4 \cdot 1 & 2 \cdot 1 + 3 \cdot 1 + 4 \cdot 0 & 2 \cdot 2 + 3 \cdot 0 + 4 \cdot 1 \end{pmatrix}$$

$$= \begin{pmatrix} 3 + 4 + 3 & 1 + 2 + 0 & 2 + 0 + 3 \\ 6 + 6 + 4 & 2 + 3 + 0 & 4 + 0 + 4 \end{pmatrix}$$

$$= \begin{pmatrix} 10 & 3 & 5 \\ 16 & 5 & 8 \end{pmatrix}$$

Notice that if A is a $m \times k$ matrix and B is a $k \times n$ matrix, we can find the product $A \cdot B$, but the product $B \cdot A$ is not defined. In fact, only when we have square matrices A and B of the same dimensions can we find both $A \cdot B$ and $B \cdot A$.

Example 2: Given $A = \begin{pmatrix} 3 & 4 \\ -1 & 2 \end{pmatrix}$ and $B = \begin{pmatrix} 2 & -1 \\ 3 & -2 \end{pmatrix}$

find (a) $A \cdot B$; and (b) $B \cdot A$.

Solution:

(a)

$$A \cdot B = \begin{pmatrix} 3 & 4 \\ -1 & 2 \end{pmatrix} \cdot \begin{pmatrix} 2 & -1 \\ 3 & -2 \end{pmatrix}$$

$$= \begin{pmatrix} 3 \cdot 2 + 4 \cdot 3 & 3(-1) + 4(-2) \\ -1 \cdot 2 + 2 \cdot 3 & (-1)(-1) + 2(-2) \end{pmatrix}$$

$$= \begin{pmatrix} 6 + 12 & -3 - 8 \\ -2 + 6 & 1 - 4 \end{pmatrix}$$

$$= \begin{pmatrix} 18 & -11 \\ 4 & -3 \end{pmatrix}$$

(b)

$$B \cdot A = \begin{pmatrix} 2 & -1 \\ 3 & -2 \end{pmatrix} \cdot \begin{pmatrix} 3 & 4 \\ -1 & 2 \end{pmatrix}$$

$$= \begin{pmatrix} 2 \cdot 3 + (-1)(-1) & 2 \cdot 4 + (-1)(2) \\ 3 \cdot 3 + (-2)(-1) & 3 \cdot 4 + (-2)(2) \end{pmatrix}$$

$$= \begin{pmatrix} 6 + 1 & 8 - 2 \\ 9 + 2 & 12 - 4 \end{pmatrix}$$

$$= \begin{pmatrix} 7 & 6 \\ 11 & 8 \end{pmatrix}$$

Observe in the above example $A \cdot B \neq B \cdot A$, hence matrix multiplication is not commutative.

5.
MORE ABOUT
MATRIX MULTIPLICATION

If we have two matrices A and B we can find their product $A \cdot B$ if and only if the number of columns of A is the same as the number of rows of B. If $A = (a_{ij})_{m \times k}$ and $B = (b_{ij})_{k \times n}$ then the product $A \cdot B = C$ is a $m \times n$ matrix. We cannot find the product $B \cdot A$.

If A and B are both square matrices with dimensions $n \times n$, we can find $A \cdot B$ and $B \cdot A$. From Example 2 in Section 4 we see that although we can find both $A \cdot B$ and $B \cdot A$ in the case of two square matrices, $A \cdot B$ is not necessarily equal to $B \cdot A$. Thus we see that matrix multiplication is not commutative.

We now ask whether or not matrix multiplication is associative. Let us first consider a specific case. Let

$$A = \begin{pmatrix} 1 & 2 & -1 \\ 2 & -2 & 3 \end{pmatrix} \quad B = \begin{pmatrix} 0 & -2 \\ -1 & -1 \\ 3 & 4 \end{pmatrix} \quad C = \begin{pmatrix} -5 & 2 \\ -2 & -3 \end{pmatrix}$$

then

$$(A \cdot B) \cdot C = \left[\begin{pmatrix} 1 & 2 & -1 \\ 2 & -2 & 3 \end{pmatrix} \cdot \begin{pmatrix} 0 & -2 \\ -1 & -1 \\ 3 & 4 \end{pmatrix} \right] \cdot \begin{pmatrix} -5 & 2 \\ -2 & -3 \end{pmatrix}$$

$$= \begin{pmatrix} -5 & -8 \\ 11 & 10 \end{pmatrix} \cdot \begin{pmatrix} -5 & 2 \\ -2 & -3 \end{pmatrix}$$

$$= \begin{pmatrix} 41 & 14 \\ -75 & -8 \end{pmatrix}$$

and

$$A \cdot (B \cdot C) = \begin{pmatrix} 1 & 2 & -1 \\ 2 & -2 & 3 \end{pmatrix} \left[\begin{pmatrix} 0 & -2 \\ -1 & -1 \\ 3 & 4 \end{pmatrix} \cdot \begin{pmatrix} -5 & 2 \\ -2 & -3 \end{pmatrix} \right]$$

$$= \begin{pmatrix} 1 & 2 & -1 \\ 2 & -2 & 3 \end{pmatrix} \begin{pmatrix} 4 & 6 \\ 7 & 1 \\ -23 & -6 \end{pmatrix}$$

$$= \begin{pmatrix} 41 & 14 \\ -75 & -8 \end{pmatrix}$$

We see in this case, $(A \cdot B) \cdot C = A \cdot (B \cdot C)$. Although we shall not prove it here, it can be shown that if $A = (a_{ij})_{m \times k}$, $B = (b_{ij})_{k \times s}$, and $C = (c_{ij})_{s \times n}$, then $(A \cdot B) \cdot C = A \cdot (B \cdot C)$. That is, if the products are defined, matrix multiplication is associative. The reader should carry out the computation for $A = (a_{ij})_{2 \times 2}$, $B = (b_{ij})_{2 \times 2}$, and $C = (c_{ij})_{2 \times 2}$ to show that $(A \cdot B) \cdot C = A \cdot (B \cdot C)$.

Now let us consider the set $M_{n \times n}$ of all $n \times n$ square matrices. If we multiply two $n \times n$ matrices we certainly get a $n \times n$ matrix, hence $M_{n \times n}$ is **closed** under matrix multiplication.

If we consider two matrices A and B in this set $M_{n \times n}$, we see that both $A \cdot B$ and $B \cdot A$ are defined. That $A \cdot B$ is not necessarily equal to $B \cdot A$, see Example 2 in Section 4 of this chapter. We see that for this system of $n \times n$ matrices matrix multiplication is not commutative.

MATRICES

Let us consider the $n \times n$ matrix

$$I_n = \begin{pmatrix} 1 & 0 & 0 & \cdots & 0 \\ 0 & 1 & 0 & \cdots & 0 \\ 0 & 0 & 1 & \cdots & 0 \\ \vdots & & & & \\ 0 & 0 & 0 & \cdots & 1 \end{pmatrix}_{n \times n}$$

This is a very special $n \times n$ matrix. It has $a_{ii} = 1$ and $a_{ij} = 0$, $i \neq j$. Now let us find the product

$$A \cdot I_n = (a_{ij})_{n \times n} \cdot I_n$$

We find the element in the ith row and the jth column of this product. We shall call this element c_{ij}. The element c_{ij} is obtained by considering the ith row

$$a_{i1}, a_{i2}, a_{i3}, \ldots, a_{in}$$

of A and the jth column of I_n. Now every element of the jth column of I_n is 0 except one, the element in the jth row. Then

$$c_{ij} = a_{i1} \cdot 0 + a_{i2} \cdot 0 + \cdots + a_{ij} \cdot 1 + \cdots + a_{in} \cdot 0$$

We see that $c_{ij} = a_{ij}$ and

$$A \cdot I_n = A$$

Similarly we can show that $I_n \cdot A = A$.

The I_n matrix has the same properties in $M_{n \times n}$ as 1 does in the real number system. I_n is called the **identity element** for $M_{n \times n}$.

Example 1:   Given

$$A = \begin{pmatrix} 1 & 2 & 3 \\ 1 & 3 & -1 \\ 0 & -2 & -3 \end{pmatrix}$$

find $A \cdot I_3$.

Solution:

$A \cdot I_3$

$$= \begin{pmatrix} 1 & 2 & 3 \\ 1 & 3 & -1 \\ 0 & -2 & -3 \end{pmatrix} \cdot \begin{pmatrix} 1 & 0 & 0 \\ 0 & 1 & 0 \\ 0 & 0 & 1 \end{pmatrix}$$

$$= \begin{pmatrix} 1\cdot1+2\cdot0+3\cdot0 & 1\cdot0+2\cdot1+3\cdot0 & 1\cdot0+2\cdot0+3\cdot1 \\ 1\cdot1+3\cdot0+(-1)\cdot0 & 1\cdot0+3\cdot1+(-1)\cdot0 & 1\cdot0+3\cdot0+(-1)\cdot1 \\ 0\cdot1+(-2)\cdot0+(-3)\cdot0 & 0\cdot0+(-2)\cdot1+(-3)\cdot0 & 0\cdot0+(-2)\cdot0+(-3)\cdot1 \end{pmatrix}$$

$$= \begin{pmatrix} 1 & 2 & 3 \\ 1 & 3 & -1 \\ 0 & -2 & -3 \end{pmatrix}$$

Example 2: Given

$$A = \begin{pmatrix} 4 & 0 & 5 & -1 \\ 3 & 1 & 2 & 0 \\ 1 & 2 & 4 & -5 \\ 1 & 2 & 1 & 2 \end{pmatrix}$$

find $I_4 \cdot A$.

Solution:

$$I_4 \cdot A = \begin{pmatrix} 1 & 0 & 0 & 0 \\ 0 & 1 & 0 & 0 \\ 0 & 0 & 1 & 0 \\ 0 & 0 & 0 & 1 \end{pmatrix} \cdot \begin{pmatrix} 4 & 0 & 5 & -1 \\ 3 & 1 & 2 & 0 \\ 1 & 2 & 4 & -5 \\ 1 & 2 & 1 & 2 \end{pmatrix}$$

$$= \begin{pmatrix} 4 & 0 & 5 & -1 \\ 3 & 1 & 2 & 0 \\ 1 & 2 & 4 & -5 \\ 1 & 2 & 1 & 2 \end{pmatrix}$$

Exercise 3

1. Find the products (1) $A \cdot B$ and (2) $B \cdot A$.

(a) $A = \begin{pmatrix} 3 & 4 \\ -2 & 3 \end{pmatrix}$ $B = \begin{pmatrix} -5 & 0 \\ -1 & -2 \end{pmatrix}$

MATRICES

(b) $A = \begin{pmatrix} 1 & 2 \\ 6 & 4 \end{pmatrix}$ $B = \begin{pmatrix} 2 & 9 \\ -1 & 11 \end{pmatrix}$

2. Find the products (1) $A \cdot B$ and (2) $B \cdot A$.

(a) $A = \begin{pmatrix} 1 & 0 \\ 0 & 1 \end{pmatrix}$ $B = \begin{pmatrix} -3 & -4 \\ 9 & -6 \end{pmatrix}$

(b) $A = \begin{pmatrix} 2 & 4 \\ -1 & -2 \end{pmatrix}$ $B = \begin{pmatrix} 0 & 2 \\ 0 & -1 \end{pmatrix}$

3. Find the products (1) $A \cdot B$; (2) $B \cdot A$; (3) $A^2 = A \cdot A$; (4) $B^2 = B \cdot B$.

(a) $A = \begin{pmatrix} -3 & -1 \\ 0 & 1 \end{pmatrix}$ $B = \begin{pmatrix} 4 & 2 \\ -4 & -2 \end{pmatrix}$

(b) $A = \begin{pmatrix} -2 & -3 \\ 4 & 2 \end{pmatrix}$ $B = \begin{pmatrix} 3 & -2 \\ 4 & -5 \end{pmatrix}$

4. Given

$$A = \begin{pmatrix} 1 & 2 & 3 \\ 1 & -1 & 0 \\ -2 & 0 & -2 \end{pmatrix}$$

find

(a) $A^2 = A \cdot A$
(b) $A^3 = A^2 \cdot A$

5. Use the following matrices to find the products (a) through (i) below.

$$A = \begin{pmatrix} 1 & 3 \\ 2 & 1 \end{pmatrix} \quad B = \begin{pmatrix} -1 & 3 \\ -4 & -2 \end{pmatrix} \quad C = \begin{pmatrix} -2 & -3 \\ 1 & 4 \end{pmatrix}$$

$$D = \begin{pmatrix} 4 & 3 \\ -5 & 7 \end{pmatrix} \quad E = \begin{pmatrix} -2 & -1 \\ 0 & 1 \end{pmatrix} \quad F = \begin{pmatrix} 4 & 2 \\ 2 & 4 \end{pmatrix}$$

(a) $(A \cdot B) \cdot C$ (d) $(A \cdot C) \cdot F$ (g) $E \cdot (C \cdot F)$
(b) $(D \cdot C) \cdot A$ (e) $(B \cdot E) \cdot D$ (h) $(B \cdot E) \cdot F$
(c) $E \cdot (F \cdot C)$ (f) $(C \cdot A) \cdot F$ (i) $(B \cdot D) \cdot E$

6. Given

$$A = \begin{pmatrix} 2 & 4 \\ -1 & -2 \end{pmatrix} \quad B = \begin{pmatrix} 0 & 2 \\ 0 & -1 \end{pmatrix}$$

find $A \cdot B$. What is unusual about this product?

7. Using

$$A = \begin{pmatrix} a_{11} & a_{12} \\ a_{21} & a_{22} \end{pmatrix} \quad B = \begin{pmatrix} b_{11} & b_{12} \\ b_{21} & b_{22} \end{pmatrix} \quad C = \begin{pmatrix} c_{11} & c_{12} \\ c_{21} & c_{22} \end{pmatrix}$$

show that $(A \cdot B) \cdot C = A \cdot (B \cdot C)$.

8. Given

$$P = \begin{pmatrix} -2 & 3 & 4 & 5 \\ 1 & -2 & 0 & -1 \end{pmatrix} \qquad Q = \begin{pmatrix} 3 & 1 & 0 \\ 1 & 0 & 1 \\ -1 & -2 & 0 \\ -1 & -1 & -1 \end{pmatrix}$$

find $P \cdot Q$.

9. Given

$$A = \begin{pmatrix} 3 & -1 & 2 \\ 2 & -3 & 4 \end{pmatrix} \qquad B = \begin{pmatrix} -1 & 3 \\ 2 & -1 \\ 4 & -5 \end{pmatrix} \qquad C = \begin{pmatrix} 0 & -3 \\ 4 & 2 \\ 1 & 6 \end{pmatrix}$$

verify that $A \cdot (B + C) = A \cdot B + A \cdot C$.

10. If $A = (a_{ij})_{4 \times 3}$ and $B = (b_{ij})_{3 \times 5}$, what are the dimensions of $A \cdot B$?

11. Find the product

$$\begin{pmatrix} -2 & 3 \\ -1 & 4 \\ 3 & -4 \end{pmatrix} \cdot \begin{pmatrix} 2 & 1 & -1 & 3 & 2 \\ -3 & 4 & 0 & 1 & 1 \end{pmatrix}$$

12. If $P = (p_{ij})_{k \times r}$, $Q = (q_{ij})_{r \times t}$, and $S = (s_{ij})_{t \times z}$, what are the dimensions of $(P \cdot Q) \cdot S$?

13. Perform the indicated operations. If the operations are impossible to perform so state.

(a)
$$\begin{pmatrix} 1 & 2 & 1 \\ 2 & 1 & 2 \\ 1 & 2 & 1 \end{pmatrix} \cdot \begin{pmatrix} 1 & 0 & 1 \\ 0 & 1 & 0 \\ 0 & 0 & 1 \end{pmatrix}$$

(b)
$$\left[\begin{pmatrix} 3 & 2 \\ 1 & 0 \\ -1 & -2 \end{pmatrix} \cdot \begin{pmatrix} -4 & 3 \\ -5 & -2 \end{pmatrix} \right] + \begin{pmatrix} 3 & 2 \\ 4 & -1 \end{pmatrix}$$

(c)
$$\begin{pmatrix} 4 & 1 \\ -1 & 2 \\ 2 & -1 \\ 4 & 2 \end{pmatrix} \cdot \begin{pmatrix} 1 & 0 & 1 \\ -1 & -1 & -1 \end{pmatrix}$$

(d)
$$\begin{pmatrix} 3 \\ -1 \\ 2 \end{pmatrix} \cdot (3 \quad -1 \quad 2)$$

14. Given

$$A = \begin{pmatrix} 2 & -1 & 0 \\ 3 & 5 & -3 \end{pmatrix} \qquad B = \begin{pmatrix} 2 & 3 & 1 & 0 \\ 1 & -1 & 1 & 0 \\ 0 & 0 & 1 & 0 \end{pmatrix}$$

(a) Find the element in the second row and the third column of $A \cdot B$.
(b) Find the element in the first row and the third column of $A \cdot B$.
(c) Find the elements in the fourth column of $A \cdot B$.

15. Given

$$A = \begin{pmatrix} 2 & 0 & 0 \\ 0 & 3 & 0 \\ 0 & 0 & 4 \end{pmatrix} \qquad B = \begin{pmatrix} 0 & 0 & 1 \\ 0 & 1 & 0 \\ 1 & 0 & 0 \end{pmatrix}$$

find

(a) $A \cdot B$.
(b) $B \cdot A$.

6.
MULTIPLICATIVE INVERSES OF 2 × 2 MATRICES

Now let us consider the set $M_{2 \times 2}$ of all 2 × 2 matrices. Since every matrix has an additive inverse, we are assured that every 2 × 2 matrix has an additive inverse. We now ask: Does every 2 × 2 matrix have a multiplicative inverse? If a 2 × 2 matrix A has a multiplicative inverse, denoted by A^{-1} (read: A inverse), then

$$A \cdot A^{-1} = A^{-1} \cdot A = I_2$$

where I_2 is the identity matrix

$$\begin{pmatrix} 1 & 0 \\ 0 & 1 \end{pmatrix}$$

Let us consider a specific case and try to find the multiplicative inverse of the 2 × 2 matrix

$$\begin{pmatrix} 3 & 1 \\ 2 & 4 \end{pmatrix}$$

6. MULTIPLICATIVE INVERSES OF 2 × 2 MATRICES

We are looking for a 2 × 2 matrix

$$\begin{pmatrix} x & y \\ z & w \end{pmatrix}$$

such that

(a)
$$\begin{pmatrix} 3 & 1 \\ 2 & 4 \end{pmatrix} \cdot \begin{pmatrix} x & y \\ z & w \end{pmatrix} = \begin{pmatrix} 1 & 0 \\ 0 & 1 \end{pmatrix}$$

and

(b)
$$\begin{pmatrix} x & y \\ z & w \end{pmatrix} \cdot \begin{pmatrix} 3 & 1 \\ 2 & 4 \end{pmatrix} = \begin{pmatrix} 1 & 0 \\ 0 & 1 \end{pmatrix}$$

First we shall consider equation (a), called a **matrix equation.** We see that

$$\begin{pmatrix} 3 & 1 \\ 2 & 4 \end{pmatrix} \cdot \begin{pmatrix} x & y \\ z & w \end{pmatrix} = \begin{pmatrix} 3x + z & 3y + w \\ 2x + 4z & 2y + 4w \end{pmatrix}$$

We want this product to equal the identity matrix I_2, and therefore we have

$$\begin{pmatrix} 3x + z & 3y + w \\ 2x + 4z & 2y + 4w \end{pmatrix} = \begin{pmatrix} 1 & 0 \\ 0 & 1 \end{pmatrix}$$

Since two matrices are equal if and only if their corresponding elements are equal we have

(1) $\qquad\qquad\qquad 3x + z = 1$
(2) $\qquad\qquad\qquad 3y + w = 0$
(3) $\qquad\qquad\qquad 2x + 4z = 0$
(4) $\qquad\qquad\qquad 2y + 4w = 1$

Multiplying the members of equation (2) by 4 and subtracting equation (4) member by member from it we have

$$12y + 4w = 0$$
$$\underline{2y + 4w = 1}$$
$$10y \qquad\quad = -1$$
$$y = -\tfrac{1}{10}$$

Replacing y by $-\frac{1}{10}$ in equation (2) we have

$$-\tfrac{3}{10} + w = 0$$
$$w = \tfrac{3}{10}$$

Multiplying each member of equation (1) by 4 and subtracting equation (3) member by member from it we have

$$12x + 4z = 4$$
$$\underline{2x + 4z = 0}$$
$$10x \qquad = 4$$
$$x = \tfrac{2}{5}$$

Replacing x by $\frac{2}{5}$ in equation (1) we have

$$\tfrac{6}{5} + z = 1$$
$$z = -\tfrac{1}{5}$$

Using the values $\frac{2}{5}$, $-\frac{1}{10}$, $-\frac{1}{5}$, and $\frac{3}{10}$ for x, y, z, and w respectively yield the matrix

$$\begin{pmatrix} \tfrac{2}{5} & -\tfrac{1}{10} \\ -\tfrac{1}{5} & \tfrac{3}{10} \end{pmatrix}$$

We must now check to see whether this matrix is the multiplicative inverse of the given matrix. We do this by checking the following multiplications:

$$\begin{pmatrix} 3 & 1 \\ 2 & 4 \end{pmatrix} \cdot \begin{pmatrix} \tfrac{2}{5} & -\tfrac{1}{10} \\ -\tfrac{1}{5} & \tfrac{3}{10} \end{pmatrix} = \begin{pmatrix} \tfrac{6}{5} - \tfrac{1}{5} & -\tfrac{3}{10} + \tfrac{3}{10} \\ \tfrac{4}{5} - \tfrac{4}{5} & -\tfrac{2}{10} + \tfrac{12}{10} \end{pmatrix}$$

$$= \begin{pmatrix} 1 & 0 \\ 0 & 1 \end{pmatrix}$$

$$\begin{pmatrix} \tfrac{2}{5} & -\tfrac{1}{10} \\ -\tfrac{1}{5} & \tfrac{3}{10} \end{pmatrix} \cdot \begin{pmatrix} 3 & 1 \\ 2 & 4 \end{pmatrix} = \begin{pmatrix} \tfrac{6}{5} - \tfrac{2}{10} & \tfrac{2}{5} - \tfrac{4}{10} \\ -\tfrac{3}{5} + \tfrac{6}{10} & -\tfrac{1}{5} + \tfrac{12}{10} \end{pmatrix}$$

$$= \begin{pmatrix} 1 & 0 \\ 0 & 1 \end{pmatrix}$$

The fact that one 2×2 matrix has a multiplicative inverse does not mean that every 2×2 matrix has an inverse. Consider the 2×2 matrix

$$\begin{pmatrix} 2 & 1 \\ 4 & 2 \end{pmatrix}$$

6. MULTIPLICATIVE INVERSES OF 2 × 2 MATRICES

Again we want to find a matrix

$$\begin{pmatrix} x & y \\ z & w \end{pmatrix}$$

such that

$$\begin{pmatrix} 2 & 1 \\ 4 & 2 \end{pmatrix} \cdot \begin{pmatrix} x & y \\ z & w \end{pmatrix} = \begin{pmatrix} x & y \\ z & w \end{pmatrix} \cdot \begin{pmatrix} 2 & 1 \\ 4 & 2 \end{pmatrix} = \begin{pmatrix} 1 & 0 \\ 0 & 1 \end{pmatrix}$$

Assuming that we can find the desired matrix, we use the same method as previously used. We find that we must solve the following system of equations:

(5) $\qquad\qquad\qquad 2x + z = 1$
(6) $\qquad\qquad\qquad 2y + w = 0$
(7) $\qquad\qquad\qquad 4x + 2z = 0$
(8) $\qquad\qquad\qquad 4y + 2w = 1$

Multiplying the members of equation (5) by 2 and subtracting it member by member from equation (7) we have the impossible result $0 = 2$. This tells us our assumption that the given matrix has an inverse is false. We see then that we cannot say that every 2×2 has a multiplicative inverse.

It can be proved, although we shall accept it without proof, that every 2×2 matrix

$$A = (a_{ij})_{2\times2} = \begin{pmatrix} a_{11} & a_{12} \\ a_{21} & a_{22} \end{pmatrix}$$

has a multiplicative inverse provided $a_{11}a_{22} - a_{21}a_{12} \neq 0$. If $a_{11}a_{22} - a_{21}a_{12} = 0$ then the matrix does not have a multiplicative inverse.

Exercise 4

1. Find the multiplicative inverse of each of the following.

(a) $\begin{pmatrix} 2 & 2 \\ 2 & 4 \end{pmatrix}$
   (b) $\begin{pmatrix} 3 & 1 \\ 1 & 2 \end{pmatrix}$

2. Find the multiplicative inverse of each of the following.

(a) $\begin{pmatrix} 1 & 2 \\ 1 & 1 \end{pmatrix}$
   (b) $\begin{pmatrix} 2 & 0 \\ 3 & 4 \end{pmatrix}$

3. Find the multiplicative inverse of each of the following.

(a) $\begin{pmatrix} 0 & 1 \\ 2 & 1 \end{pmatrix}$

(b) $\begin{pmatrix} 3 & -2 \\ -1 & 2 \end{pmatrix}$

4. Show that the following matrices do not have multiplicative inverses.

(a) $\begin{pmatrix} 8 & 1 \\ 16 & 2 \end{pmatrix}$

(b) $\begin{pmatrix} 2 & 5 \\ 4 & 10 \end{pmatrix}$

(c) $\begin{pmatrix} 3 & 2 \\ 6 & 4 \end{pmatrix}$

5. Show that the following matrices do not have multiplicative inverses.

(a) $\begin{pmatrix} 9 & 6 \\ 3 & 2 \end{pmatrix}$

(b) $\begin{pmatrix} 2 & 4 \\ 4 & 8 \end{pmatrix}$

(c) $\begin{pmatrix} 3 & 6 \\ 4 & 8 \end{pmatrix}$

6. Which of the following matrices have multiplicative inverses? Which do not?

(a) $\begin{pmatrix} 3 & 4 \\ 1 & -2 \end{pmatrix}$

(c) $\begin{pmatrix} 4 & 8 \\ 8 & 16 \end{pmatrix}$

(b) $\begin{pmatrix} 6 & -4 \\ -12 & 8 \end{pmatrix}$

(d) $\begin{pmatrix} 3 & -2 \\ 1 & -7 \end{pmatrix}$

7. Which of the following matrices have multiplicative inverses? Which do not?

(a) $\begin{pmatrix} 7 & 0 \\ 0 & 1 \end{pmatrix}$

(c) $\begin{pmatrix} 5 & 6 \\ 10 & 12 \end{pmatrix}$

(b) $\begin{pmatrix} -9 & -6 \\ -6 & -4 \end{pmatrix}$

(d) $\begin{pmatrix} -1 & 3 \\ -2 & 7 \end{pmatrix}$

8. Using

$$A = \begin{pmatrix} 3 & 6 \\ 4 & 9 \end{pmatrix} \qquad B = \begin{pmatrix} 3 & 2 \\ 4 & 3 \end{pmatrix}$$

show that

(a) $(A^{-1})^{-1} = A$

(b) $(A \cdot B)^{-1} = B^{-1} \cdot A^{-1}$

9. Sometimes the multiplicative inverse of a 2×2 matrix can be written down using the following rule: Interchange a_{11} and a_{22} and replace a_{12} and a_{21} by their additive inverses. For example, the multiplicative inverse of

$$\begin{pmatrix} 9 & 7 \\ 5 & 4 \end{pmatrix} \quad \text{is} \quad \begin{pmatrix} 4 & -7 \\ -5 & 9 \end{pmatrix}$$

Some matrices on which you can use this rule are

$$\begin{pmatrix} 8 & 3 \\ 5 & 2 \end{pmatrix} \quad \begin{pmatrix} 7 & 11 \\ 5 & 8 \end{pmatrix} \quad \begin{pmatrix} 13 & 8 \\ 8 & 5 \end{pmatrix} \quad \begin{pmatrix} 5 & 11 \\ 4 & 9 \end{pmatrix}$$

Try to find other 2×2 matrices for which this rule works. In general when does this rule work for finding the multiplicative inverse of a 2×2 matrix?

10. The rule for finding the multiplicative inverse of a 2×2 matrix given in problem 9 does not work for all 2×2 matrices that have inverses. For example, performing the rule on

$$\begin{pmatrix} 5 & 4 \\ 2 & 3 \end{pmatrix} \quad \text{gives} \quad \begin{pmatrix} 3 & -4 \\ -2 & 5 \end{pmatrix} \quad \text{but the inverse is} \quad \begin{pmatrix} \frac{3}{7} & -\frac{4}{7} \\ -\frac{2}{7} & \frac{5}{7} \end{pmatrix}$$

We see then that to obtain the inverse of the given matrix we can use the rule given in problem 9 but in addition we divide each element by some number. Find the inverses of several matrices and try to discover the number by which you divide each element.

7.
MATRICES AND
SYSTEMS OF EQUATIONS

Let us consider the equations

(1)
$$a_1 x + b_1 y = c_1$$
(2)
$$a_2 x + b_2 y = c_2$$

These equations form a **system of two linear equations in two variables.** We can use the following matrix equation to denote the above system:

(3)
$$\begin{pmatrix} a_1 & b_1 \\ a_2 & b_2 \end{pmatrix} \begin{pmatrix} x \\ y \end{pmatrix} = \begin{pmatrix} c_1 \\ c_2 \end{pmatrix}$$

since

$$\begin{pmatrix} a_1 & b_1 \\ a_2 & b_2 \end{pmatrix} \begin{pmatrix} x \\ y \end{pmatrix} = \begin{pmatrix} a_1 x + b_1 y \\ a_2 x + b_2 y \end{pmatrix}$$

and $a_1 x + b_1 y = c_1$ and $a_2 x + b_2 y = c_2$.

Letting

$$A = \begin{pmatrix} a_1 & b_1 \\ a_2 & b_2 \end{pmatrix} \quad X = \begin{pmatrix} x \\ y \end{pmatrix} \quad \text{and} \quad C = \begin{pmatrix} c_1 \\ c_2 \end{pmatrix}$$

equation (3) may be written

(4)
$$AX = C$$

Since A, X, and C are matrices, equation (4) is called a **matrix equation.**
If matrix A has an inverse, A^{-1}, we can multiply each member of equation (4) by A^{-1} (on the left) to obtain

$$A^{-1}(AX) = A^{-1}C$$

Since $A^{-1}A = I_2$, we have

$$X = A^{-1}C$$

Using the definition of equality of matrices we can find the values of x and y that satisfy equations (1) and (2) of our system of equations.
For example, consider the system

$$\begin{aligned} 3x + y &= 2 \\ 5x + 2y &= 3 \end{aligned}$$

Denoting this system by a matrix equation we have

(5)
$$\begin{pmatrix} 3 & 1 \\ 5 & 2 \end{pmatrix}\begin{pmatrix} x \\ y \end{pmatrix} = \begin{pmatrix} 2 \\ 3 \end{pmatrix}$$

Since

$$A = \begin{pmatrix} 3 & 1 \\ 5 & 2 \end{pmatrix}$$

has an inverse

$$A^{-1} = \begin{pmatrix} 2 & -1 \\ -5 & 3 \end{pmatrix}$$

we can multiply each member of equation (5) by A^{-1} (on the left) to obtain

$$\begin{pmatrix} 2 & -1 \\ -5 & 3 \end{pmatrix}\begin{pmatrix} 3 & 1 \\ 5 & 2 \end{pmatrix}\begin{pmatrix} x \\ y \end{pmatrix} = \begin{pmatrix} 2 & -1 \\ -5 & 3 \end{pmatrix}\begin{pmatrix} 2 \\ 3 \end{pmatrix}$$

Then

$$\begin{pmatrix} 1 & 0 \\ 0 & 1 \end{pmatrix}\begin{pmatrix} x \\ y \end{pmatrix} = \begin{pmatrix} 2 & -1 \\ -5 & 3 \end{pmatrix}\begin{pmatrix} 2 \\ 3 \end{pmatrix}$$

$$\begin{pmatrix} x \\ y \end{pmatrix} = \begin{pmatrix} 1 \\ -1 \end{pmatrix}$$

Hence $x = 1$ and $y = -1$. Checking these values of x and y in our system we have

$$3 \cdot 1 + (-1) = 3 - 1 = 2$$
$$5 \cdot 1 + 2(-1) = 5 - 2 = 3$$

Exercise 5

Using matrices find the values of x and y that satisfy the following systems.

1. $5x + 3y = 13$
   $8x + 5y = 21$
2. $9x - 7y = 12$
   $5x - 4y = 7$
3. $2x + 3y = 8$
   $3x - 4y = -22$
4. $4x + 3y = 12$
   $x - 5y = 26$
5. $x + y = 8$
   $x - y = -2$
6. $2y - 8x = 8$
   $3x - 2y = -8$
7. $8x - 16y = 12$
   $10x + 4y = 3$
8. $8x - 9y = -4$
   $-16x + 15y = 6$
9. $9x - 10y = 9$
   $-6x - 25y = 13$
10. $\frac{1}{3}x + \frac{2}{5}y = 0$
    $\frac{3}{8}x - \frac{2}{3}y = 0$

8.
DETERMINANTS OF SQUARE MATRICES

In finding the inverse of the matrix

$$A = \begin{pmatrix} a_{11} & a_{12} \\ a_{21} & a_{22} \end{pmatrix}$$

we found the expression $a_{11}a_{22} - a_{21}a_{12}$ playing a critical role. If

$a_{11}a_{22} - a_{21}a_{12} \neq 0$, then matrix A has an inverse; if $a_{11}a_{22} - a_{21}a_{12} = 0$, matrix A does not have an inverse.

The expression $a_{11}a_{22} - a_{21}a_{12}$ is called the **determinant** of matrix A and is denoted by

$$\det A = \begin{vmatrix} a_{11} & a_{12} \\ a_{21} & a_{22} \end{vmatrix} = a_{11}a_{22} - a_{21}a_{12}$$

Notice that the elements of a matrix are enclosed in parentheses; the elements of the determinant of a matrix are enclosed between parallel lines. Although a matrix has no number value, its determinant is a single number called the **value** of the determinant. It is important to remember that a matrix is an array of numbers and a determinant is a single number associated with the matrix.

Example 1: Given

$$A = \begin{pmatrix} 3 & -2 \\ 4 & 5 \end{pmatrix}$$

find $\det A$.

Solution:

$$\det A = \begin{vmatrix} 3 & -2 \\ 4 & 5 \end{vmatrix} = 3 \cdot 5 - 4(-2) = 15 + 8 = 23$$

Example 2: Prove that the determinant of a 2×2 matrix is zero† if the elements of the two rows are equal.

Solution: Since the elements of the two rows of the matrix are equal we have

$$A = \begin{pmatrix} a_{11} & a_{12} \\ a_{11} & a_{12} \end{pmatrix}$$

Then

$$\det A = \begin{vmatrix} a_{11} & a_{12} \\ a_{11} & a_{12} \end{vmatrix} = a_{11}a_{12} - a_{11}a_{12} = 0$$

† Saying that a determinant is zero is an abbreviated way of saying that the value of the determinant is zero.

Now let us consider a square matrix of order 3:

$$A = \begin{pmatrix} a_{11} & a_{12} & a_{13} \\ a_{21} & a_{22} & a_{23} \\ a_{31} & a_{32} & a_{33} \end{pmatrix}$$

We define the value of the determinant associated with A, denoted by

$$\det A = \begin{vmatrix} a_{11} & a_{12} & a_{13} \\ a_{21} & a_{22} & a_{23} \\ a_{31} & a_{32} & a_{33} \end{vmatrix}$$

in the following fashion.

(1) $\det A$
$$= a_{11}(a_{22}a_{33} - a_{32}a_{23}) - a_{12}(a_{21}a_{33} - a_{31}a_{23}) + a_{13}(a_{21}a_{32} - a_{31}a_{22})$$

Note that if we scratch out the row and column containing the element a_{11}, we are left with a 2×2 matrix whose determinant has been defined.

$$\begin{pmatrix} \cancel{a_{11}} & \cancel{a_{12}} & \cancel{a_{13}} \\ \cancel{a_{21}} & a_{22} & a_{23} \\ \cancel{a_{31}} & a_{32} & a_{33} \end{pmatrix}$$

This determinant is called the **minor** of element a_{11} and is denoted by A_{11}. Observe that the minor of a_{11} is $a_{22}a_{33} - a_{32}a_{23}$.

In a similar manner we can find the minor of any element of $\det A$. Table 9.1 gives some of the elements of $\det A$ and their minors.

Note that the value of $\det A$ (given by the right member of equation (1)) is

$$a_{11}A_{11} - a_{12}A_{12} + a_{13}A_{13}$$

TABLE 9.1

Element	Minor
a_{11}	$A_{11} = a_{22}a_{33} - a_{32}a_{23}$
a_{12}	$A_{12} = a_{21}a_{33} - a_{31}a_{23}$
a_{13}	$A_{13} = a_{21}a_{32} - a_{31}a_{23}$
a_{22}	$A_{22} = a_{11}a_{33} - a_{31}a_{13}$
a_{33}	$A_{33} = a_{11}a_{22} - a_{21}a_{12}$

In defining det A in equation (1) we used the elements of the first row and their minors. Let us now consider the value of det A

$$a_{11}(a_{22}a_{33}) - a_{12}(a_{21}a_{33} - a_{31}a_{23}) + a_{13}(a_{22}a_{32} - a_{31}a_{22})$$
$$= a_{11}a_{22}a_{33} - a_{11}a_{32}a_{23} - a_{12}a_{21}a_{33} + a_{12}a_{31}a_{23} + a_{13}a_{22}a_{32} - a_{13}a_{31}a_{22}$$
$$= a_{13}(a_{21}a_{32} - a_{31}a_{22}) - (a_{23}(a_{11}a_{32} - a_{31}a_{12}) + a_{33}(a_{11}a_{22} - a_{21}a_{12})$$
$$= a_{13}A_{13} - a_{23}A_{23} + a_{33}A_{33}$$

In a similar fashion we can show that

$$\det A = a_{11}A_{11} - a_{21}A_{21} + a_{31}A_{31}$$
$$= -a_{12}A_{12} + a_{22}A_{22} - a_{32}A_{32}$$
$$= a_{31}A_{31} - a_{32}A_{32} + a_{33}A_{33}$$
$$= -a_{21}A_{21} + a_{22}A_{22} - a_{23}A_{23}$$

We see from the above that the value of the determinant of a 3×3 matrix is found by multiplying each element of any row or column by its minor, attaching the proper sign to these products, and adding them. We find the proper sign in the following manner. Let us consider the element a_{ij}. The minor of a_{ij} is A_{ij}. If $i + j$ is an even number, the sign of the product $a_{ij}A_{ij}$ is $+$; if $i + j$ is an odd number, the sign of the product $a_{ij}A_{ij}$ is $-$. We see that the product $a_{ij}A_{ij}$ has the sign $(-1)^{i+j}$ assigned to it. The minor A_{ij} together with the proper sign is called the **cofactor** of a_{ij} and is denoted by C_{ij}. Thus $C_{ij} = (-1)^{i+j}A_{ij}$.

Suppose we wish to find the value of

$$\det A = \begin{vmatrix} a_{11} & a_{12} & a_{13} \\ a_{21} & a_{22} & a_{23} \\ a_{31} & a_{32} & a_{33} \end{vmatrix}$$

using the elements of the second column and their minors. The elements of the second column are a_{12}, a_{22}, and a_{32}. The minors of these elements are A_{12}, A_{22} and A_{32} respectively. Since

$1 + 2 = 3$, an odd number, we use a minus sign with the product $a_{12}A_{12}$
$2 + 2 = 4$, an even number, we use a plus sign with the product $a_{22}A_{22}$
$3 + 2 = 5$, an odd number, we use a minus sign with the product $a_{32}A_{32}$

Then

$$\det A = -a_{12}A_{12} + a_{22}A_{22} - a_{32}A_{32}$$

8. DETERMINANTS OF SQUARE MATRICES

Example 3: Find the value of

$$\det\begin{pmatrix} 1 & 2 & 3 \\ 1 & 0 & 1 \\ 2 & 1 & -1 \end{pmatrix} = \begin{vmatrix} 1 & 2 & 3 \\ 1 & 0 & 1 \\ 2 & 1 & -1 \end{vmatrix}$$

using the elements of the second row and their minors.

Solution: The elements of the second row are $a_{12} = 1$, $a_{22} = 0$, and $a_{23} = 1$. The minors of these elements are

$$A_{12} = \begin{vmatrix} 2 & 3 \\ 1 & -1 \end{vmatrix} = 2(-1) - 3 \cdot 1 = -2 - 3 = -5$$

$$A_{22} = \begin{vmatrix} 1 & 3 \\ 2 & -1 \end{vmatrix} = 1(-1) - 2 \cdot 3 = -1 - 6 = -7$$

$$A_{23} = \begin{vmatrix} 1 & 2 \\ 2 & 1 \end{vmatrix} = 1 \cdot 1 - 2 \cdot 2 = 1 - 4 = -3$$

Finding the products $a_{12}A_{12}$, $a_{22}A_{22}$, and $a_{23}A_{23}$ and attaching the proper signs and finding the sums we have

$$\begin{vmatrix} 1 & 2 & 3 \\ 1 & 0 & 1 \\ 2 & 1 & -1 \end{vmatrix} = -1(-5) + 0(-7) - 1(-3)$$

$$= 5 + 0 + 3$$
$$= 8$$

Exercise 6

1. Find the values of the determinants of the following matrices.

(a) $\begin{pmatrix} 3 & 2 \\ -1 & 4 \end{pmatrix}$

(b) $\begin{pmatrix} 0 & 3 \\ 1 & -2 \end{pmatrix}$

(c) $\begin{pmatrix} 4 & -2 \\ -3 & -2 \end{pmatrix}$

(d) $\begin{pmatrix} 0 & 1 \\ 2 & 3 \end{pmatrix}$

MATRICES

Find the values of the following determinants (Exercises 2–11).

2. $\begin{vmatrix} 4 & -2 \\ 3 & 4 \end{vmatrix}$

7. $\begin{vmatrix} 3\sqrt{2} & -\sqrt{5} \\ 5\sqrt{5} & 4\sqrt{2} \end{vmatrix}$

3. $\begin{vmatrix} 6 & -1 \\ 0 & 1 \end{vmatrix}$

8. $\begin{vmatrix} -3 & -4 \\ 4 & 3 \end{vmatrix}$

4. $\begin{vmatrix} 0 & 1 \\ 1 & 0 \end{vmatrix}$

9. $\begin{vmatrix} \frac{2}{3} & \frac{3}{4} \\ \frac{8}{3} & 9 \end{vmatrix}$

5. $\begin{vmatrix} 0 & 0 \\ 0 & 0 \end{vmatrix}$

10. $\begin{vmatrix} a & b \\ b & a \end{vmatrix}$

6. $\begin{vmatrix} \sqrt{2} & \sqrt{2} \\ -\sqrt{2} & \sqrt{2} \end{vmatrix}$

11. $\begin{vmatrix} a^2 & ab \\ ab & b^2 \end{vmatrix}$

Find the values of the determinants of the following matrices (Exercises 12–15).

12. $\begin{pmatrix} 1 & 0 & 2 \\ 1 & -1 & 0 \\ 1 & 1 & 1 \end{pmatrix}$

14. $\begin{pmatrix} 2 & 3 & 1 \\ -2 & -3 & -1 \\ 1 & -1 & 0 \end{pmatrix}$

13. $\begin{pmatrix} -1 & 5 & 2 \\ 0 & 3 & -2 \\ 1 & 2 & 0 \end{pmatrix}$

15. $\begin{pmatrix} 3 & 4 & 5 \\ 2 & 3 & 4 \\ 3 & 4 & 5 \end{pmatrix}$

16. Prove that the determinant of a 2×2 matrix is zero if two of the columns are equal.

17. Prove that

$$\begin{vmatrix} a & b \\ c & d \end{vmatrix} = - \begin{vmatrix} c & d \\ a & b \end{vmatrix}$$

18. Prove that

$$\begin{vmatrix} a & b \\ ka & kb \end{vmatrix} = 0$$

19. Prove

$$\begin{vmatrix} a & b & c \\ d & e & f \\ g & h & i \end{vmatrix} = \begin{vmatrix} a & d & g \\ b & e & h \\ c & f & i \end{vmatrix}$$

20. Show that

$$\begin{vmatrix} 0 & 0 & 0 \\ a & b & c \\ e & f & g \end{vmatrix} = 0$$

Oswald Veblen (1880–1960), one of America's leading mathematicians, was born in Iowa. He studied at Harvard and the University of Chicago. He taught at the University of Chicago and at Princeton, and in 1932 he became professor at the Princeton Institute for Advanced Study. Veblen's most notable contributions involved differential and projective geometry, the foundations of geometry and topology. Finite geometries were not brought into prominence until 1906 when Veblen and W. H. Bussey made a study of them.

FINITE
GEOMETRY

10

1.
PHYSICAL GEOMETRY VERSUS ABSTRACT GEOMETRY

Most persons who studied Euclidean geometry in high school think of geometry as physical geometry. That is, they think of a geometry which is concerned with the shapes and sizes of actual physical objects. Although Euclid attempted to develop his geometry as an abstract mathematical system, as we read *The Elements* we feel that he associates points and lines with physical points and lines rather than thinking of them as undefined mathematical objects.

In the modern development of geometry, the mathematician thinks of point and line as undefined terms. He then develops geometry as an abstract mathematical system. We recall that a mathematical system consists of a set of elements. In geometry these elements are usually the undefined words "point" and "line." We are also given an undefined relation. In the mathematical systems developed in Chapter 6 the undefined relation was "equals." Although formal definitions of the elements are not given, they are subject to the restrictions imposed upon them by the postulates of the system.

The best way to demonstrate how a modern mathematician develops an abstract geometry is by considering a **finite geometry,** that is a geometry which, by virtue of the postulates selected, deals with a finite number of points and lines.

It was G. Fano, who, in 1892, first considered a finite geometry. Fano's finite geometry was a three-dimensional geometry containing fifteen points, thirty-five lines and fifteen planes. Each plane contained seven points and seven lines. Finite geometries were not brought into prominence until 1906, when O. Veblen and W. H. Bussey made a study of them. Since then, the study of finite geometries has grown considerably.

2.
A FINITE GEOMETRY

Our first example of a finite geometry consists of a set S of undefined elements called "points" and "lines." The undefined relation is "on." In English it is more convenient to say a line "contains" a point rather than a line is "on" a point and a pair of lines "has" one point in common rather than a pair of lines is "on" one common point. We shall assume that in such a context "on," "contain" and "has" have the same meaning. We state the following five postulates for our system. We must keep in mind that whenever two or more points or lines are mentioned, it will be understood that we mean distinct points and lines.

POSTULATE 1: *Each pair of lines in S has at least one point in common.*

POSTULATE 2: *Each pair of lines in S has not more than one point in common.*

POSTULATE 3: *Every point in S is on at least two lines.*

POSTULATE 4: *Every point in S is on not more than two lines.*

POSTULATE 5: *The total number of lines in S is four.*

These postulates were selected arbitrarily. We could have had an entirely different set if we had so desired. There are some rules that must be observed in constructing a set of postulates for a mathematical system such as a finite geometry. The most important rule is that it must be possible for all the postulates to be true at the same time. This is called **consistency.** If it is possible for all the postulates to be true at the same time, then the set of postulates is said to be **consistent.**

Let us look at our set of postulates and determine whether or not they are consistent. There are several ways to do this, but one of the easiest is to see if we can set up a model. What do we mean by a model? A **model** simply means an example of some kind in which all the postulates are true. A model is sometimes called an **interpretation.** We will now give an interpretation of the undefined terms in our postulates. The postulates as they are written are *neither* true nor false, but when we assign meanings

327

to the undefined terms then the postulates are *either* true or false. If we can find just one interpretation of the undefined terms for which all of the postulates are true, then we have shown that it is possible for all to be true at the same time and so be consistent.

We now attempt to find a model for the five postulates of our finite geometry. Let us think of the set S as a set of tin soldiers arranged in rows. Then the soldiers represent the points and the rows represent the lines of our postulates. We have now given an interpretation or meaning to the undefined terms of the postulates. Remember that the lines (rows) are just a collection of points (soldiers). They are completely empty between the points (soldiers).

How can we decide how many tin soldiers are in our set S? Postulates 1 and 2 tell us that there is one and only one soldier in each pair of rows. Postulates 3 and 4 tell us that every soldier is in two and only two rows. Postulate 5 tells that there are four rows of soldiers. Since there are four rows and every pair of rows has one and only one soldier in it, to find the number of soldiers in the set we must find the number of combinations of four things (rows) taken two (a pair) at a time. Since

$$\binom{4}{2} = \frac{4!}{2!2!} = 6$$

there are six soldiers in set S.

Suppose we let I, II, III, and IV represent the rows of soldiers, and A, B, C, D, E, and F represent the soldiers. The six pairs of rows are

I–II	II–III
I–III	II–IV
I–IV	III–IV

Each of these six pairs of rows have one soldier common to each pair. Let us call the soldier in rows I and II, A; the soldier in rows I and III, B; the soldier in rows I and IV, C; the soldier in rows II and III, D; the soldier in rows II and IV, E; and the soldier in rows III and IV, F.

I–II	A
I–III	B
I–IV	C
II–III	D
II–IV	E
III–IV	F

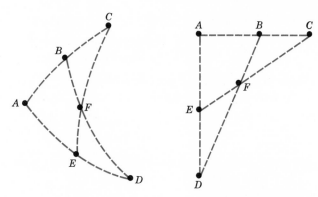

FIGURE 10.1

From the discussion above we see that soldiers *A*, *B*, and *C* are in row I; soldiers *A*, *D*, and *E* are in row II; soldiers *B*, *D*, and *F* are in row III; and soldiers *C, E,* and *F* are in row IV. We see then that there are exactly three soldiers in each row.

Now let us draw a diagram showing the placement of the soldiers (points) in rows (lines). Figure 10.1 shows two possibilities for placing the six soldiers in four rows. In the figure, soldiers (points) are represented by black dots and rows (lines) are represented by broken lines.

Since row I consists of soldiers *A*, *B*, and *C*, we shall call it line *ABC*. Similarly, row II is called line *ADE*; row III is called line *BDF*; and row IV is called line *CEF*.

We now return to the postulates and determine whether they are all true for our arrangement of tin soldiers in rows. Postulate 1 says that each pair of rows must have at least one soldier in common. Rows I and II have soldier *A* in common; rows I and II, soldier *B*; rows I and IV, soldier *C*; rows II and III, soldier *D*; rows II and IV, soldier *E*; and rows III and IV, soldier *F*. No pair of rows has more than one soldier in common, hence Postulate 2 is satisfied. Looking at the diagram we see that every soldier is in at least two rows, satisfying Postulate 3, and that no soldier is in more than two rows, satisfying Postulate 4. There are four rows in all, satisfying Postulate 5.

Exercise 1

1. With the interpretation:

> "point" as bead
> "line" as wire

restate the postulates of our finite geometry.

2. With the interpretation:

> "point" as member
> "line" as committee

restate the postulates of our finite geometry.

3. With the interpretation:

> "point" as tree
> "line" as row

restate the postulates of our finite geometry.

4. With the interpretation:

> "point" as stock
> "line" as mutual fund

restate the postulates of our finite geometry.

5. With the interpretation:

> "point" as student
> "line" as class

restate the postulates of our finite geometry.

6. Is the following diagram a correct model for the five postulates of our finite geometry? If not, which of the postulates are not satisfied? The lines are *AB*, *AD*, *BE*, and *CE*.

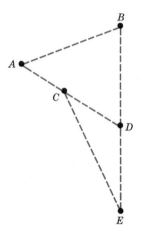

7. Interpreting the points in Figure 10.1 as San Diego, Los Angeles, Oakland, San Francisco, Portland, and Seattle, construct a permissible interpretation of our finite geometry by choosing appropriate airline flights.

8. With the interpretation:

> a point is one of the digits from the set {1, 2, 3, 4}
> a line is a triple of digits: the lines are 123, 124, 134, 234

Check to see whether the postulates of our finite geometry are true for this model.

3.
THEOREMS FOR
THE FINITE GEOMETRY

Just what have we shown by setting up the model in Section 2? We have shown that in at least one instance it is possible for all five of the postulates to be true at the same time. We now say that the five postulates are **consistent.**

Can we prove any other statements that are true for this system? If so, these statements are called **theorems.** In order for the theorems we prove to be true for all interpretations of the system, that is, that they are true in general and not for some particular interpretation, they must be stated in terms of the undefined elements of the system. Although we may use a diagram or an interpretation to help us understand relationships between the elements of the system, our proofs cannot use anything that depends upon the particular model that we use. Keeping this in mind, we look at our model of tin soldiers in rows.

When we set up this model we discovered two statements that were not mentioned in the postulates. We discovered that there are exactly six points in S and that there are exactly three points on each line. Let us state these two statements as theorems and attempt to prove them.

THEOREM 10.1:   *There are exactly six points in S.*

      Proof: By Postulate 5 there are exactly four lines in S. Let us call them I, II, III, and IV. By Postulate 1 each pair of these lines has at least one point of S in common. Naming these points by the two lines on which they lie, we have I–II, I–III, I–IV, II–III, II–IV, and III–IV. These six points are the only possible points by Postulate 2. No two of these points coincide by Postulate 4. Therefore, there are exactly six points in S.

THEOREM 10.2:   *There are exactly three points on each line.*

      Proof: By Theorem 10.1 there are exactly six points in S. They are I–II, I–III, I–IV, II–III, II–IV, and III–IV. We see that on line I we have three points, I–II, I–III, and I–IV; on line II we have three points, I–II, II–III,

331

and II–IV; on line III we have three points, I–III, II–III, and II–IV; and on line IV we have three points, I–IV, II–IV, and III–IV. Since this list contains all the points in *S*, there are exactly three points on each line.

We now define **parallel points** for this finite geometry.

DEFINITION 10.1: *Two points that have no lines in common are called parallel points.*

Looking at the diagram in Figure 10.1 we see that each point in *S* has one and only one point parallel to it. Let us prove this statement for our system.

THEOREM 10.3: *Each point in S has one and only one point parallel to it.*

     Proof: By Postulates 1 and 2, any given point of *S* is determined by exactly two lines. By Postulate 5, there are exactly two lines in *S* on which any given point does not lie. By Postulates 1 and 2, there is one and only one point determined by these two lines. Therefore, there is one and only one point in *S* parallel to any given point.

In looking at Theorem 10.3 we see that another statement is true because of this theorem. This statement, since it follows directly from Theorem 10.3, is called a corollary of 10.3.

COROLLARY 10.1: *On a given line in S, not containing a given point, there is one and only one point parallel to a given point.*

The proof of this corollary is left to the reader.

Since there are exactly six points in the finite geometry just developed, in the following discussion it shall be referred to as **six-point geometry.**

Exercise 2

Prove the theorems in Exercises 1 and 2 for six-point geometry.
  1. Two distinct lines have exactly one point in common.
  2. Every point is on exactly two lines.
  3. In the following set of postulates the set *S* consists of undefined elements "points" and "lines." The undefined relation is "contains."

A-1. Every pair of points in S has at least one line in common.
A-2. Each pair of points in S has not more than one line in common.
A-3. Every line in S contains at least two points.
A-4. Every line in S contains not more than two points.
A-5. The total number of points in S is four.

Construct a model for this system to show that the postulates are consistent.

4. With the following interpretation:

"point" as musical note
"line" as chord

restate the postulates in problem 3.

5. Draw a diagram that could represent the postulates of problem 3.

6. Which of the postulates in problem 3 would not be true in the following diagram?

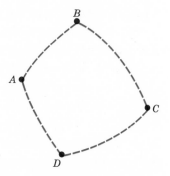

7. Letting the cities New York, Boston, Philadelphia, and Washington represent points for the finite geometry in problem 3, choose flights to represent the lines.

Prove the following theorems for the finite geometry in problem 3 (Exercises 8–10).

8. There are exactly six lines in S.

9. There are exactly three lines passing through each point of S.

10. Two points are on exactly one line.

4.
INDEPENDENCE
OF POSTULATES

To construct a finite geometry we select some words which we call **undefined** (we make no attempt to define them), an **undefined relation,** and some statements, called **postulates,** about the undefined terms. These statements are assumed to be true. We now set up a model to show that our postulates are consistent. Next we define new terms using our undefined terms as a basis of the definitions. We now conjecture new statements which appear to be true for our system. We attempt to prove these conjectures to be true using our definitions, postulates, and deductive reasoning. If we can prove these conjectures true, we call them **theorems.**

Another property of a set of postulates for a mathematical system such as finite geometry is **independence.** We call a set of postulates **independent** if no one postulate can be deduced from the remaining ones. If any one of the postulates of a system can be deduced from the remaining ones, it is said to be **dependent.** If one postulate is dependent, there is no need to include it in our set of postulates, as we can prove it as a theorem using the other postulates of the system.

For example, in stating the postulates for the set of integers in Chapter 2, there was no need to include the closure property of subtraction, since that property can be proved using the other postulates of the system and the definition of subtraction.

To prove the independence of any postulate we only have to show that the other postulates of the system, together with some contradiction of the postulate in question, form a consistent set of postulates. Returning to the postulates of six-point geometry it is not difficult to supply independence proofs for each of the postulates. For convenience, let us restate the postulates.

POSTULATE 1: *Each pair of lines in S has at least one point in common.*

POSTULATE 2: *Each pair of lines in S has not more than one point in common.*

POSTULATE 3: *Every point in S is on at least two lines.*

POSTULATE 4: *Every point in S is on not more than two lines.*

POSTULATE 5: *The total number of lines in S is four.*

To show the independence of Postulate 1, let us interpret S to consist of five points, A, B, C, D, and E, distributed among four lines, I, II, III, and IV, as follows:

I	*AB*
II	*ACD*
III	*BDE*
IV	*CE*

In this interpretation Postulate 1 fails because pairs of lines I and IV have no point in common. On the other hand, no pair of lines has more than one point in common, so Postulate 2 is satisfied. Each point is on at least two lines and not more than two lines, so Postulates 3 and 4 are satisfied. We have four lines in S so Postulate 5 is satisfied. In the same way, independence proofs can be supplied for Postulates 2, 3, 4, and 5.

Exercise 3

1. Prove the independence of Postulate 2 of six-point geometry. (*Hint:* Interpret S to consist of seven points.)

2. Prove the independence of Postulate 4 of six-point geometry. (*Hint:* Interpret S to consist of four points.)

3. Prove the independence of Postulate 3 of six-point geometry. (*Hint:* Interpret S to consist of seven points.)

4. Prove the independence of Postulate 5 of six-point geometry.

5.
ANOTHER
FINITE GEOMETRY

The finite geometry we studied in the previous sections of this chapter had exactly six points. We shall now state the postulates of another finite geometry. In this system the set S consists of undefined elements called "points" and "lines." The undefined relation is "on." In English it is more convenient to say a line "contains" a point rather than a line is "on" a point. We shall assume that in such a context "on" and "contain" have the same meaning.

POSTULATE 1: *If P_1 and P_2 are any two points in S, there is at least one line containing both P_1 and P_2.*

POSTULATE 2: *If P_1 and P_2 are any two points in S, there is at most one line containing both P_1 and P_2.*

POSTULATE 3: *If L_1 and L_2 are any two lines in S, there is at least one point that lies on both L_1 and L_2.*

POSTULATE 4: *There are exactly three points on each line.*

POSTULATE 5: *If L is any line in S, there is at least one point that does not lie on L.*

POSTULATE 6: *There exists at least one line.*

Now let us construct a model to test the above postulates for consistency. Let us think of the points as men and the lines as clubs to which the men belong. There exists at least one club having exactly three members by Postulates 4 and 6. Let us designate men by symbols $A, B, C, D, \ldots$, and the club by ABC. By Postulate 5 there exists a man D who is not a member of club ABC. Now D must belong to the same club as A by Postulate 1, and neither B nor C can belong to this club by Postulate 2. Thus there exists another club, ADE, and still another, BDF, where E and F are other men.

At present we have three clubs (ABC, ADE, BDF) and six men, A, B, C, D, E, and F. Every pair of men, however, must belong to exactly one club. Man A now belongs to clubs with B, C, D, and E. Thus A and F cannot belong to clubs with B, C, D, or E. (Why?) Hence there must be a seventh man in a club with A and F. Let us call this club AFG. If there were an eighth man, H, then the club containing A and H could not have a member in common with BDF since A is already in a club with B, D, and F. Thus the set consists of exactly seven men. By continuing this same type of reasoning, clubs CEF, BEG, and CDG can be found. There are then exactly seven men and seven clubs, as indicated by the seven columns below:

$$
\begin{array}{ccccccc}
A & A & B & A & C & B & C \\
B & D & D & F & E & E & D \\
C & E & F & G & F & G & G
\end{array}
$$

With this interpretation it is easy to verify that each of the six postulates is fulfilled. Postulates 4, 5, and 6 obviously are justified. In order to verify Postulates 1 and 2, we must consider every possible pair of men and verify that there is one and only one club to which both men belong. For example, A and B belong to club ABC, but to no other club. The pairs and their common clubs are shown in Table 10.1.

Finally, to verify Postulate 3, we must consider every possible pair of clubs and verify that in each case there is at least one man who belongs

TABLE 10.1

Club	Pairs of men	Club	Pairs of men
ABC	A, B	BDF	D, F
ABC	A, C	BEG	B, E
ABC	B, C	BEG	B, G
ADE	A, D	BEG	E, G
ADE	A, E	CDG	C, D
ADE	D, E	CDG	C, G
AFG	A, F	CDG	D, G
AFG	A, G	CEF	C, E
AFG	F, G	CEF	C, F
BDF	B, D	CEF	E, F
BDF	B, F		

to both clubs. The pairs of clubs and their common members are shown in Table 10.2.

We have now shown that our set of postulates is consistent by exhibiting a concrete model for this finite geometry.

We see from the above that there are seven points in this geometry and seven lines; so we shall call this geometry **seven-point geometry.**

We can now prove some simple theorems that follow from the postulates of the seven-point geometry.

TABLE 10.2

Member	Pairs of clubs	Member	Pairs of clubs
A	ABC, ADE	F	AFG, BDF
A	ABC, AFG	G	AFG, BEG
B	ABC, BDF	F	AFG, CDG
B	ABC, BEG	F	AFG, CEF
C	ABC, CDG	B	BDF, BEG
C	ABC, CEF	D	BDG, CDG
A	ADE, AFG	F	BDF, CEF
D	ADE, BDF	D	BEG, CDG
E	ADE, BEG	E	BEG, CEF
D	ADE, CDG	C	CDG, CEF
E	ADE, CEF		

THEOREM 10.4: *There exists at least one point.*

Proof: By Postulate 6 there exists at least one line, and by Postulate 4 every line contains exactly three points. Hence the assertion that at least one point exists is surely true.

THEOREM 10.5: *If L_1 and L_2 are any two lines, there is at most one point that lies on both L_1 and L_2.*

Proof: To prove this theorem, let us assume the contrary and suppose it is possible for two lines, say L_1 and L_2, to have two points, say P_1 and P_2, in common. This leads at once to a contradiction since Postulate 2 asserts that there is at most one line that contains each of two given points. Hence it follows that the two lines can have at most one point in common.

THEOREM 10.6: *Two points determine exactly one line.*

Proof: This follows immediately from Postulates 1 and 2.

THEOREM 10.7: *Two lines have exactly one point in common.*

Proof: This follows immediately from Postulate 3 and Theorem 10.5.

THEOREM 10.8: *If P is any point, there is at least one line that does not contain P.*

Proof: By Postulate 6 there exists at least one line. If this line, L, does not contain P, our proof is complete. Suppose therefore, that L passes through P. By Postulate 4, L contains two points besides P. Call one of them P'. By Postulate 5 there is at least one point, say P'', that does not lie on L. By Theorem 10.6 there is a unique line, say L', that contains P' and P''. Moreover, L and L' are distinct, since L' contains P'' and L does not. Hence by Theorem 10.6, L and L' have exactly one point in common, and this point is P'. Therefore P, which lies on L, cannot lie on L'. In other words, L' is a line that does not contain P.

THEOREM 10.9: *Every point lies on at least three lines.*

Proof: Let P be an arbitrary point. By Theorem 10.8 there is at least one line, L, that does not pass through P, and by Postulate 4 this line contains three points, P_1, P_2, and P_3. By Theorem 10.6 each of these points determines with P a unique line. Moreover, all these lines are distinct, for if two of them coincided, that line would have two points in common with L, which is impossible by Theorem 10.7. Hence there are at least three lines passing through an arbitrary point P as asserted.

Exercise 4

1. Draw a diagram that represents the seven-point geometry. (*Hint:* Remember that the lines need not be straight; one line may be a circle.)

2. Give another interpretation for the postulates of seven-point geometry.

3. Consider the system consisting of seven triples of the form $P = (u, v, w)$. Let seven points be represented by:

$$P_1 = (1, 0, 0) \qquad P_2 = (0, 1, 0) \qquad P_3 = (0, 0, 1)$$
$$P_4 = (0, 1, 1) \qquad P_5 = (1, 0, 1) \qquad P_6 = (1, 1, 0)$$
$$P_7 = (1, 1, 1)$$

Let seven lines, $L_1, L_2, \ldots, L_7$, be those sets of points that satisfy the following equations:

$$L_1: u = 0 \qquad L_2: v = 0 \qquad L_3: w = 0$$
$$L_4: v + w = 0 \qquad L_5: u + w = 0 \qquad L_6: u + v = 0$$
$$L_7: u + v + w = 0$$

where all the calculations are computed modulo 2. For example P_2, P_3, and P_4 are on L_1 because the first component, u, of these three points is 0. Verify that in this system the postulates of seven-point geometry are satisfied.

4. Use the interpretation:

> point as politician
> line as committee

Restate the postulates of seven-point geometry.

For seven-point geometry prove the following theorems (Exercises 5–6).

5. Any two points are on exactly one line.

6. There exist three points that are not on the same line.

7. In the following set of postulates the set S consists of undefined elements "points" and "lines". The undefined relation is "on".

> S-1. If L_1 and L_2 are any two lines in S, there is at least one point on both L_1 and L_2.

S-2. If L_1 and L_2 are any two lines in S, there is at most one point on both L_1 and L_2.

S-3. If P_1 and P_2 are any two points in S, there is at least one line that contains both P_1 and P_2.

S-4. There are exactly three lines containing each point.

S-5. If P is any point in S, there is at least one line that does not contain P.

S-6. There exists at least one point.

Construct a model for this system to show that the postulates are consistent.

8. Using the postulates in problem 7 prove that two lines contain exactly one point.

9. Using the postulates in problem 7 prove that there exists at least one line.

Rene Descartes (1596–1650) was a seventeenth century French genius who revolutionized mathematical concepts and ushered in modern mathematics with his creation of plane analytic (coordinate) geometry. Descartes was a very delicate child and, because of this, while attending a Jesuit boarding school he was allowed to lie in bed as late as he pleased in the morning. This habit stayed with him all his life. He slept ten hours each night and never let any one disturb him before noon. Descartes did not agree with the philosophy taught by the Jesuits; as soon as he was old enough he developed his own views of philosophy that were revolutionary in their break with the past and caused him to be known as the father of modern philosophy. In 1637 he published a book outlining his views on philosophy. In the appendix of this book he introduced analytic geometry for which he is famous today. It is amazing that Descartes accomplished so much because he was never a strenuous worker. He claimed that he spent only a few hours daily in thoughts that occupy the imagination and very few hours yearly in those thoughts that require understanding.

ANALYTIC GEOMETRY: THE STRAIGHT LINE AND THE CIRCLE

11

1.
INTRODUCTION

In 1637, Réné Descartes (1596–1650) published a book called *La Géométrie* in which he introduced the subject of analytic geometry. **Analytic geometry** is the study of geometry through the use of a coordinate system and an associated algebra. In (plane) analytic geometry we solve two types of problems. In the first type we are given an equation in two variables and are asked to draw the graph of this equation; in the second type we are given a set of points defined by certain geometric conditions and asked to find an equation whose graph consists of this set of points.

In Chapter 5, Section 4, we discussed Cartesian products and their graphs. At that time we considered the graph of the Cartesian product $R \times R$ (R the set of real numbers) as the set of all points in the Cartesian plane. Thus we introduced a coordinate system that is the basis of analytic geometry.

In Euclidean geometry the words "point," "line," and "plane" are undefined terms. In analytic geometry we identify a point with an ordered pair of real numbers, we call the set $R \times R$ the Cartesian plane and identify a line with a first degree equation in two variables.

2.
MIDPOINT OF
A LINE SEGMENT

In Chapter 7, Section 7, we derived the formula for finding the distance between two points when the coordinates of these points were given. We found that if P_1 and P_2 are two points with coordinates (x_1, y_1) and (x_2, y_2), respectively, then the distance, d, of the line segment P_1P_2 is given by the formula

$$d = \sqrt{(x_2 - x_1)^2 + (y_2 - y_1)^2}$$

This is called the **distance formula.**

Now let us consider two points P_1 and P_2 with coordinates (x_1, y_1) and (x_2, y_2) respectively (Figure 11.1). We would like to find the coordinates (x, y) of the midpoint, M, of line segment $\overline{P_1P_2}$.

Using Figure 11.1 we see that triangle P_1P_2S is similar to triangle P_1MR, hence, since the ratios of the lengths of corresponding sides of similar triangles are proportional,

$$(1) \qquad \frac{x_2 - x_1}{P_1P_2} = \frac{x - x_1}{P_1M}$$

Since M is the midpoint of $\overline{P_1P_2}$, $P_1P_2 = 2P_1M$, and equation (1) can be written

$$\frac{x_2 - x_1}{2P_1M} = \frac{x - x_1}{P_1M}$$

From this we obtain the x-coordinate of point M:

$$P_1M(x_2 - x_1) = 2P_1M(x - x_1)$$
$$x_2 - x_1 = 2(x - x_1)$$
$$x_2 - x_1 = 2x - 2x_1$$
$$x = \frac{x_2 + x_1}{2}$$

In a similar fashion we obtain the y-coordinate of point M:

$$y = \frac{y_2 + y_1}{2}$$

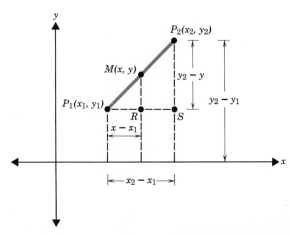

FIGURE 11.1

The coordinates of the midpoint, M, of line segment $\overline{P_1P_2}$ are

$$\left(\frac{x_1 + x_2}{2}, \frac{y_1 + y_2}{2}\right)$$

Example 1: Find the coordinates of the midpoint of the line segment whose endpoints have coordinates $(5, -4)$ and $(-3, 6)$.

Solution:

$$x = \frac{5 + (-3)}{2} = \frac{2}{2} = 1$$

$$y = \frac{-4 + 6}{2} = \frac{2}{2} = 1$$

The coordinates of the midpoint are $(1, 1)$.

Example 2: Find the coordinates of the midpoints of the sides of the triangle whose vertices have coordinates $(4, 3)$, $(6, -7)$, and $(-8, -1)$.

Solution: Let us call $(4, 3)$ point A†, $(6, -7)$ point B, and $(-8, -1)$ point C. The coordinates of the midpoint of $\overline{AB}$ are $(5, -2)$ since

$$x = \frac{4 + 6}{2} = 5 \quad \text{and} \quad y = \frac{3 + (-7)}{2} = -2$$

The coordinates of the midpoint of $\overline{BC}$ are $(-1, -4)$ since

$$x = \frac{6 + (-8)}{2} = -1 \quad \text{and} \quad y = \frac{-7 + (-1)}{2} = -4$$

The coordinates of the midpoint of $\overline{AC}$ are $(-2, 1)$ since

$$x = \frac{4 + (-8)}{2} = -2 \quad \text{and} \quad y = \frac{3 + (-1)}{2} = 1$$

---

† Instead of saying the point A whose coordinates are $(4, 3)$ we often say the point $(4, 3)$. When we write $A(4, 3)$ we mean the point A whose coordinates are $(4, 3)$.

Exercise 1

1. Determine the distance between the following pairs of points.

(a) (1, 3) and (3, −2)

(b) (−3, −4) and (0, 5)

(c) (−2, 5) and (7, 1)

(d) (0, 0) and (−2, 3)

(e) (−3, −1) and (−3, −4)

(f) (−5, −4) and (4, −5)

(g) (−1, −1) and (4, −6)

(h) (0, −1) and (7, −2)

2. Find the coordinates of the midpoints of the line segments joining the points with coordinates given below.

(a) (3, 4) and (−5, 12)

(b) (−2, −3) and (6, −9)

(c) (−6, −4) and (2, 0)

(d) (3, −7) and (−8, −5)

3. Given the triangle the coordinates of whose vertices are (2, −3), (−7, −5), and (−6, 1). Find the coordinates of the midpoints of the sides of the triangle.

4. The coordinates of the vertices of an isosceles triangle are (−1, −3), (9, 5), and (−4, 11). Find the coordinates of the midpoints of the sides of equal length.

5. The coordinates of the vertices of a right triangle are (−1, 4), (2, −1), and (12, 5). Find the coordinates of the midpoint of the hypotenuse.

6. The coordinates of the vertices of a square are (0, −2), (2, 0), (2, −2), and (0, 0). Find the coordinates of the midpoints of the two diagonals of the square.

7. Locate a point on the x-axis that is equidistant from the points (−3, 5) and (2, −4). (*Hint:* Use the distance formula and denote the desired point on the x-axis by (x, 0) and find x.)

8. The endpoints of a line segment are (−8, 6) and (4, 12). Find three points on the segment that divide it into four segments of equal length.

9. Given the points $A(−1, 2)$, $B(−3, 4)$, and $C(2, 5)$. Do these points lie on a straight line or are they the vertices of a triangle? Why?

10. An isosceles triangle has vertices (4, 3), (3, 1), and (1, 2). Find the area of this triangle.

3.

THE EQUATION OF A LINE

Let us consider a line not parallel to the y-axis. The **slope,** m, of this line is the ratio of the rise to the run of any segment of the line (Figure 11.2).

Let P_1 and P_2 with coordinates (x_1, y_1) and (x_2, y_2), respectively, be any two points on the line. Then the slope, m, of the line is given by

$$m = \frac{P_2 R}{P_1 R} = \frac{y_2 - y_1}{x_2 - x_1}$$

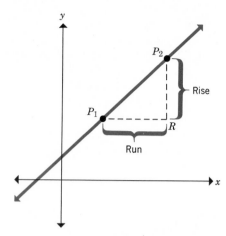

FIGURE 11.2

A line parallel to the y-axis has no slope since in this case $x_2 = x_1$ and $\dfrac{y_2 - y_1}{x_2 - x_1}$ is not defined because division by 0 is impossible in the field of real numbers.

Suppose we know that a line, L, passes through point P_1 with coordinates (x_1, y_1) and has slope m. Then the point P with coordinates (x, y) is on L if and only if

(1)
$$m = \frac{y - y_1}{x - x_1}$$

or

(2)
$$y - y_1 = m(x - x_1)$$

That is, a line with slope m and passing through the point with coordinates (x, y) is the graph of the relation

$$\{(x, y) \mid y - y_1 = m(x - x_1)\}$$

We call the equation $y - y_1 = m(x - x_1)$ the **point-slope equation** of a line.

For any value of x other than x_1, equations (1) and (2) give equal values of (y). Replacing x by x_1 in equation (2) gives $y = y_1$ showing that point P is a point on the graph of equation (2). Hence equation (2) is the equation of the given line.

Since line L is not parallel to the y-axis it must intersect the y-axis at

348

some point. This point has coordinates of the form $(0, b)$, b a real number. Since $(0, b)$ is a point on the given line its coordinates must satisfy equation (2). Replacing y_1 by b and x_1 by 0 in equation (2) we obtain

$$y - b = m(x - 0)$$

which simplifies to

(3) $$y = mx + b$$

This equation is called the **slope-intercept** form of the equation of the line; m is the **slope** and b is called the **y-intercept** of the line.

That is, a line with slope m and y-intercept b is the graph of the relation

$$\{(x, y)\,|\,y = mx + b\}$$

Notice that both equation (2) and equation (3) are first degree equations in two variables x and y.

Now let us consider a line parallel to the y-axis and k units from it. If the line is k units to the right of the y-axis, every point on the line has coordinates (k, y), hence the equation of the line is

(4) $$x = k$$

If, on the other hand, the line is k units to the left of the y-axis, every point on the line has coordinates $(-k, y)$ and the equation of the line is

(5) $$x = -k$$

We see then, that a line parallel to the y-axis and k units from it is the graph of the relation

$$\{(x, y)\,|\,x = k\}$$

We observe that both equations (4) and (5) are first degree equations.

We conclude from the above that every line is represented by a first degree equation in two variables.

In discussing the linear function in Chapter 5, Section 10, we stated that the graph of a first degree equation is a line. We shall now show that this is true.

Let us consider the linear equation

(6) $$Ax + By + C = 0$$

where A, B, and C are real numbers and A and B are not both zero. If $B = 0$, we have

$$Ax + C = 0$$

Since $A \neq 0$, this equation may be written

$$x = -\frac{C}{A}$$

From the above discussion we see that the graph of this equation is a line parallel to the y-axis.

If $B \neq 0$, equation (6) may be written

$$y = -\frac{A}{B}x - \frac{C}{B}$$

Comparing this equation with equation (3) we see that its graph is a line with slope $-\frac{A}{B}$ and passing through the point $(0, -\frac{C}{B})$ and that its y-intercept is $-\frac{C}{B}$.

We see then that every linear equation $Ax + By + C = 0$, A and B not both zero, has a graph that is a line. This form of the equation of a line is called the **general form.**

Example 1: Find the equation of a line with slope 3 and passing through the point with coordinates $(-2, 5)$.

Solution: Using equation (2) with $m = 3$, $y_1 = 5$, and $x_1 = -2$ we obtain

$$y - 5 = 3[x - (-2)]$$

which may be written

$$y - 3x - 11 = 0$$

Example 2: Find the equation of a line through points with coordinates $(-3, 4)$ and $(0, 5)$.

Solution: The slope of the given line is

$$m = \frac{4 - 5}{-3 - 0} = \frac{1}{3}$$

Using equation (2) with $x_1 = -3$, and $y_1 = 4$ we obtain

$$y - 4 = \frac{1}{3}[x - (-3)]$$

which may be written as

$$3y - x - 15 = 0$$

The coordinates of either of the given points could have been used in equation (2). The reader should use the coordinates (0, 5) to convince himself that the same equation results in either case.

Example 3: What is the slope of the line that is the graph of $3x + 2y - 6 = 0$?

Solution: Solving $3x + 2y - 6 = 0$ explicitly for y, that is, putting the general form of the equation of the line into slope-intercept form we obtain

$$y = -\frac{3}{2}x + 3$$

The slope of the line is $-\frac{3}{2}$.

4.
PARALLEL AND PERPENDICULAR LINES

Two nonvertical lines are parallel if and only if they have the same slope. We can easily verify that if two lines are parallel they have the same slope by using Figure 11.3a. Let us consider two parallel lines L_1 and L_2. We choose two points, P_1 and P_2, on L_1. Now we draw lines through P_1 and P_2 perpendicular to the x-axis and intersecting lines L_2 in points P_3 and P_4, respectively. We also draw P_1R and P_3S parallel to the x-axis. Since $P_1P_2P_4P_3$ is a parallelogram, line segments P_1P_2 and P_3P_4 are equal in length. Since P_1RSP_3 is a rectangle, P_1R and P_3S are also equal in length. Now triangle P_1P_2R and triangle P_3P_4S are congruent right triangles, hence P_2R and P_4S are equal in length and

$$\frac{P_2R}{P_1R} = \frac{P_4S}{P_3S}$$

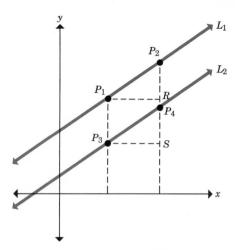

FIGURE 11.3a

But the slope, m_1, of L_1 is $\dfrac{P_2R}{P_1R}$ and the slope, m_2, of L_2 is $\dfrac{P_4S}{P_3S}$, hence $m_1 = m_2$.

We now prove that if $m_1 = m_2$, then L_1 is parallel to L_2. If $m_1 = m_2 = 0$, then L_1 and L_2 are each parallel to the x-axis and hence are parallel to each other. Now consider $m_1 = m_2 \neq 0$. Using Figure 11.3b, we see that the slope, m_1, of L_1 is $\frac{BC}{AC}$; the slope, m_2, of L_2 is $\frac{B'C'}{A'C'}$. Since $m_1 = m_2$,

$$\frac{BC}{AC} = \frac{B'C'}{A'C'}$$

Since angles BCA and $B'C'A'$ are right angles, triangle ABC is similar to triangle $A'B'C'$ and hence angles BAC is congruent to angle $B'A'C'$. Therefore L_1 is parallel to L_2 because a pair of corresponding angles made by a transversal (x-axis) are congruent.

Now let us consider two lines with equations

$$A_1x + B_1y + C_1 = 0$$
$$A_2x + B_2y + C_2 = 0$$

If these two lines are parallel, their slopes, $-\dfrac{A_1}{B_1}$ and $-\dfrac{A_2}{B_2}$ are equal, and we see that

$$\frac{A_1}{A_2} = \frac{B_1}{B_2}$$

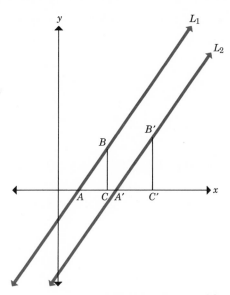

FIGURE 11.3b

Now let us consider two distinct lines with equal slope, $-\frac{A}{B}$. We can write the equations of these two lines as

$$y = -\frac{A}{B}x + C_1$$

$$y = -\frac{A}{B}x + C_2$$

where C_1 and C_2 are the y-intercepts of the two lines and $C_1 \neq C_2$ (why?). These equations may be simplified to

$$Ax + By + C_1 = 0$$
$$Ax + By + C_2 = 0$$

Example 1: Given $3x + 4y + 5 = 0$ and $6x + 8y + 3 = 0$. Are these two lines parallel?

Solution: Since

$$\frac{3}{6} = \frac{4}{8}$$

the two given lines are parallel.

In Section 3 of this chapter we showed that the slope of a line whose equation is $Ax + By + C = 0$ is $-\frac{A}{B}$. In other words, the lines whose equations are

$$Ax + By + C_1 = 0$$

and

$$Ax + By + C_2 = 0$$

are parallel.

Example 2: Find the equation of a line parallel to $2x + 3y + 1 = 0$ and passing through the point whose coordinates are $(0, 5)$.

Solution: We write the desired equation as

$$2x + 3y + C = 0$$

Replacing x by 0 and y by 5 we obtain

$$2 \cdot 0 + 3 \cdot 5 + C = 0$$

and

$$C = -15$$

The required equation is $2x + 3y - 15 = 0$.

Example 3: Given right triangle ABC, with vertices $A(2, -1)$, $B(-7, 6)$, and $C(3, 4)$. (a) Find the equation of the line through points A and C; (b) find the equation of the line through points B and C.

Solution: (a) The slope of the line $\overleftrightarrow{AC}$ is

$$\frac{4 + 1}{3 - 2} = \frac{5}{1}$$

The equation of $\overleftrightarrow{AC}$ is

$$\frac{y - 4}{x - 3} = \frac{5}{1}$$
$$y - 4 = 5x - 15$$
$$5x - y - 11 = 0$$

(b) The slope of the line $\overleftrightarrow{BC}$ is

$$\frac{4 - 6}{3 + 7} = -\frac{2}{10} = -\frac{1}{5}$$

4. PARALLEL AND PERPENDICULAR LINES

The equation of $\overleftrightarrow{BC}$ is

$$\frac{y - 4}{x - 3} = -\frac{1}{5}$$

$$5y - 20 = -x + 3$$

$$x + 5y - 23 = 0$$

In Example 3 above $\overline{AC}$ and $\overline{BC}$ are the legs of right triangle ABC and hence are perpendicular. Observe the equations of these two perpendicular lines:

(1) $\qquad\qquad\qquad\qquad 5x - y - 11 = 0$

(2) $\qquad\qquad\qquad\qquad x + 5y - 23 = 0$

It is no coincidence that the coefficient of x in equation (1) is the coefficient of y in equation (2) and that the coefficient of y in equation (1) is the opposite of the coefficient of x in equation (2). It can be proved, although we shall not do it here, that a line whose equation is $Bx - Ay + C_1 = 0$ is perpendicular to a line whose equation is $Ax + By + C_2 = 0$. That is, a line whose slope is $\frac{B}{A}$ is perpendicular to a line whose slope is $-\frac{A}{B}$.

Example 4: Find the equation of a line perpendicular to a line whose equation is $2x - 4y + 7 = 0$ and that passes through the point whose coordinates are $(-3, 5)$.

Solution: We write the required equation as $-4x - 2y + C_1 = 0$. Replacing x by -3 and y by 5 we obtain

$$-4(-3) - 2(5) + C_1 = 0$$

Solving for C_1 we obtain $C_1 = -2$. The required equation is $-4x - 2y - 2 = 0$ which may be written $4x + 2y + 2 = 0$.

Exercise 2

1. Find the equations of the lines passing through points with the following coordinates.

(a) $(-2, 4)$ and $(5, 6)$

(b) $(-3, 5)$ and $(6, 3)$

(c) $(3, -4)$ and $(0, 0)$

(d) $(-2, -7)$ and $(3, -6)$

(e) $(0, 7)$ and $(8, -5)$

(f) $(-7, 9)$, and $(12, -5)$

2. In each of the following the coordinates of a point on a line and the slope of the line are given. Find the equations of the lines.

    (a) $(2, -1)$; slope 2   (d) $(-5, -6)$; slope $-\frac{2}{3}$

    (b) $(3, 4)$; slope -4 (e) $(-7, 2)$; slope $\frac{5}{8}$

    (c) $(0, 0)$; slope $\frac{1}{2}$ (f) $(-12, -3)$; slope 2

3. Which of the following pairs of lines are parallel?

    (a) $x - y = 0$; $x + y = 0$

    (b) $3x + 2y = 6$; $3x - 2y + 4 = 0$

    (c) $5x - 2y + 4 = 0$; $5x - 2y + 1 = 0$

    (d) $y - 3x + 2 = 0$; $3x - y + 1 = 0$

    (e) $2x - 3y + 1 = 0$; $3y + 2x - 1 = 0$

4. Which of the pairs of lines in problem 3 are perpendicular?

5. The vertices of a triangle have coordinates $(2, 3)$, $(-4, 5)$, and $(4, -3)$. Find the equations of the three altitudes of the triangle.

6. Find the equation of the perpendicular bisector of the line segment whose endpoints have coordinates $(4, 3)$ and $(2, 5)$.

7. Prove that the points $(3, 2)$, $(5, 1)$, $(0, -3)$, and $(2, -4)$ are vertices of a parallelogram.

8. Determine the equations of two lines parallel to $5x + 3y + 4 = 0$, one through $(2, 3)$ and the other through the origin.

9. Two lines have equations $3x + 2y + 4 = 0$ and $3x + 2y - 1 = 0$.

    (a) Are the two lines parallel?

    (b) Consider the equation $3x + 2y + 4 + k(3x + 2y - 1) = 0$. Is this the equation of a line when k is a real number? Is it parallel to the two given lines?

10. The vertices of a triangle have coordinates $(7, -2)$, $(3, 6)$, and $(-5, 4)$. Find the equations of the three medians of the triangle.

11. Three vertices of a parallelogram are $(10, 4)$, $(5, -3)$, and $(0, 2)$. Find the fourth vertex.

5.
PROVING GEOMETRIC
THEOREMS ANALYTICALLY

Many theorems of Euclidean geometry can be proved using a coordinate system and algebraic methods. Since the properties of geometric figures depend upon the relation of their parts and not upon the position of the figures with reference to the plane, that is, with reference to the coordinate axis, choosing a suitable position of the axes with reference to the figure in question often simplifies the work considerably.

5. PROVING GEOMETRIC THEOREMS ANALYTICALLY

The coordinates of the points which determine the figure in question must be expressed in general terms so that the proof is for the general case not a specific case.

Study the examples below.

Example 1: Prove: The diagonals of a rectangle are equal in length.

  Solution: Let $ABCD$ be a rectangle. The vertices A, B, C, D have coordinates $(0, 0)$, $(a, 0)$, (a, b), and $(0, b)$ respectively (Figure 11.4). Using the distance formula we find the length of diagonal $\overline{AC}$ is

$$AC = \sqrt{(a - 0)^2 + (b - 0)^2} = \sqrt{a^2 + b^2}$$

  and the length of diagonal $\overline{BD}$ is

$$BD = \sqrt{(a - 0)^2 + (0 - b)^2} = \sqrt{a^2 + b^2}$$

  From these results the theorem follows.

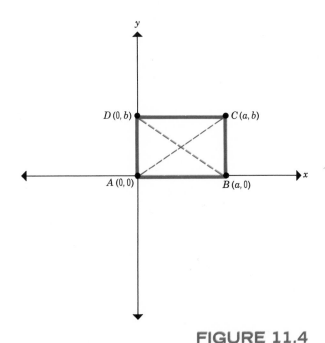

FIGURE 11.4

Example 2: Prove: The diagonals of a square are perpendicular to each other.

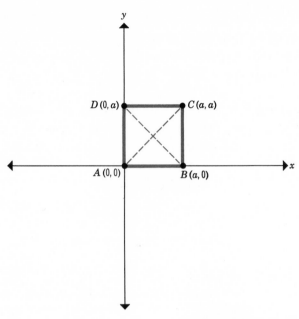

FIGURE 11.5

Solution: Let $ABCD$ be a square. The axes are chosen so that the vertices A, B, C, and D have coordinates $(0, 0)$, $(a, 0)$, (a, a), and $(0, a)$ respectively (Figure 11.5). The slope of diagonal $\overline{AC}$ is

$$\frac{a - 0}{a - 0} = 1$$

The slope of diagonal $\overline{BD}$ is

$$\frac{a - 0}{0 - a} = -1$$

The equation of $\overline{AC}$ is

$$y - a = 1(x - a)$$

which can be written $x - y = 0$.
The equation of $\overline{BD}$ is

$$y - a = -1(x - 0)$$

which may be written $x + y - a = 0$.

From the discussion in Section 4 we see that these two equations are the equations of two lines that are perpendicular.

Exercise 3

Prove each of the following theorems analytically. In the first four exercises the placement of the geometric figures on the axes are given in the diagrams to aid you in your work.

1. Two medians of an isosceles triangle are equal in length (Figure 11.6).

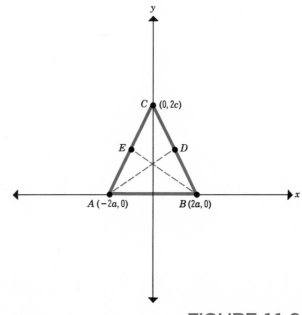

FIGURE 11.6

2. The diagonals of a parallelogram bisect each other (Figure 11.7).
3. The diagonals of a rectangle are equal in length (Figure 11.8).
4. The line segment joining the midpoints of two sides of a triangle are parallel to the third side and equal to one-half its length (Figure 11.9).
5. The diagonals of a rhombus are perpendicular to each other. (A rhombus is a parallelogram with four sides equal in length.)
6. The midpoint of the hypotenuses of a right triangle is equidistant from the three vertices.
7. The diagonals of an isosceles trapezoid are equal in length. (An isosceles trapezoid is a trapezoid whose nonparallel sides are equal in length.)

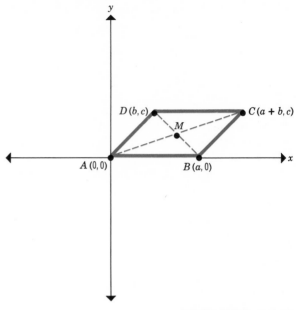

FIGURE 11.7

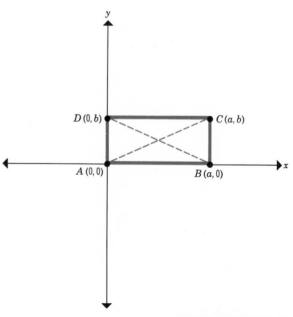

FIGURE 11.8

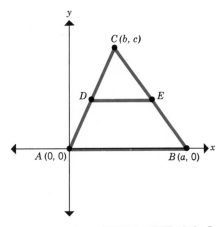

FIGURE 11.9

8. The sum of the squares of the lengths of the sides of a parallelogram equal the sum of the squares of the lengths of the diagonals.

9. The distance between the midpoints of the nonparallel sides of a trapezoid is one-half the sum of the lengths of the parallel sides.

10. If the median of a triangle is equal in length to one-half of the length of the side upon which it is drawn, the triangle is a right triangle.

6.
THE CIRCLE

A **circle** is the locus (path) of a point that moves in a plane so that it is always at a constant distance from a fixed point. The fixed point is called the **center** of the circle and the constant distance is called the **radius.**

To derive the equation of a circle we let point C with coordinates (h, k) be the center and denote the radius by r. Let P with coordinates (x, y) be any point on the circle. Since the distance between point C and point P is always r (see Figure 11.10), we have

$$\sqrt{(x - h)^2 + (y - k)^2} = r$$

or

(1) $$(x - h)^2 + (y - k)^2 = r^2$$

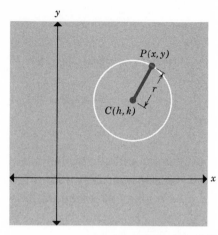

FIGURE 11.10

This is the equation of a circle since it is satisfied by the coordinates of every point on the circle and by the coordinates of no other point.

If the center of the circle is at the origin, then $h = 0$ and $k = 0$ and equation (1) becomes

$$x^2 + y^2 = r^2$$

Example 1: Find the equation of the circle whose center is the point with coordinates $(-3, 5)$ and whose radius is 4.

Solution: In this case $h = -3$, $k = 5$, and $r = 4$. Using equation (1) we obtain

$$[x - (-3)]^2 + (y - 5)^2 = 4^2$$

This equation may be simplified and written

$$x^2 + 6x + y^2 - 10y + 18 = 0$$

Example 2: Find the equation of a circle having a diameter whose endpoints have coordinates $(4, 3)$ and $(-4, -3)$.

Solution: The length of a diameter of this circle is found using the distance formula:

$$\sqrt{[4 - (-4)]^2 + [3 - (-3)]^2} = \sqrt{8^2 + 6^2}$$
$$= \sqrt{64 + 36}$$
$$= 10$$

Since the length of a diameter of a circle is twice the length of the radius of the circle, the radius of the given circle is 5. The midpoint of the diameter of a circle is the center of the circle. Hence the center of the desired circle has coordinates $(0, 0)$. The equation of the circle is

$$(x - 0)^2 + (y - 0)^2 = 5^2$$

which may be written

$$x^2 + y^2 = 25$$

Equation (1) above may be written

(2) $$x^2 + y^2 - 2hx - 2ky + h^2 + k^2 - r^2 = 0$$

If we let $g = -h, f = -k$, and $c = h^2 + k^2 - r^2$, this equation becomes

(3) $$x^2 + y^2 + 2gx + 2fy + c = 0$$

We now show that any equation of the form of equation (3) represents a circle, if it represents any real curve at all. We do this by a process called **completing the square** as shown below.

$$x^2 + 2gx + y^2 + 2fy = -c$$
$$x^2 + 2gx + g^2 + y^2 + 2fy + f^2 = -c + g^2 + f^2$$
$$(x + g)^2 + (y + f)^2 = g^2 + f^2 - c$$

If $g^2 + f^2 - c > 0$, equation (4) has the form of equation (1). It therefore represents a circle with center at the point whose coordinates are $(-g, -f)$, and whose radius is $g^2 + f^2 - c$.
If $g^2 + f^2 - c = 0$, the only real values that satisfy equation (4) are $-g$ and $-f$ respectively. The graph of equation (4) in this case is a point whose coordinates are $(-g, -f)$ and is called a **point circle.**
If $g^2 + f^2 - c < 0$, no real values of x and y satisfy equation (4). In this case the equation has no real graph.

Example 3: Find the center and radius of the circle

$$4x^2 + 4y^2 - 4x - 8y + 1 = 0$$

Solution: Dividing each member of the given equation by 4 we obtain

$$x^2 + y^2 - x - 2y + \frac{1}{4} = 0$$

By completing the square we obtain

$$\left(x^2 - x + \frac{1}{4}\right) + (y^2 - 2y + 1) = -\frac{1}{4} + \frac{1}{4} + 1$$

$$\left(x - \frac{1}{2}\right)^2 + (y - 1)^2 = 1$$

The center of the circle has coordinates $(\frac{1}{2}, 1)$ and its radius is 1.

Exercise 4

Find the equation of each circle (Exercises 1–10).
  1. Center at (2, 4); radius 2.
  2. Center at (−3, 1); radius 3.
  3. Center at (0, 0); radius $\frac{1}{4}$.
  4. Center at (−4, −5); radius 6.
  5. Center at (1, 0); and passing through the origin.
  6. Center at (−2, −2); and passing through (1, −5).
  7. Center at (−5, 12) and passing through (2, 8).
  8. Having a diameter with endpoints (0, 0) and (3, 4).
  9. Having a diameter with endpoints (−1, 5) and (−2, −3).
  10. Having a diameter with endpoints (−7, −4) and (9, 8).

Find the center and radius of each of the following circles.
  11. $x^2 + y^2 - 16 = 0$
  12. $x^2 + y^2 - 2x + 4y = 0$
  13. $x^2 + y^2 + 6y = 0$
  14. $x^2 + y^2 - 20x + 18y + 100 = 0$
  15. $4x^2 + 4y^2 - 4x + 12y + 1 = 0$
  16. $2x^2 + 2y^2 - 2x + 8y + 3 = 0$
  17. $36x^2 + 36y^2 - 36x + 24y + 1 = 0$
  18. $3x^2 + 3y^2 - 8x - 12y + 6 = 0$
  19. $36x^3 + 36y^2 - 36x - 36y + 17 = 0$
  20. $9x^2 + 9y^2 + 12x - 6y - 11 = 0$

Il fut dans l'Univers connu par ses Ouvrages,
Et dans son Païs même, il se fit respecter;
Il instruisit les Rois, il éclaira les Sages,
Plus sage qu'eux il sut douter.

Gottfried Wilhelm von Leibnitz (1646–1716), a genius in many fields including mathematics, advocated the binary system of notation and he attached mystic significance to this system, believing it was the "image of Creation." He imagined that unity (one) represented God and zero the void from which the Supreme Being drew all things, just as one and zero are the only symbols needed to express all numbers in the binary system. In 1672 Leibnitz invented a calculating machine that could add, subtract, multiply, and divide. He made only a few of these machines, one of which, still in working condition, was in the Leibnitz museum in Hanover prior to World War II.

COMPUTERS
AND
NUMERATION
SYSTEMS

12

1.
THE BASIC PRINCIPLES
OF A COMPUTER

A day seldom passes that we are not reminded in some way of the existence of digital computers. This reminder may be a cartoon, a newspaper article, a television program, or even a bill. The fact is, the extent of our everyday association with computers is still in its infant stage. Since all of us will become more involved with these so-called "thinking machines," a basic understanding of how they operate will help us intelligently assess stories and claims we hear about them, both good and bad.

We can understand the rudiments of a digital computer's operation without delving into its complicated switching logic and electronic design. The most important thing to remember is a digital computer does not think. It is a slave that is told *what* to do and *when* to do it. In the computers we refer to here, the "what" and "when" is supplied by a **program.** The extent of the "what," that is the manipulations the computer is capable of performing, is built into the machine's **circuitry.**

The machine's circuitry is a system of sophisticated electronic devices designed to send or receive electric currents. Each component is built to react to the energy—or lack of energy—received and, therefore, may transmit a pulse to other devices. The circuitry in a digital computer is designed, developed, and built by men and women. A simple example of such circuitry can be seen in an electrical dimmer switch. As your hand controls the switch (the transmitting device) the light bulb (the receiving device) responds.

A program is a set of instructions that a digital computer's circuitry is built to receive. The computer responds to these instructions by setting up unique electromagnetic patterns that cause the components to react as they were designed. Just as the machine was developed by men and women, the programs too are designed and written by men and women.

A computer retains information by using tiny ferromagnetic rings threaded on wires for its data storage—called its core storage or "memory." When a current is passed through these wires, a magnetic field is set up in the rings involved. Passing current in one direction sets up a

"+" field representing a "1"; changing the direction of the current changes the polarity of the ring to a "−" field representing a "0." This can be interpreted as an on-off scheme whose simple principle provides the digital computer with its greatest asset, speed. Also, the on-off principle lends itself well to scientific advancements and will continue in use when present day computer storage concepts are replaced by those predicted for the future.

Because a "+" polarity is represented by a "1" and a "−" polarity is represented by a "0", everything retained in a computer's memory can be expressed in the **binary system of numeration.** Therefore, numbers, symbols, and words must be reduced to patterns using only these two digits. It is easy to see that using binary numerals in everyday computations would be clumsy. Each numeral would contain many digits and therefore make it hard to see at a glance what number is represented. Because of this drawback, people in the computer field use other systems of numeration, based on powers of two, for shorthand representations of the binary system. These shorthand systems ease the problem of translating from what is stored in the computer's memory in binary form, to our everyday decimal system of numeration. When it becomes necessary to investigate all of the information retained in a computer's memory, it is possible to print out the entire contents. This print is in the shorthand system, based on a power of two, not in binary (the printout is called a **"core dump"**).

2.
THE DECIMAL
SYSTEM OF NUMERATION

The **Hindu-Arabic system of numeration** is a place-value system with symbols for the numbers zero, one, two, three, four, five, six, seven, eight, and nine. These symbols, called **digits,** are 0, 1, 2, 3, 4, 5, 6, 7, 8, and 9. The next natural number, ten, is the compounding point in the system and is called the **base** of the system. This system is commonly called the **decimal system** of numeration from the Latin word "decem" meaning ten.

COMPUTERS AND NUMERATION SYSTEMS

To write numerals for numbers other than zero, one, two, three, four, five, six, seven, eight, and nine, we use a combination of these digits formed according to a pattern determined by our system of **place-value.**

Let us look at the numeral 624.839 and analyze what is meant by a place-value system.

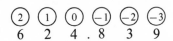

Each digit in the numeral is designated a position as indicated by the numerals in the circles above the digits. Starting at the decimal point, as we move to the left in the numeral, are the 0-position, the 1-position, and the 2-position; as we move to the right there are the -1-position, the -2-position, and the -3-position. To each position we assign a number that is the place-value of that position. The place-values of the decimal system are shown in Table 12.1.

TABLE 12.1
Place-value chart for base ten

Position	Decimal numeral	Power of ten		Name
⋮	⋮	⋮	⋮	⋮
-5	.00001	$\dfrac{1}{100,000}$	$= 10^{-5}$	One hundred-thousandth
-4	.0001	$\dfrac{1}{10,000}$	$= 10^{-4}$	One ten-thousandth
-3	.001	$\dfrac{1}{1,000}$	$= 10^{-3}$	One-thousandth
-2	.01	$\dfrac{1}{100}$	$= 10^{-2}$	One-hundredth
-1	.1	$\dfrac{1}{10}$	$= 10^{-1}$	One-tenth
0	1.0		$1 = 10^0$	One
1	10.0		$10 \times 1 = 10^1$	Ten
2	100.0		$10 \times 10 = 10^2$	One hundred
3	1,000.0		$10 \times 100 = 10^3$	One thousand
4	10,000.0		$10 \times 1,000 = 10^4$	Ten thousand
5	100,000.0		$10 \times 10,000 = 10^5$	Hundred thousand
⋮	⋮		⋮	⋮

2. THE DECIMAL SYSTEM OF NUMERATION

Now let us look again at the numeral 624.839. The place-value of

the "6" is 10^2
the "2" is 10^1
the "4" is 10^0
the "8" is 10^{-1}
the "3" is 10^{-2}
the "9" is 10^{-3}

Notice that the position of a digit in a numeral indicates two things:

(1) Its position and direction in relation to the units place (10^0)
(2) The exponent of the power of the base (ten) of its place-value.

For example the position of the "2" is ①, telling us that it is one place to the left of the units place and that its place-value is 10^1; the position of the "9" is ⊖₃ telling us that it is three places to the right of the units place and that its place-value is 10^{-3}.

In any decimal numeral, the number represented by a digit, called the **value** of the digit, is a product. This product is the number named by the digit and the place-value assigned to the position the digit occupies. In the numeral 624.839

The value of "6" is 6×10^2
The value of "2" is 2×10^1
The value of "4" is 4×10^0
The value of "8" is 8×10^{-1}
The value of "3" is 3×10^{-2}
The value of "9" is 9×10^{-3}

The number named by the numeral is the sum of these products. Thus

624.839
$$= (6 \cdot 10^2) + (2 \cdot 10^1) + (4 \cdot 10^0) + (8 \cdot 10^{-1}) + (3 \cdot 10^{-2}) + (9 \cdot 10^{-3})$$

When a numeral is written as a sum of products involving powers of the base we say that it is written in **expanded notation.**

Example 1: Write the decimal numeral 20024.7 in expanded notation.

Solution: $20024.7 = (2 \cdot 10^4) + (0 \cdot 10^3) + (0 \cdot 10^2)$
$+ (2 \cdot 10^1) + (4 \cdot 10^0) + (7 \cdot 10^{-1})$

Example 2: What is the place-value and the value of the "5" in the numeral 143.659?

Solution: The place-value of the "5" is 10^{-2}. The value of the "5" is

$$(5 \cdot 10^{-2}) = 5 \cdot \frac{1}{100} = 0.05.$$

Example 3: Write a standard decimal numeral for

$$(3 \cdot 10^7) + (4 \cdot 10^3) + (1 \cdot 10^{-1}) + (6 \cdot 10^{-5})$$

Solution: 30,004,000.10006

Exercise 1

1. Write the following decimal numerals in expanded notation.
   (a) 78.1   (c) 827.6   (e) 82,000.0
   (b) 163.22   (d) 7619.32   (f) 130,999.9
2. Write the following decimal numerals in expanded notation.
   (a) 93,000,001.0   (c) 132.1   (e) 876.007
   (b) 18.2   (d) 1,671.321   (f) 96.1041
3. What is the place value of the "4" in each of the following?
   (a) 426.13   (e) 4.001
   (b) 4,000,000   (f) 6.0394
   (c) 0.004   (g) 146.872
   (d) 36.142   (h) 5648.01
4. What is the value of the "6" in each of the following?
   (a) 68.32   (e) 64
   (b) 6,000,000   (f) 1.067
   (c) 346.91   (g) 32.1069
   (d) 0.0006   (h) 368,912.8
5. Write a standard decimal numeral for each of the following
   (a) $(1 \cdot 10^6) + (2 \cdot 10^3) + (3 \cdot 10^0) + (5 \cdot 10^{-2})$
   (b) $7 \cdot 10^7$
   (c) $(9 \cdot 10^6) + (1 \cdot 10^5) + (4 \cdot 10^1) + (3 \cdot 10^{-1}) + (6 \cdot 10^{-2})$
   (d) $9 \cdot 10^4$
   (e) $(4 \cdot 10^{-3}) + (1 \cdot 10^{-4}) + (1 \cdot 10^{-6})$

3.
THE BINARY NUMERATION SYSTEM

The representation of a **bit** (a contraction of binary digit) of data by either a 1 or a 0 is the basis of the **binary numeration system.** As we shall see later, many binary digits are required to represent equivalent decimal numerals and thus the core storage of a computer must contain many magnetic rings. (One version of the Control Data Corporation's 6400 computer has 65,535 words, each word contains 60 bits; thus, the core storage contains $60 \times 65,535 = 3,932,100$ magnetic rings.)

Before we investigate how information is handled by a computer, we must be able to translate our language into something equivalent in the binary system.

The **binary system** of numeration (base two) is a place-value system with two digits, 0 and 1, which represent the numbers zero and one, respectively. All numbers, symbols, and alphabetic characters used internally by digital computers must be combinations of these two symbols.

In base ten, the place-values are powers of ten. In base two, the place-values are powers of two:

$$\vdots$$
$$\text{two}^{-3} = \text{one-eighth}$$
$$\text{two}^{-2} = \text{one-fourth}$$
$$\text{two}^{-1} = \text{one-half}$$
$$\text{two}^{0} = \text{one}$$
$$\text{two}^{1} = \text{two}$$
$$\text{two}^{2} = \text{four}$$
$$\text{two}^{3} = \text{eight}$$
$$\vdots$$

The binary numeral for two is $10 = (1 \cdot \text{two}^{1}) + (0 \cdot \text{two}^{0})$. The first few binary numerals and their decimal equivalents are:

	Binary Notation	Decimal Notation
0		0
1		1
$10 = 1$ two $+ 0$ ones		2
$11 = 1$ two $+ 1$ one		3
$100 = 1$ four $+ 0$ twos $+ 0$ ones		4
$\vdots$		$\vdots$
$1000 = 1$ eight $+ 0$ fours $+ 0$ twos $+ 0$ ones		8
$1001 = 1$ eight $+ 0$ fours $+ 0$ twos $+ 1$ one		9
$\vdots$		$\vdots$

Table 12.2 is the place-value chart for the binary system.

TABLE 12.2

Place-value chart for the binary system of notation

Position	Binary Power of two	Binary numeral	Decimal equivalents
$\vdots$	$\vdots$	$\vdots$	$\vdots \quad \vdots \quad \vdots \quad \vdots$
-5	two^{-5}	.00001	$2^{-5} = \dfrac{1}{2^5} = \dfrac{1}{32} = 0.03125$
-4	two^{-4}	.0001	$2^{-4} = \dfrac{1}{2^4} = \dfrac{1}{16} = 0.0625$
-3	two^{-3}	.001	$2^{-3} = \dfrac{1}{2^3} = \dfrac{1}{8} = 0.125$
-2	two^{-2}	.01	$2^{-2} = \dfrac{1}{2^2} = \dfrac{1}{4} = 0.25$
-1	two^{-1}	.1	$2^{-1} = \dfrac{1}{2^1} = \dfrac{1}{2} = 0.5$
0	two^0	1.	$2^0 = \dfrac{1}{2^0} = \dfrac{1}{1} = 1.$
1	two^1	10.	$2^1 = 2^1 = 2 = 2.$
2	two^2	100.	$2^2 = 2^2 = 4 = 4.$
3	two^3	1000.	$2^3 = 2^3 = 8 = 8.$
4	two^4	10000.	$2^4 = 2^4 = 16 = 16.$
5	two^5	100000.	$2^5 = 2^5 = 32 = 32.$
$\vdots$	$\vdots$	$\vdots$	$\vdots \quad \vdots \quad \vdots \quad \vdots$

Now let us examine some binary numerals and find their decimal equivalents.

Example 1: Find the decimal equivalent of the binary numeral 1011.1

Solution:

$$
\begin{aligned}
1011.1 &= (1 \cdot \text{two}^3) + (0 \cdot \text{two}^2) + (1 \cdot \text{two}^1) \\
&\quad + (1 \cdot \text{two}^0) + (1 \cdot \text{two}^{-1}) \\
&= (1 \cdot 2^3) + (0 \cdot 2^2) + (1 \cdot 2^1) + (1 \cdot 2^0) \\
&\quad + (1 \cdot 2^{-1}) \\
&= 8 + 0 + 2 + 1 + 0.5 \\
&= 11.5
\end{aligned}
$$

The first line of the expansion is the binary numeral written in expanded notation. Here we spell out "two" since the binary system has no numeral "2." The second, third, and fourth lines are decimal equivalents. Notice an expansion of a binary numeral is the same as an expansion of a decimal numeral except the place values are powers of two instead of powers of ten.

We could have written the first line of the expansion as:

$$
\begin{aligned}
1011.1 &= (1 \cdot 1000_{(\text{two})}) + (0 \cdot 100_{(\text{two})}) + (1 \cdot 10_{(\text{two})}) \\
&\quad + (1 \cdot 1_{(\text{two})}) + (1 \cdot .1_{(\text{two})})
\end{aligned}
$$

The word "two" in the subscript indicates the numerals are binary numerals.

Example 2: Find the decimal equivalent of the binary numeral 11.001.

Solution:

$$
\begin{aligned}
11.001 &= (1 \cdot \text{two}^1) + (1 \cdot \text{two}^0) + (0 \cdot \text{two}^{-1}) \\
&\quad + (0 \cdot \text{two}^{-2}) + (1 \cdot \text{two}^{-3}) \\
&= (1 \cdot 2^1) + (1 \cdot 2^0) + (0 \cdot 2^{-1}) + (0 \cdot 2^{-2}) \\
&\quad + (1 \cdot 2^{-3}) \\
&= 2 + 1 + 0 + 0 + 0.125 \\
&= 3.125
\end{aligned}
$$

Example 3: Find the decimal equivalent of the binary numeral 11001.01.

Solution:

$$
\begin{aligned}
11001.01 &= (1 \cdot \text{two}^4) + (1 \cdot \text{two}^3) + (0 \cdot \text{two}^2) + (0 \cdot \text{two}^1) \\
&\quad + (1 \cdot \text{two}^0) + (0 \cdot \text{two}^{-1}) + (1 \cdot \text{two}^{-2}) \\
&= (1 \cdot 2^4) + (1 \cdot 2^3) + (0 \cdot 2^2) + (0 \cdot 2^1) + (1 \cdot 2^0) \\
&\quad + (0 \cdot 2^{-1}) + (1 \cdot 2^{-2}) \\
&= 16 + 8 + 0 + 0 + 1 + 0 + 0.25 \\
&= 25.25
\end{aligned}
$$

COMPUTERS AND NUMERATION SYSTEMS

If we wish to find a binary numeral equivalent to a given decimal numeral, we use the concept of place-value.

Example 4: Find the binary numeral equivalent to the decimal numeral 85.

Solution: We ask ourselves: What is the largest power of 2 less than 85? We see that the largest power of 2 less than 85 is $2^6 = 64$. Since we have one set of $2^6 = 64$, we put a "1" in position 6.

6	5	4	3	2	1	0
1						

To find the digit in position 5 we determine whether we can take a set of $2^5 = 32$ from $85 - 64 = 21$. Since we cannot take a set of 32 from 21, we place a "0" in position 5.

6	5	4	3	2	1	0
1	0					

To find the digit in position 4 we determine whether we can take a set of $2^4 = 16$ from 21. Since we can take a set of $2^4 = 16$ from 21, we put a "1" in position 4.

6	5	4	3	2	1	0
1	0	1				

Now we ask whether we can take a set of $2^3 = 8$ from $21 - 16 = 5$? Since we cannot take a set of $2^3 = 8$ from 5, we put a "0" in position 3.

6	5	4	3	2	1	0
1	0	1	0			

Can we take a set of $2^2 = 4$ from 5? Since we can take a set of $2^2 = 4$ from 5, we put a "1" in position 2.

6	5	4	3	2	1	0
1	0	1	0	1		

We see that we cannot take a set of $2^1 = 2$ from $5 - 4 = 1$; so we put a "0" in position 1.

6	5	4	3	2	1	0
1	0	1	0	1	0	

We can take a set of $2^0 = 1$ from 1, hence we put a "1" in the 0 position.

6	5	4	3	2	1	0
1	0	1	0	1	0	1

The binary equivalent of 85 is

$$1,010,101_{(two)}$$

The above discussion is shown below in a shortened form.

$$85 = 64 + 0 + 16 + 0 + 4 + 0 + 1$$
$$= (1 \cdot 64) + (0 \cdot 32) + (1 \cdot 16) + (0 \cdot 8) + (1 \cdot 4) + (0 \cdot 2)$$
$$+ (1 \cdot 1)$$
$$= (1 \cdot two^6) + (0 \cdot two^5) + (1 \cdot two^4) + (0 \cdot two^3) + (1 \cdot two^2)$$
$$+ (0 \cdot two^1) + (1 \cdot two^0)$$
$$= 1,010,101_{(two)}$$

The binary digits need not be separated by commas, but we use them for clarity.

Example 5: Convert the decimal numeral 369 to an equivalent binary numeral.

377

Solution: $369 = 256 + 0 + 64 + 32 + 16 + 0 + 0 + 0 + 1$
$= (1 \cdot 256) + (0 \cdot 128) + (1 \cdot 64) + (1 \cdot 32) + (1 \cdot 16)$
$+ (0 \cdot 8) + (0 \cdot 4) + (0 \cdot 2) + (1 \cdot 1)$
$= (1 \cdot \text{two}^8) + (0 \cdot \text{two}^7) + (1 \cdot \text{two}^6) + (1 \cdot \text{two}^5)$
$+ (1 \cdot \text{two}^4) + (0 \cdot \text{two}^3) + (0 \cdot \text{two}^2)$
$+ (0 \cdot \text{two}^1) + (1 \cdot \text{two}^0)$
$= 101,110,001_{(\text{two})}$

Another method for converting whole numbers from decimal to binary notation involves repeated divisions. Each subsequent division is the subtraction of the next higher power of the base and the remainder is the digit for that particular place.

Example 6: Use repeated division to find the binary equivalent to the decimal numeral 33.

Solution:

$$
\begin{array}{r|ll}
2)\,33 & & \\
2)\,16 & R = 1 & 1 \times 2^0 \\
2)\,8 & R = 0 & 0 \times 2^1 \\
2)\,4 & R = 0 & 0 \times 2^2 \\
2)\,2 & R = 0 & 0 \times 2^3 \\
2)\,1 & R = 0 & 0 \times 2^4 \\
0 & R = 1 & 1 \times 2^5 \\
\end{array}
$$

To find the answer we read up; thus

$$33 \text{ (base ten)} = 100,001_{(\text{two})}$$

Example 7: Find the binary numeral equivalent to the decimal numeral 77.5.

Solution: $77.5 = 64 + 0 + 0 + 8 + 4 + 0 + 1 + .5$
$= (1 \cdot 64) + (0 \cdot 32) + (0 \cdot 16) + (1 \cdot 8) + (1 \cdot 4)$
$+ (0 \cdot 2) + (1 \cdot 1) + (1 \cdot \tfrac{1}{2})$
$= (1 \cdot 64) + (0 \cdot 32) + (0 \cdot 16) + (1 \cdot 8) + (1 \cdot 4)$
$+ (0 \cdot 2) + (1 \cdot 1) + (1 \cdot 2^{-1})$
$= (1 \cdot \text{two}^6) + (0 \cdot \text{two}^5) + (0 \cdot \text{two}^4) + (1 \cdot \text{two}^3)$
$+ (1 \cdot \text{two}^2) + (0 \cdot \text{two}^1) + (1 \cdot \text{two}^0)$
$+ (1 \cdot \text{two}^{-1})$
$= 1,001,101.1_{(\text{two})}$

3. THE BINARY NUMERATION SYSTEM

In the decimal numeral 77.5, the fractional part is separated from the whole number part by a **decimal point,** ".". In the binary numeral 1,001,101.1, the fractional part and the whole number part are separated by a point called a **binary point.** In any place-value system of numeration to separate the fractional part from the whole number part we use a point that takes its name from the base of the numeration system used.

Example 8: Find the binary numeral equivalent to 0.90625.

   Solution: Again we begin by separating the numeral into powers of two.

$$0.90625 = 0.5 + 0.25 + 0.125 + 0.03125$$
$$= (1 \cdot 0.5) + (1 \cdot 0.25) + (1 \cdot 0.125)$$
$$+ (0 \cdot 0.0625) + (1 \cdot .03125)$$
$$= \left(1 \cdot \frac{1}{2}\right) + \left(1 \cdot \frac{1}{4}\right) + \left(1 \cdot \frac{1}{8}\right)$$
$$+ \left(0 \cdot \frac{1}{16}\right) + \left(1 \cdot \frac{1}{32}\right)$$
$$= \left(1 \cdot \frac{1}{2^1}\right) + \left(1 \cdot \frac{1}{2^2}\right) + \left(1 \cdot \frac{1}{2^3}\right)$$
$$+ \left(0 \cdot \frac{1}{2^4}\right) + \left(1 \cdot \frac{1}{2^5}\right)$$
$$= (1 \cdot 2^{-1}) + (1 \cdot 2^{-2}) + (1 \cdot 2^{-3}) + (0 \cdot 2^{-4})$$
$$+ (1 \cdot 2^{-5})$$
$$= 0.11101_{\text{(two)}}$$

We recall that some rational numbers may be denoted by terminating decimal fractions while others may not. For example

$$\frac{1}{2} = 0.5$$

$$\frac{1}{8} = 0.125$$

$$\frac{1}{25} = 0.04$$

The only rational numbers that can be represented by terminating decimals are those which when represented by common fractions in simplest form (that is fractions whose numerator and denominator are relatively prime) have denominators whose complete factorization contains only powers of 2 or powers of 5. Thus $\frac{3}{250}$ may be represented as a terminating decimal since $(3, 250) = 1$ and $250 = 2 \cdot 5^3$.

Rational numbers named by (decimal) common fractions in simplest form whose denominators have prime factors other than 2 and 5 cannot be represented by terminating decimals. Thus $\frac{1}{3}$ may not be represented by a terminating decimal. These rational numbers may be approximated by decimal fractions.

Rational numbers represented by (decimal) common fractions with denominators that are powers of two may be expressed as terminating binary numerals; those represented by (decimal) common fractions with denominators with prime factors other than powers of two cannot be represented by terminating binary numerals, but they may be approximated by binary numerals.

For example

$$0.125 = \frac{125}{1000} = \frac{1}{8} = \frac{1}{2^3} = 1 \cdot 2^{-3} = 0.001_{\text{(two)}}$$

$$0.5 = \frac{5}{10} = \frac{1}{2} = 1 \cdot 2^{-1} = 0.1_{\text{(two)}}$$

$$1.25 = \frac{125}{100} = \frac{5}{4} = 1 + \frac{1}{4} = (1 \cdot 2^0) + (1 \cdot 2^{-2}) = 1.01_{\text{(two)}}$$

$$14.0625 = \frac{140625}{10000} = \frac{225}{16} = 14 + \frac{1}{16} = 14 + \frac{1}{2^4}$$

$$= 8 + 4 + 2 + \frac{1}{2^4}$$

$$= (1 \cdot 2^3) + (1 \cdot 2^2) + (1 \cdot 2^1) + (1 \cdot 2^{-4})$$

$$= 1110.0001_{\text{(two)}}$$

There is a method for converting decimal fractions naming numbers between 0 and 1 to equivalent binary numerals. This method involves repeated multiplications by 2. The integral parts of these products are the successive digits, reading from left to right, to the right of the binary point in the binary numeral. If the decimal fraction has an exact binary equivalent, the fractional part in one of the successive products will be zero. If the decimal fraction has no exact binary equivalent, no matter how many successive multiplications are completed, the fractional part of the product will never be zero. In this case, the approximate binary numeral may be found by carrying out as many multiplications as required.

We present this method without proof by a series of examples.

Example 9: Use repeated multiplication to convert 0.25 to a binary numeral.

Solution: $0.25 \times 2 = 0.50$ The integral part of the product is 0. This means that the digit with place value 2^{-1} is 0. We now multiply the fractional part of the product, .50, by 2.

$0.50 \times 2 = 1.00$ The integral part of the product is 1. The digit with place value 2^{-2} is 1.

The decimal numeral converts to a terminating binary numeral since the fractional part of the product on the second multiplication is 0.

$$0.25 = 0.01_{(two)}$$

Example 10: Use the repeated multiplication method to convert the decimal fraction 0.625 to an equivalent binary numeral.

Solution:
$$0.625 \cdot 2 = 1.250$$
$$.250 \cdot 2 = 0.500$$
$$.500 \cdot 2 = 1.000$$

.625 converts to an exact binary numeral since on the third multiplication the fractional part of the product is zero. The digits in the binary numeral to the right of the binary point are the digits in the integral parts of the products reading from top to bottom.

$$.625 = .101_{(two)}$$

We can check our result by converting $0.101_{(two)}$ to decimal notation. Using Table 12.2 we see that

$$0.101 = (1 \cdot two^{-1}) + (0 \cdot two^{-2}) + (1 \cdot two^{-3})$$
$$= (1 \cdot .5) + (0 \cdot .250) + (1 \cdot .125)$$
$$= .500 + .125$$
$$= .625$$

Example 11: Use repeated multiplication to convert .7144 to an equivalent binary numeral correct to six binary places.

Solution:
$$.7144 \cdot 2 = 1.4288$$
$$.4288 \cdot 2 = 0.8576$$
$$.8576 \cdot 2 = 1.7152$$
$$.7152 \cdot 2 = 1.4304$$
$$.4304 \cdot 2 = 0.8608$$
$$.8608 \cdot 2 = 1.7216$$

This decimal numeral does not convert exactly to binary.

$$.7144 \doteq .101101_{\text{(two)}}$$

The symbol $\doteq$ means "is approximately equal to." If we convert $.101101_{\text{(two)}}$ back to decimal notation, we can see the accuracy lost.

$$
\begin{aligned}
.101101 &= (1 \cdot \text{two}^{-1}) + (0 \cdot \text{two}^{-2}) + (1 \cdot \text{two}^{-3}) \\
&\quad + (1 \cdot \text{two}^{-4}) + (0 \cdot \text{two}^{-5}) + (1 \cdot \text{two}^{-6}) \\
&= (1 \cdot 0.5) + (0 \cdot 0.25) + (1 \cdot 0.125) \\
&\quad + (1 \cdot 0.0625) + (0 \cdot 0.03125) + (1 \cdot 0.015625) \\
&= 0.5 + 0.125 + 0.0625 + 0.015625 \\
&= 0.703125
\end{aligned}
$$

Exercise 2

1. Convert the following binary numerals to equivalent decimal numerals.
  (a) $111_{\text{(two)}}$ (d) $101,101_{\text{(two)}}$
  (b) $1,010_{\text{(two)}}$ (e) $111,111,111_{\text{(two)}}$
  (c) $11,001_{\text{(two)}}$ (f) $101,000,101_{\text{(two)}}$

2. Convert the following decimal numerals to equivalent binary numerals.
  (a) 7   (d) 129
  (b) 12   (e) 306
  (c) 38   (f) 5000

3. Convert the following binary numerals to equivalent decimal numerals.
  (a) $100.1_{\text{(two)}}$ (d) $1000.011_{\text{(two)}}$
  (b) $111.001_{\text{(two)}}$ (e) $1111.111_{\text{(two)}}$
  (c) $1010.10101_{\text{(two)}}$ (f) $10,000,000.0_{\text{(two)}}$

4. Convert the following binary numerals to equivalent decimal numerals.
  (a) 1.0111 (d) $100000.0001_{\text{(two)}}$
  (b) $101,101.101_{\text{(two)}}$ (e) $11100.111_{\text{(two)}}$
  (c) $1110011.0_{\text{(two)}}$ (f) $11011011.0_{\text{(two)}}$

5. Convert the following decimal numerals to equivalent binary numerals.

(a) 625.75

(b) 69.0625

(c) 99.3125

(d) 106.28125

(e) 986.375

(f) 33.5625

6. Convert the following decimal numerals to equivalent binary numerals using repeated division (on the whole number part) and repeated multiplication (on the fractional part). If the fractional parts do not convert exactly, carry the conversion to five binary places, that is five places to the right of the binary point.

(a) 81.1875

(b) 126.1305

(c) 512.187

(d) 34.03105

(e) 65.111

(f) 1024.3751

7. When does the fractional portion of a decimal numeral convert exactly to a binary numeral?

8. In looking at the following powers of two one might conjecture that a decimal fraction (naming a number less than 1) ending in 0 or 5 converts exactly to a binary numeral.

$$2^{-1} = 0.500000$$
$$2^{-2} = 0.250000$$
$$2^{-3} = 0.125000$$
$$2^{-4} = 0.062500$$
$$2^{-5} = 0.031250$$
$$2^{-6} = 0.015625$$
$$\cdots$$

Is this conjecture true? Convert 0.35 to binary notation before you answer yes or no.

9. When does a binary numeral name (a) an odd number; (b) an even number?

10. Can you think of any problem arising from the fact that some decimal fractions cannot have an exact binary equivalent in the storage of a computer? As an example, suppose you were to convert $1.23 into binary notation and then back into decimal notation. What would you have?

4.
THE OCTAL
NUMERATION SYSTEM

Numerals written in the **octal** (base eight) system are represented by a combination of eight symbols 0, 1, 2, 3, 4, 5, 6, 7. The octal system is also a place-value system. A place-value chart for base eight is given in Table 12.3.

TABLE 12.3
Place-value chart for base eight

Base eight			Decimal equivalent			
Position	Power of Eight	Octal Numeral				
$\vdots$	$\vdots$	$\vdots$	$\vdots$	$\vdots$	$\vdots$	$\vdots$
-4	eight^{-4}	.0001	$8^{-4} = \dfrac{1}{8^4}$	$= \dfrac{1}{4096}$	$\doteq$	.000244
-3	eight^{-3}	.001	$8^{-3} = \dfrac{1}{8^3}$	$= \dfrac{1}{512}$	$\doteq$	.001953
-2	eight^{-2}	.01	$8^{-2} = \dfrac{1}{8^2}$	$= \dfrac{1}{64}$	$=$	.015625
-1	eight^{-1}	.1	$8^{-1} = \dfrac{1}{8^1}$	$= \dfrac{1}{8}$	$=$	.125
0	eight^0	1.	$8^0 = \dfrac{1}{8^0}$	$= \dfrac{1}{1}$	$=$	1.
1	eight^1	10.	$8^1 = \dfrac{8^1}{1}$	$= \dfrac{8}{1}$	$=$	8.
2	eight^2	100.	$8^2 = \dfrac{8^2}{1}$	$= \dfrac{64}{1}$	$=$	64.
3	eight^3	1000.	$8^3 = \dfrac{8^3}{1}$	$= \dfrac{512}{1}$	$=$	512.
4	eight^4	10000.	$8^4 = \dfrac{8^4}{1}$	$= \dfrac{4096}{1}$	$=$	4096.
$\vdots$	$\vdots$	$\vdots$	$\vdots$	$\vdots$	$\vdots$	$\vdots$

When we write **octal numerals** we think of objects grouped in sets of eight. Just as in base ten the representation of ten or more objects results in a regrouping or "carry" into the next higher position, so does the representation of eight or more objects result in a "carry" in base eight. In the following examples, the first lines represent the expanded notation in base eight; lines two, three, and four are the decimal equivalents. We use the notation $73_{(eight)}$ to signify that the numeral 73 is written in base-eight notation.

Example 1: Convert $73_{(eight)}$ to an equivalent decimal numeral.

Solution:
$$73_{(eight)} = (7 \cdot eight^1) + (3 \cdot eight^0)$$
$$= (7 \cdot 8^1) + (3 \cdot 8^0)$$
$$= 56 + 3$$
$$= 59$$

Example 2: Convert $165.35_{(eight)}$ to an equivalent decimal numeral.

Solution:
$$165.35_{(eight)} = (1 \cdot eight^2) + (6 \cdot eight^1)$$
$$+ (5 \cdot eight^0) + (3 \cdot eight^{-1}) + (5 \cdot eight^{-2})$$
$$= (1 \cdot 8^2) + (6 \cdot 8^1) + (5 \cdot 8^0)$$
$$+ \left(3 \cdot \frac{1}{8^1}\right) + \left(5 \cdot \frac{1}{8^2}\right)$$
$$= 64 + 48 + 5 + .375 + .078125$$
$$= 117.453125$$

From the examples above we see that the value of a number written in octal notation is the sum of the products formed by each digit multiplied by its place-value.

To change a decimal numeral naming a whole number to an equivalent octal numeral, we must think of the number of objects grouped in sets of $eight^0$, $eight^1$, $eight^2$, and so forth instead of sets of ten^0, ten^1, ten^2, and so forth.

Example 3: Find the base eight numeral equivalent to the decimal numeral 123.

Solution: We ask ourselves what is the largest power of 8 less than 123. Since $8^3 = 512$ we cannot take a set of 8^3 from 123; $8^2 = 64$ is less than 123 so to find the digit in position 2, we determine the greatest multiple of 64 less than 123.

$$\begin{array}{r} 1 \\ 64\overline{)123} \\ \underline{64} \\ 59 \end{array}$$

We see that we can subtract one $8^2 = 64$ from 123, hence we write "1" in position 2.

2	1	0
1		

To find the digit for position 1 we determine the largest multiple of $8^1 = 8$ less than $123 - 64 = 59$.

$$\begin{array}{r} 7 \\ 8\overline{)59} \\ 56 \\ \hline 3 \end{array}$$

We put a "7" in position 1.

2	1	0
1	7	

Since there is a remainder of 3 when we subtract $8 \times 7 = 56$ from 59 we put a "3" in position 0.

2	1	0
1	7	3

Hence

$$123_{(ten)} = 173_{(eight)}$$

The above discussion is shown below in shortened form.

$$\begin{aligned} 123 &= 64 + 56 + 3 \\ &= (1 \cdot 64) + (7 \cdot 8) + (3 \cdot 1) \\ &= (1 \cdot \text{eight}^2) + (7 \cdot \text{eight}^1) + (3 \cdot \text{eight}^0) \\ &= 173_{(eight)} \end{aligned}$$

Example 4: Convert 1096 to a base eight numeral.

Solution:
$$\begin{aligned} 1096 &= 1024 + 64 + 8 \\ &= (2 \cdot 512) + (1 \cdot 64) + (1 \cdot 8) + (0 \cdot 1) \\ &= (2 \cdot \text{eight}^3) + (1 \cdot \text{eight}^2) + (1 \cdot \text{eight}^1) \\ &\quad + (0 \cdot \text{eight}^0) \\ &= 2110_{(eight)} \end{aligned}$$

Example 5: Change 975 to its base eight equivalent.

Solution:
$$975 = 512 + 448 + 8 + 7$$
$$= (1 \cdot 512) + (7 \cdot 64) + (1 \cdot 8) + (7 \cdot 1)$$
$$= (1 \cdot \text{eight}^3) + (7 \cdot \text{eight}^2) + (1 \cdot \text{eight}^1)$$
$$+ (7 \cdot \text{eight}^0)$$
$$= 1717_{(\text{eight})}$$

A base ten numeral may be converted to an equivalent base eight numeral by repeated division by eight. Each successive division is the extraction of the next higher power of the base.

Example 6: Convert 123 to its equivalent base eight numeral using the method of repeated division.

Solution:

$$
\begin{array}{r|l|l}
8\overline{)123} & \text{Remainder} & \\
8\overline{)\,15} & R = 3 \uparrow & 3 \times \text{eight}^0 \\
8\overline{)\,\,1} & R = 7 & 7 \times \text{eight}^1 \\
0 & R = 1 & 1 \times \text{eight}^2 \\
\end{array}
$$

The answer is read up, thus

$$123 = 173_{(\text{eight})}$$

Decimal fractions naming numbers between 0 and 1 may be converted to equivalent octal fractions by repeated multiplications by eight. The integer part of the products are the digits to the right of the octal point in the octal numeral. If the fractional part of any product is zero, the process is complete and the decimal fraction has an exact representation in the octal system. If there is no exact conversion, we continue the multiplication process until a sufficient number of octal digits have been generated.

Example 7: Convert 0.140625 to an equivalent octal fraction.

Solution:

$$
\begin{array}{c|r}
 & .140625 \\
 & \times \quad 8 \\
1 & .125000 \\
 & \times \quad 8 \\
1 \downarrow & .000000 \\
\end{array}
$$

The answer is read down, thus

$$.140625 = .11_{(\text{eight})}$$

Example 8:   Convert .1585 to an octal fraction.

Solution:

$$
\begin{array}{r|l}
 & .1585 \\
 & \times\ 8 \\
\hline
1 & .2680 \\
 & \times\ 8 \\
\hline
2 & .1440 \\
 & \times\ 8 \\
\hline
1 & .1520 \\
 & \times\ 8 \\
\hline
1 \downarrow & .2160 \\
\end{array}
$$

$$.1585 \doteq .1211_{(eight)}$$

By converting $.1211_{(eight)}$ back to a decimal fraction we can check the amount of accuracy lost. We shall do this using expanded notation with the aid of Table 12.3.

$$
\begin{aligned}
.1211_{(eight)} &\doteq (1 \cdot .125) + (2 \cdot .015625) + (1 \cdot .001953) + (1 \cdot 000244) \\
&\doteq .125 + .031250 + .001953 + .000244 \\
&\doteq .158447
\end{aligned}
$$

Octal fractions may be converted to equivalent decimal fractions by division. To use this method we divide the lowest order octal digit by eight and add the quotient to the next highest octal digit. We repeat this process until all the digits of the octal fraction are used. The last quotient is the decimal answer. This method of conversion is demonstrated in the examples below.

Example 9:   Use repeated division to convert $.25_{(eight)}$ to a decimal fraction.

Solution: 5 is the lowest order digit (that is the digit on the extreme right of the numeral)

$$
\begin{array}{r}
.625 \\
8\overline{)5.000} \\
\underline{4\ 8} \\
20 \\
\underline{16} \\
40 \\
\underline{40} \\
\end{array}
$$

Add the quotient to the next higher digit, 2.

$$2 + .625 = 2.625$$

$$
\begin{array}{r}
.328125 \\
8\overline{)2.625000} \\
2\,4 \\
\hline
22 \\
16 \\
\hline
65 \\
64 \\
\hline
10 \\
8 \\
\hline
20 \\
16 \\
\hline
40 \\
40 \\
\hline
\end{array}
$$

$$.25_{(eight)} = .328125$$

Check: $\quad 0.25_{(eight)} = (2 \times 8^{-1}) + (5 \times 8^{-2}) = \frac{2}{8} + \frac{5}{64} = \frac{21}{64}$
$$= 0.328125$$

Exercise 3

1. Convert the following base eight numerals to equivalent base ten numerals using expanded notation.

(a) $2047.51_{(eight)}$ (c) $64.02_{(eight)}$ (e) $77.07_{(eight)}$

(b) $527.3_{(eight)}$ (d) $100.77_{(eight)}$ (f) $72,000.4_{(eight)}$

2. Convert the following base eight numerals to their equivalent base ten numerals.

(a) $6.15_{(eight)}$ (c) $4056.005_{(eight)}$

(b) $22.012_{(eight)}$ (d) $33.333_{(eight)}$

3. Convert the following base ten numerals to base eight numerals. Use the method of repeated division for the whole number part and repeated multiplication for the fractional part. If the fractional part has no exact octal representation, carry out the process until there is agreement through and including the third fractional digit.

(a) 4056.140625   (c) 932.1478

(b) 32,768.100   (d) 64.93277

4. Convert the following decimal numerals to octal numerals using any method discussed. Use expanded notation to check for at least two digits of agreement in the fractional parts.

(a) 634.125   (c) 4096.875

(b) 997.375   (d) 0.126953125

5. How many digits are used in the octal system?

6. What is the decimal value of "5" in each of the following octal numerals?

(a) $562_{(eight)}$ (c) $5637_{(eight)}$

(b) $53_{(eight)}$ (d) $51,677_{(eight)}$

7. What is the place value of "4" in each of the following?

(a) $416_{(eight)}$ (c) $35.124_{(eight)}$

(b) $724_{(eight)}$ (d) $100.0412_{(eight)}$

5.
BINARY
OCTAL RELATION

Binary numerals have many digits and are clumsy. Since they contain so many digits it is hard to see at a glance what they represent.

Binary numerals convert very easily to octal numerals which contain fewer digits. Study the following examples.

Example 1: Convert the binary numeral 1,011,110 to an equivalent octal numeral.

Solution:
$$1,011,110_{(two)} = [(1 \cdot two^6)] + [(0 \cdot two^5) + (1 \cdot two^4) \\ + (1 \cdot two^3)] + [(1 \cdot two^2) \\ + (1 \cdot two^1) + (0 \cdot two^0)]$$

Each portion of the binary numeral separated by commas has been enclosed in brackets in the expanded notation.

Notice that the first set of brackets encloses the number represented by "1"; the second, the number represented by "011"; and the third, the number represented by "110."

Observe the three enclosed expressions.

$$[(1 \cdot two^6)] = \text{sixty-four}$$
$$= 1 \cdot eight^2$$
$$[(0 \cdot two^5) + (1 \cdot two^4) + (1 \cdot two^3)] = \text{zero} + \text{sixteen} + \text{eight}$$
$$= 0 + 16 + 8$$
$$= 24$$
$$= 3 \cdot eight^1$$
$$[(1 \cdot two^2) + (1 \cdot two^1) + (0 \cdot two^0)] = \text{four} + \text{two} + 0$$
$$= 4 + 2 + 0$$
$$= 6$$
$$= 6 \cdot eight^0$$

Combining these, we have

$$1,011,110_{(two)} = 136_{(eight)}$$

Example 2: Convert $11,111,001_{(two)}$ to an equivalent octal numeral.
Solution:

$$11,111,001_{(two)} = [(1 \cdot two^7) + (1 \cdot two^6)]$$
$$+ [(1 \cdot two^5) + (1 \cdot two^4)$$
$$+ (1 \cdot two^3)] + [(0 \cdot two^2)$$
$$+ (0 \cdot two^1) + (1 \cdot two^0)]$$
$$[(1 \cdot two^7) + (1 \cdot two^6)] = \text{one hundred twenty-eight}$$
$$+ \text{sixty-four}$$
$$= 128 + 64$$
$$= 192$$
$$= 3 \cdot 64$$
$$= 3 \cdot eight^2$$
$$[(1 \cdot two^5) + (1 \cdot two^4) + (1 \cdot two^3)] = \text{thirty-two} + \text{sixteen} + \text{eight}$$
$$= 32 + 16 + 8$$
$$= 56$$
$$= 7 \cdot eight^1$$
$$[(0 \cdot two^2) + (0 \cdot two^1) + (1 \cdot two^0)] = 0 + 0 + 1$$
$$= 1 \cdot eight^0$$
$$11,111,001_{(two)} = 371_{(eight)}$$

Examples 1 and 2 above suggest a convenient method for converting binary numerals to equivalent octal numerals. Notice the two binary numerals in the above examples and consider each group of digits separated by commas separately:

$$1,011,110_{(two)} = 136_{(eight)}$$

names 1   names 3   names 6

$$11,111,001_{(two)} = 371_{(eight)}$$

names 3   names 7   names 1

Observing the pattern demonstrated above we see that to convert a binary numeral to an equivalent octal numeral:

(a) We separate the binary numeral into groups of three digits by commas starting at the right in the usual fashion.

(b) We determine the decimal numeral equivalent to each group of binary digits.

(c) These digits, in order, found in (b) are the digits of the equivalent octal numeral.

We reverse this process to convert an octal numeral to an equivalent binary numeral. In converting an octal numeral to an equivalent binary numeral, each digit in the octal numeral is represented by three digits in the binary numeral.

Example 3: Convert $4,371_{(eight)}$ to an equivalent binary numeral.

Solution: Octal numeral   4   3   7   1

Binary numeral   100   011   111   001

$$4,371_{(eight)} = 100,011,111,001_{(two)}$$

Exercise 4

1. Write the first twenty binary numerals starting with 1.
2. Write the first twenty octal numerals starting with 1.
3. Convert the following base-two numerals to base-eight numerals.

    (a) $1,001_{(two)}$ (d) $100,100,100_{(two)}$ (g) $10,000,111_{(two)}$
    (b) $101,010_{(two)}$ (e) $11,101,011_{(two)}$ (h) $110,011,110_{(two)}$
    (c) $111,111_{(two)}$ (f) $1,001,101_{(two)}$ (i) $101,111,010_{(two)}$

4. Use the binary-octal relation to convert each of the following binary numerals to base-eight numerals.

    (a) $110,111_{(two)}$ (c) $11,110,100_{(two)}$ (e) $10,101,011_{(two)}$

    (b) $101,101,101_{(two)}$ (d) $111,001,010_{(two)}$ (f) $100,001,011_{(two)}$

5. Use the binary-octal relation to convert each of the following octal numerals to binary numerals.

    (a) $1234_{(eight)}$ (d) $36,245_{(eight)}$ (g) $56,200,063_{(eight)}$

    (b) $7462_{(eight)}$ (e) $77_{(eight)}$ (h) $44,766_{(eight)}$

    (c) $5731_{(eight)}$ (f) $100_{(eight)}$ (i) $2361_{(eight)}$

6. Use the symbols "$=$," "$<$," "$>$" to compare the following pairs of numerals.

    (a) $1,010_{(eight)}, 268_{(eight)}$ (d) $63_{(eight)}, 1,011_{(ten)}$

    (b) $100_{(ten)}, 101_{(eight)}$ (e) $111,000_{(two)}, 100_{(ten)}$

    (c) $11,111_{(two)}, 25_{(eight)}$ (f) $42_{(eight)}, 101,010_{(two)}$

6.
THE HEXADECIMAL SYSTEM OF NUMERATION

To write numerals in the hexadecimal (base sixteen) system, we need sixteen digits. As we know, the symbols used in a numeration system are arbitrary. For the sixteen in the hexadecimal system, we shall use the familiar 0, 1, 2, 3, 4, 5, 6, 7, 8, 9 to represent, zero, one, two, three, four, five, six, seven, eight and nine, respectively, and A, B, C, D, E, F to represent ten, eleven, twelve, thirteen, fourteen, and fifteen, respectively. Although we could have chosen any set of sixteen symbols, these are the ones most often used in the computer field. In base ten, we group in sets of ten; in base eight, we group in sets of eight; in base sixteen we group in sets of sixteen. Examine Table 12.4 which relates the first few hexadecimal numerals to their decimal equivalent.

Notice that the representation of the grouping of sixteen or more objects produces a "carry" into the next higher place. A place-value chart for base sixteen is given in Table 12.5.

As we have observed in the conversions of binary, octal, and decimal numerals, the fractional part is converted separately from its integer part. In fact, the process of converting a fractional part is the inverse of the process used in converting the integer part. If the integer part is converted

TABLE 12.4
Hexadecimal and decimal numerals

Hexadecimal numeral	Decimal equivalent	Hexadecimal numeral	Decimal equivalent
1	1	11	17
2	2	12	18
3	3	13	19
4	4	14	20
5	5	15	21
6	6	16	22
7	7	17	23
8	8	18	24
9	9	19	25
A	10	1A	26
B	11	1B	27
C	12	1C	28
D	13	1D	29
E	14	1E	30
F	15	1F	31
10	16	20	32

by repeated division, then the fractional part is converted by repeated multiplication and vice versa. Base sixteen conversions follow the same scheme as base eight conversions.

In the following examples, we will work only with the integer portions of hexadecimal numerals. In actual practice, people in the computer field use tables to convert from one system to another relieving much tedious work.

Example 1: Convert $A4F_{(sixteen)}$ to its decimal equivalent.

Solution: Using expanded notation:

$$A4F = (A \cdot sixteen^2) + (4 \cdot sixteen^1) + (F \cdot sixteen^0)$$
$$= (10 \cdot 16^2) + (4 \cdot 16^1) + (15 \cdot 16^0)$$
$$= (10 \cdot 256) + (4 \cdot 16) + (15 \cdot 1)$$
$$= 2560 + 64 + 15$$
$$= 2639$$

Example 2: Convert $26D3_{(sixteen)}$ to an equivalent decimal numeral.

TABLE 12.5
Place-value chart for base sixteen

Base	Sixteen			Decimal equivalent			
⋮	⋮	⋮	⋮	⋮	⋮	⋮	⋮
−4	sixteen^{-4}	.0001	$16^{-4} = \dfrac{1}{16^4} = \dfrac{1}{65536} \doteq$				.00001526
−3	sixteen^{-3}	.001	$16^{-3} = \dfrac{1}{16^3} = \dfrac{1}{4096} \doteq$				.00024414
−2	sixteen^{-2}	.01	$16^{-2} = \dfrac{1}{16^2} = \dfrac{1}{256} =$				.00390625
−1	sixteen^{-1}	.1	$16^{-1} = \dfrac{1}{16^1} = \dfrac{1}{16} =$				.0625
0	sixteen^{0}	1.	$16^{0} = \dfrac{1}{16^0} = \dfrac{1}{1} =$				1.
1	sixteen^{1}	10.	$16^{1} = 16^1 =$	16	$=$	16.	
2	sixteen^{2}	100.	$16^{2} = 16^2 =$	256	$=$	256.	
3	sixteen^{3}	1000.	$16^{3} = 16^3 =$	4096	$=$	4096.	
4	sixteen^{4}	10000.	$16^{4} = 16^4 =$	65536	$=$	65536.	
⋮	⋮	⋮	⋮	⋮	⋮	⋮	⋮

Solution: Using expanded notation:

$$26D3 = (2 \cdot \text{sixteen}^3) + (6 \cdot \text{sixteen}^2) + (D \cdot \text{sixteen}^1) + (3 \cdot \text{sixteen}^0)$$
$$= (2 \cdot 16^3) + (6 \cdot 16^2) + (13 \cdot 16^1) + (3 \cdot 16^0)$$
$$= (2 \cdot 4096) + (6 \cdot 256) + (13 \cdot 16) + (3 \cdot 1)$$
$$= 8192 + 1536 + 208 + 3$$
$$= 9939$$

Example 3: Convert 8837 to its hexadecimal equivalent.
Solution: Here we will use repeated division by 16.

$$16\overline{)8837}$$
$$16\overline{)552} \quad R = 5 \uparrow \quad 5 \cdot \text{sixteen}^0$$
$$16\overline{)34} \quad R = 8 \quad 8 \cdot \text{sixteen}^1$$
$$16\overline{)2} \quad R = 2 \quad 2 \cdot \text{sixteen}^2$$
$$0 \quad R = 2 \quad 2 \cdot \text{sixteen}^3$$

The answer is read up, thus

$$8837 = 2285_{(sixteen)}$$

We must be careful in using the repeated division method of converting decimal numerals to hexadecimal numerals and always write the remainders in the hexadecimal system. Notice the example below.

Example 4: Convert 1711 to its hexadecimal equivalent using repeated division.

Solution:

$$16 \overline{)1711}$$

$16 \overline{)106}$	$R = 15$	$\uparrow$	$F \cdot sixteen^0$	(the symbol for fifteen is F)
$16 \overline{)6}$	$R = 10$		$A \cdot sixteen^1$	(the symbol for ten is A)
0	$R = 6$		$6 \cdot sixteen^2$	

$$1711 = 6AF_{(sixteen)}$$

Exercise 5

1. Convert the following hexadecimal numerals to decimal numerals.
   (a) $1234_{(sixteen)}$ (c) $BAD_{(sixteen)}$ (e) $333A_{(sixteen)}$
   (b) $E2B_{(sixteen)}$ (d) $1FED_{(sixteen)}$ (f) $690C_{(sixteen)}$

2. Convert the following hexadecimal numerals to decimal numerals.
   (a) $ABC_{(sixteen)}$ (c) $BAAO_{(sixteen)}$ (e) $FAD_{(sixteen)}$
   (b) $DEAD_{(sixteen)}$ (d) $AOOD_{(sixteen)}$ (f) $B1D_{(sixteen)}$

3. Use repeated division to convert the following decimal numerals to hexadecimal.
   (a) 1708   (c) 652   (e) 2989
   (b) 2730   (d) 777   (f) 4096

4. Use repeated division to convert the following decimal numerals to hexadecimal.
   (a) 100   (c) 3410   (e) 43981
   (b) 10000   (d) 36864   (f) 3150

5. In the last two sections, several methods of conversion from binary or octal fractions to decimal fractions were given. Using the same ideas, can you convert the following hexadecimal fractions to decimal fractions?
   (a) $.E_{(sixteen)}$ (b) $.04_{(sixteen)}$ (c) $.D8_{(sixteen)}$

6. In section 12.3 and 12.4, methods of repeated multiplications are given to convert a decimal fraction to its equivalent binary and octal fractions, respectively.

6. THE HEXADECIMAL SYSTEM OF NUMERATION

Study this method and see if you can discover how to convert the following decimal fractions to their hexadecimal equivalents.

(a) .5   (b) .9375   (c) .625

7. If you had a table of hexadecimal numerals, ranging from 000 to FFF, giving their decimal equivalents, what would be the range of the table in decimal notation?

Blaise Pascal (1623–1662) at the age of twelve discovered many geometric facts by drawing circles and triangles with chalk on the floor. When his father discovered this, he gave his son a copy of Euclid's *Elements* to read. Young Pascal quickly mastered Euclid and two years later began to attend weekly meetings led by Professor Boberval at the University of Paris. At sixteen he had written a book on the conic sections which Descartes refused to believe written by so young a boy. At eighteen, Pascal began plans to build a calculating machine, and within two years had built and sold fifty of these machines. A model of Pascal's calculating machine can be seen today in the Arts and Sciences Department of IBM. Together with Fermat, Pascal laid the foundation of the mathematical theory of probability by solving a question presented to him by his friend Chevalier de Mere.

Pascal from the age of seventeen until his death in 1662 suffered from acute dyspepsia and at times was partly paralyzed. We wonder what he might have accomplished had he been of robust health!

COMPUTER
HISTORY
AND
OPERATIONS

13

1.
A BRIEF HISTORY OF THE DIGITAL COMPUTER

Mathematics has been and still is fascinating to man. On the other hand, computation—adding, subtracting, multiplying, and dividing—can become extremely boring. To remedy this, man has always tried to devise computing tools to make this drudgery easier. The earliest calculating device known (other than the fingers) was the abacus. This was followed by Napier's Bones (1614), Pascal's Calculator (1642), Leibnitz's Reckoning Machine (1671), and Babbage's Difference Machine (1830). The latest device to aid man's campaign against computation is the **digital computer.**

In 1642, a French mathematician, Plaise Pascal, invented what was probably the first mechanical adding machine. This modest machine with its hand-fashioned rackets and gears, began the chain of events leading to today's digital computers.

Early in the nineteenth century, Charles Babbage, a young Englishman, began work on a machine that was to compute mathematical tables. His "difference machine" needed a human operator to start the calculation, then it proceeded automatically until all the answers had been printed. Although Babbage's machine was never fully successful, many of the ideas that he proposed were adopted approximately one hundred years later in the development of the digital computer.

Around 1890 Herman Hollerith, who worked for the United States Bureau of the Census, devised a system in which holes punched in cards represented numbers, letters, and symbols. His next project was the development of an electromechanical device that "read" the cards and tabulated the data they contained. Such punched cards are seen every day in pay checks, registration cards, bills, and the like.

At the turn of the century desk calculators had become compact and reliable, yet their new efficiency was not enough. Need was fast arising for a machine that would perform a sequence of operations automatically while saving intermediate results for later use. The first attempts at this type of device was an enlargement on the desk calculator. This resulted in a maze of wheels, shafts, gears, and the like, all requiring precision adjustment and intricate balance. This soon became impractical both in size and cost. The next step was to enlarge the capabilities of the punched-card machines. Even though these machines could do only one step at a time, it was easier and faster to pass decks of cards through many times than to do the calculations on a desk calculator.

Around 1937 George Stibitz and Howard Aiken began to work independently on a sequentially operated digital computer. The first digital computer was unveiled in 1940 by George Stibitz and the Bell Telephone Laboratories at the American Mathematical Society meeting in Hanover, New Hampshire. Howard Aiken, in collaboration with International Business Machines (IBM), designed the Mark I computer which was announced to the public in 1944. It was installed at Harvard University.

World War II was one of the major factors that hastened the development of the digital computer. The need for more sophisticated weapons demanded more extensive analyses. The computer filled this demand. More and more research was concentrated on improving its capabilities and speed. Soon after the end of the war, Remington Rand Corporation introduced the Eniac computer. In the Eniac the electrical relays, used in older computers to represent the digits, had been replaced by electron tubes. This replacement increased the speed of computation by computers as much as 100 per cent.

One of the most important inventions that furthered the development of computers was a tiny ferromagnetic ring to replace the electron tube for digit storage. These rings, or **cores** as they are sometimes called, are very reliable, generate no heat, require little current to operate, and permit a great reduction in computer storage size.

The computer era has just begun, and it is here to stay. The authors of this book hope that the brief material presented here will give the reader a better understanding of these machines. The basic fact we must remember is that computers do not think. They are capable only of performing the manipulations wired into their systems and executing the sequence of instructions supplied by men and women.

2.
COMPUTER COMPONENTS

The operation of a digital computer can be divided roughly into five phases: (1) **input,** (2) **control,** (3) **storage** and **memory,**

(4) **processing,** and (5) **output.** We shall outline these phases, omitting the technical details.

Input

Input is the **information,** which may be in the form of numbers, letters, or symbols, submitted to the machine. The input may be on punched cards, paper or magnetic tape, or entered manually by a keyboard or typewriter.

Control

Computers operate under direction of a **control** unit. This unit interprets and executes the sequence of instructions called a **program.** In today's computers the program is stored in the machine's memory.

Storage and Memory

Data are retained internally in what is called the **memory** or **core.** The memory is a multitude of tiny ferromagnetic rings arranged in groups. These groups, often called **words,** may contain a number or a program instruction. Each word, regardless of the content, can be located by the control unit.

Other storage devices are magnetic tape, magnetic disks, and magnetic drums. All three storage devices use patterns of magnetic spots to represent data. Magnetic-tape units are similar to ordinary tape recorders using heads to record or "play back" data. Magnetic-disk storage may be thought of as a stack of phonograph records. Data stored on either side of the disks may be located by sensing arms that move in and out as the disks rotate. A magnetic drum is a cylinder that revolves several thousand times per second past a series of read-write heads.

Processing

The **processing unit** performs the arithmetical operations of addition, subtraction, multiplication, and division. All complex mathematical problems must be broken down into combinations of these operations. The processing unit is also capable of logical decisions when comparing two words (greater than, less than, or equal to). It can also distinguish positive, negative, and zero values.

Output

Output is the answer to the problem. This may be obtained in printed form or on punched cards or magnetic tape, depending on the program.

3.
HOW FAST
ARE COMPUTERS?

Timing of machine operations is based on access time, called a **machine cycle.** A **cycle** is the amount of time required to transmit information between the storage area and the arithmetic unit. All instructions are timed in multiples of cycles—for example, an instruction to add may take four cycles to perform. The cycle time for the IBM 704, introduced in 1956, is 12 microseconds (0.000012 second). The IBM 7094, released in 1961, has cycle time of 2 microseconds (0.000002 second). Addition and subtraction of integers using the 7094 takes two cycles. Multiplication and division of integers on the 7094 take two to five cycles and three to eight cycles respectively, depending on the magnitude of the numbers. Nonintegral numbers require more time to manipulate.

The most recent computers, the so-called "third generation," are even faster. IBM's System/360–65 has a cycle time of 0.75 microsecond, and the CDC† 6800 can add two decimal numbers in 0.0000001 second.

The fact that the 6800 can add 10,000,000 decimal numbers in one second emphasizes the speed of computers. The machines of tomorrow certainly will surpass present-day standards and no one can predict how fast they will be in the future.

† Control Data Corporation.

4.
MAGNETIC CORE MEMORY

A digital computer retains data in its **memory** or **core.** The memory consists of many many ferrite rings, also called cores. Each ring represents one bit (binary digit) of data. When current is passed through a ring, a magnetic field is set up. If the direction of the current is changed, the polarity of the field is changed, thus providing two unique states. The states are named "1" and "0" and are represented by the binary system of numeration.

Rings are arranged in arrays, each array forming a plane as in Figure 13.1.

We can think of the elements of the plane as having X and Y coordinates. The planes are stacked upon each other giving a three-dimensional grouping of cores as depicted in Figure 13.2. There are as many planes as bits in the computer **word.**

Words are groups of cores that the computer processes together. Word size varies from computer to computer. The IBM System/360 uses a 32-bit word; whereas, CDC's series 6000 has a 60-bit word. The individual cores, that represent one data word, all have the same X and Y locations in their respective planes.

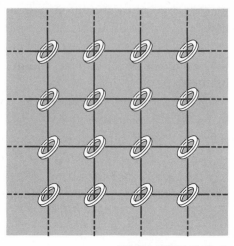

FIGURE 13.1

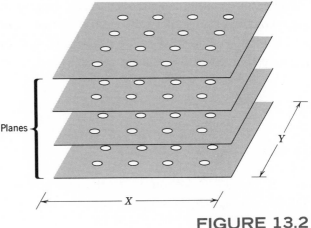

FIGURE 13.2

A matrix of X and Y wires plus a "sense" wire are threaded through each array of rings at right angles as shown in Figure 13.3.

To write a 1 in a particular core in a plane, both the X and Y wires carry one-half the magnetizing current. Only the core at the intersection of the two wires can become a 1. To change the state of the core (to a 0), the direction of the current flow is changed in both the X and Y wires.

To determine what is stored in a ring (1 or 0), the type of current to produce a "0" state is passed through the X and Y wires. If the intersected ring changes state, it is known it contained a "1." If no state change takes place, a "0" was represented. The changing of state produces a pulse which is sent down the sense wire. Since the changing of state may have produced a 0 in the ring where we once had a 1, something must be done to restore the core to its original value. The restoration process differs from machine to machine with some using a fourth wire, and others a combination of the original three.

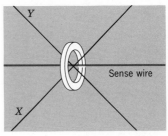

FIGURE 13.3

When the computer's circuitry needs the information stored in a particular word, the states of all the rings assigned to that word are reproduced in another word located in the processing unit and called a **register.** The states produced in the register may be represented by a binary numeral.

5.
COMPLEMENT ARITHMETIC

Addition, when the numbers to be added are named in the binary system, is simple because only four addition facts need be learned. These are:

$$0 + 0 = 0$$
$$0 + 1 = 1$$
$$1 + 0 = 1$$
$$1 + 1 = 10$$

These addition facts are shown more compactly in the addition table for the binary system in Table 13.1.

Just as a sum of ten or more in decimal addition results in a **regrouping** (usually called **carrying**) into the next higher position, a sum of two or more in binary addition results in a regrouping (called a **carry**) into the next higher position.

Study the following binary addition example. The explanation appears on the right; the carries are encircled.

①① ①①		
1 1 , 0 1 1	$1 + 1 = 10$	Write the 0 and carry 1 (two^1).
1 , 1 0 1	$1 + 1 + 0 = 10$	Write the 0 and carry 1 (two^2).
1 0 1 , 0 0 0	$1 + 0 + 1 = 10$	Write the 0 and carry 1 (two^3).
	$1 + 1 + 1 = 11$	Write the 1 and carry 1 (two^4).
	$1 + 1 = 10$	Write the 10.

TABLE 13.1

+	0	1
0	0	1
1	1	10

Example 1: Add (all numerals are binary numerals) 11,101 + 1,101.
  Solution:

$$
\begin{array}{r}
①① \quad ① \\
1\,1\ 1\,0\,1 \\
1\ 1\,0\,1 \\
\hline
1\,0\,1,0\,1\,0
\end{array}
$$

Example 2: Add (all numerals are binary numerals) 10,111 + 1010.
  Solution:

$$
\begin{array}{r}
①①① \\
1\,0\ 1\,1\,1 \\
1\ 0\,1\,0 \\
\hline
1\,0\,0,0\,0\,1
\end{array}
$$

Subtraction in the binary system is done the same way as subtraction in the decimal system. Study the following examples.

Example 3: Subtract (all numerals are binary numerals) 110 − 1.
  Solution:

$$
\begin{array}{r}
0\ 10 \\
1\,\cancel{1}\,0 \\
1 \\
\hline
1\,0\,1
\end{array}
$$

One two in *this* place becomes

two ones in *this* place.

Example 4: Subtract (all numerals are binary numerals) 10,110 − 11.
  Solution:

Step 1

$$
\begin{array}{r}
0\ 10 \\
1\,0\,1\,\cancel{1}\,0 \\
1\,1 \\
\hline
1
\end{array}
$$

Step 2

$$
\begin{array}{r}
10 \\
0\ 0\ 10 \\
1\,0\,\cancel{1}\,\cancel{1}\,0 \\
1\,1 \\
\hline
1\,0\,0\,1\,1
\end{array}
$$

407

Example 5: Subtract (all numerals are binary numerals) $10{,}001 - 1111$.
Solution:

Step 1	Step 2	Step 3
		1 10
	1 10	1 ~~10~~
10	0 ~~10~~	0 ~~10~~
~~1~~ 0 0 0 1	~~1~~ 0 0 0 1	~~1~~ 0 0 0 1
1 1 1 1	1 1 1 1	1 1 1 1
0	0	0 0 0 ~~1~~ 0

Example 6: Subtract (all numerals are binary numerals) $1000 - 110$.
Solution:

Step 1	Step 2
	1 10
0 10	0 ~~10~~
~~1~~ 0 0 0	~~1~~ 0 0 0
1 1 0	1 1 0
0	0 0 1 0

Computers subtract by using a method called "subtraction by adding the complement." The **ones' complement** of a number N named by a binary numeral of n digits is

$$\underbrace{111 \ldots 1_{\text{(two)}}}_{n \text{ 1's}} - N_{\text{(two)}}$$

For example the ones' complement of $10{,}100_{\text{(two)}}$ is

$$11111_{\text{(two)}} - 10100_{\text{(two)}} = 01011_{\text{(two)}}$$

The ones' complement of $1{,}011_{\text{(two)}}$ is

$$1111_{\text{(two)}} - 1011_{\text{(two)}} = 0100_{\text{(two)}}$$

The ones' complement of $10{,}010_{\text{(two)}}$ is

$$11111_{\text{(two)}} - 10010_{\text{(two)}} = 01101_{\text{(two)}}$$

Notice that the ones' complement of any binary numeral is found by changing the 1's in the numeral to 0's and the 0's to 1's.

The **twos' complement** of a number N named by a binary numeral of n digits is

$$10^n_{\text{(two)}} - N_{\text{(two)}}$$

The twos' complement of $10,110_{(two)}$ is

$$10^5_{(two)} - 10110_{(two)} = 100000_{(two)} - 10110_{(two)} = 1010_{(two)}$$

The twos' complement of $1111_{(two)}$ is

$$10^4_{(two)} - 1111_{(two)} = 10000_{(two)} - 1111_{(two)} = 1_{(two)}$$

The twos' complement of $10,100_{(two)}$ is

$$10^5_{(two)} - 10100_{(two)} = 100000_{(two)} - 10100_{(two)} = 1100_{(two)}$$

The twos' complement of $10010_{(two)}$ is

$$10^5_{(two)} - 10010_{(two)} = 100000_{(two)} - 10010_{(two)} = 1110_{(two)}$$

Notice that the twos' complement of a number is 1 more than the ones' complement and

$$10^n_{(two)} - N_{(two)} = \underbrace{111\ldots1}_{n\ 1\text{'s}} - N + 1$$

People in the computer field obtain the twos' complement by first finding the ones' complement and then adding 1 to the lowest order bit (the last digit to the right in the numeral). The ones' complement, as we saw above, is obtained by inverting the digits (changing the 0 to 1 and 1 to 0). This is the same as subtracting each digit in the numeral from 1 as shown below.

$$
\begin{array}{ll}
\text{binary numeral} & 1100110 \\
\text{inverted} & 0011001 \ \ (\text{ones' complement})
\end{array}
$$

The above is the same as (all numerals are binary)

$$
\begin{array}{r}
1111111 \\
-\,1100110 \\
\hline
0011001 \ \ (\text{ones' complement})
\end{array}
$$

For the twos' complement, add 1 to the lowest order bit thus (all numerals are binary)

$$
\begin{array}{r}
0011001 \ \ (\text{ones' complement}) \\
+\,1 \\
\hline
0011010 \ \ (\text{twos' complement})
\end{array}
$$

Computers subtract a number N (with n digits) by adding its twos' complement and then subtracting $10_{(two)}^n$ from this sum. Thus,

$$1101_{(two)} - 101_{(two)} = 1101_{(two)} + (1000_{(two)} - 101_{(two)}) - 1000_{(two)}$$
$$= 1000_{(two)}$$

Using ordinary subtraction, we have

$$1101_{(two)} - 101_{(two)} = 1000_{(two)}$$

Using complement arithmetic to find $1101_{(two)} - 101_{(two)}$ first we find the twos' complement by inverting and adding 1 to the least significant digit:

$$
\begin{array}{ll}
\text{numeral} & 101 \\
\text{inverted} & 010 \quad \text{(ones' complement)} \\
& \underline{+1} \\
& 011 \quad \text{(twos' complement)}
\end{array}
$$

We then add the twos' complement to $1101_{(two)}$

$$
\begin{array}{r}
1101 \\
+011 \\
\hline
10000 \\
-1000 \\
\hline
1000
\end{array}
$$

The last step is to subtract $10_{(two)}^n$ which was added in obtaining the twos' complement.

This probably seems like an inefficient way to subtract two numbers. When we perform the operations shown above, subtraction by adding a complement takes two additional steps: (1) Obtaining the complement $10_{(two)}^n - N$; (2) Subtracting $10_{(two)}^n$. In the computer, these two extra steps are eliminated by the machine's circuitry. First, a negative number is automatically converted to its twos' complement. Second, the registers where the arithmetic operations are performed are designed to eliminate the need to subtract $10_{(two)}^n$.

Suppose the machine we were using had a register named (A) that could represent six bits of data and that register (A) contained the following information,

0	0	1	0	1	1	(A)

Somewhere in memory is another six bit word (X). Word (X) contains the integer -5 which in binary notation is $-(000101)$. In storage the twos' complement of $-(000101)$ appears so that the word (X) is

| 1 | 1 | 1 | 0 | 1 | 1 | (X)

Adding word (X) to register (A) is the same as subtracting $+5$ from register (A).

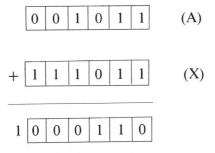

| 0 | 0 | 1 | 0 | 1 | 1 | (A)

$+$ | 1 | 1 | 1 | 0 | 1 | 1 | (X)

1 | 0 | 0 | 0 | 1 | 1 | 0 |

Notice that in the addition the first digit of the sum has place-value $10^6_{(two)}$. Thus, one place greater than the registers' capacity was generated (this is called **overflow**). The computer will discard this bit, a process that equals the final subtraction of $10^n_{(two)}$ as shown above.

For a check

$$001011_{(two)} - 000101_{(two)} = 000110_{(two)}$$

which is the same answer we obtained by adding the twos' complement of $000101_{(two)}$ to register (A).

If we were to examine the entire contents of the storage of most computers by using a core dump,† we would find each negative number represented by a twos' complement.

Complement arithmetic eliminates the need for each word and register to have one bit of information used to represent a sign. By not dedicating one bit to a sign, each word can represent a number greater in magnitude by one higher power of two. Complement arithmetic also simplifies the circuitry. The arithmetic registers need only add. Subtracting is done by adding the twos' complement; multiplication is done by adding and shifting; division is done by adding the twos' complement and shifting.

† See Chapter 12, Section 1.

Exercise 1

1. Add, all numerals are binary numerals.

(a) 1011
101

(c) 1111
111

(b) 11001
1100

(d) 10000
10011

2. Add, all numerals are binary numerals.

(a) 10111
1111

(c) 11011
10110

(b) 1010
1011

(d) 110001
100111

3. Subtract, all numerals are binary numerals.

(a) 111
10

(c) 100101
10110

(b) 1011
110

(d) 101011
11110

4. Subtract, all numerals are binary numerals.

(a) 1000
101

(c) 1000111
111001

(b) 101010
1111

(d) 1010111
101101

5. Find the ones' complement of each of the following.
(a) $110011_{(two)}$
(b) $11111_{(two)}$
(c) $10011_{(two)}$
(d) $11001_{(two)}$
(e) $10001_{(two)}$
(f) $0001_{(two)}$
(g) $0000_{(two)}$
(h) $10100_{(two)}$

6. Find the twos' complement of each of the following.

(a) $0001_{(two)}$ (e) $10101_{(two)}$

(b) $1110_{(two)}$ (f) $100001_{(two)}$

(c) $110011_{(two)}$ (g) $00011_{(two)}$

(d) $101110_{(two)}$ (h) $10000_{(two)}$

7. Perform the following binary subtractions by adding the twos' complement. If you think of these numbers in a six position register, the overflow bit may be ignored.

(a)
$$\begin{array}{r} 001110 \\ -\,000101 \\ \hline \end{array}$$

(d)
$$\begin{array}{r} 111011 \\ -\,011111 \\ \hline \end{array}$$

(b)
$$\begin{array}{r} 110011 \\ -\,001100 \\ \hline \end{array}$$

(e)
$$\begin{array}{r} 111101 \\ -\,100111 \\ \hline \end{array}$$

(c)
$$\begin{array}{r} 000101 \\ -\,000011 \\ \hline \end{array}$$

(f)
$$\begin{array}{r} 100001 \\ -\,000011 \\ \hline \end{array}$$

6.
ASSEMBLY LANGUAGE

A **word** in a computer's storage device consists of a group of ferrite rings processed together by the machine's control unit. A word may represent two types of binary information, **datum** or a **program operation.** A program is a set of instructions that directs the computer to perform operations in the solution of some problem. To help understand how a computer goes about solving a problem, we outline the basic steps below.

1. A computer is built to perform a set of operations (e.g., add, divide, transfer, complement, etc.). The actual performance of each operation is built into the circuitry.
2. Each operation has its own individual binary pattern. This binary pattern is always in the same location within a computer word. These patterns are called **operation codes** or **"opp codes."**
3. The program, which is a set of opp codes, is stored in a particular location in memory.

4. The control unit, starting with the first word of the program, translates each binary opp code—one after another—into pulses causing the circuitry to react.

5. The preparation of the program involves selecting the proper operations to solve the problem. In an assembly language, each machine operation (built into the circuitry) is represented by a mnemonic which allows the person writing the program (the programmer) to use simple symbols to represent the binary operation codes.

6. Assembly languages use what is called **symbolic programming.** Symbolic programming involves selecting the correct mnemonics (representing machine operations) in proper order that will direct the computer to solve a particular problem.

7. The assembly language instructions are punched on cards in some prescribed format.

8. The deck of punched cards is fed into the computer where an **assembly program,** provided by the manufacturer, translates the symbolic instructions into binary patterns that control the machine's operation.

Computers come in all sizes. The smaller ones are called "mini," the largest ones could be called maxi. Most of these have an assembly language. The larger the computer, the more operations it can perform and thus the assembly language will contain more opp codes. (Some mini computers for guiding missiles and airplanes have only twelve symbolic instructions, whereas the IBM System/360 has approximately one hundred and fifty.)

Let's invent a set of assembly language instructions for some fictitious computer. If we remember that each instruction, after translation into the machine's binary pattern, actually causes the operation it represents to be performed by the computer, we will understand that machines only do what *we* instruct them to do.

Our computer will be in the mini class, since our aim is the understanding of how a machine is directed to solve problems. We won't concern ourselves with things like word size, memory size, the orientation of symbols on punched cards, or the beginning or ending of a symbolic program.

Our machine will have ten symbolic instructions constructed thus:

N   O   A

N   The name of this operation: (when translated to machine language the name of a core location).

O   The operation.

A   The address this operation refers to, this can be the name of another operation or the name of a word of datum. (This will be translated to the name of a core location.) We should note that only the operations that are referenced by some "A" need have an "N." Within the computer "N" is a name for a computer word. If this word is never "addressed" by any reference by the field "A," there is no need to give it a unique name.

We shall restrict our names, operations, and addresses (except the LT operation) to two alphabetic characters. In the following definitions, **R** refers to an arithmetic register and **X** a word of core storage.

The operations and their definitions are:

N	O	A	
	CA	X	Clear register **R,** place the contents of word **X** in **R.** The contents of word **X** remains unchanged.
	AD	X	Add to the contents of register **R** the contents of word **X.** The contents of word **X** remains unchanged.
	SB	X	Subtract from the contents of register **R** the contents of word **X.** The results of the subtraction is placed in register **R,** the contents of word **X** remains unchanged.
	ML	X	Multiply the contents of register **R** by the contents of word **X.** The product is placed in register **R,** the contents of word **X** remains unchanged.
	DV	X	Divide the contents of register **R** by the contents of word **X.** The quotient is placed in register **R,** the contents of word **X** remains unchanged.
	ST	X	Store the contents of register **R** in word **X.**
	TR	N	Transfer to and execute next the operation named **N.**

N	O	A	
	CM	X	Compare the contents of register **R** to the contents of word **X**. If **R** is less than **X**, execute the next instruction. If **R** is equal to **X**, skip the next instruction and execute the second instruction. If **R** is greater than **X**, skip the next two instructions and execute the third instruction.
	WD		This reserves a word in storage for something that will be computed by the program or some external data that was read in as input.
	LT	i	LT means a "literal." A literal is a precise numerical value represented by "i" and stored in word **N**.

Example 1: Add word **AA** (=2) to word **BB** (=6), store answer in word **CC**.

N	O	A	
	CA	AA	Notice **AA** and **BB** are literals. The operation
	AD	BB	**WD** saves a storage word for **CC** which is com-
	ST	CC	puted.
	⋮		
AA	LT	2	
BB	LT	6	
CC	WD	—	

Example 2: Divide word **AA** (a computed word) by the word **XX** (=12). Store the quotient in word **ZZ**.

N	O	A	
	⋮		
	CA	AA	**XX** is the literal 12. **AA** and **ZZ** contain values
	DV	XX	that are computed by the program. In this
	ST	ZZ	example **AA** must have been computed by some
	⋮		previous steps.
AA	WD	—	
XX	LT	12	
ZZ	WD	—	

Example 3: Test the value of a number that has been computed and stored in word **DA**. If it is less than zero, go to the operation

416

named **DL**; if it is equal to zero, go to the operation named **DE**; if it is greater than zero, go to the operation named **DG**.

N	O	A	
	⋮		
	CA	DA	**DA** was computed before this test. **ZR** is the
	CM	ZR	literal zero. **DL, DE,** and **DG** are names for
	TR	DL	operations in other parts of the program. This
	TR	DE	example shows how the program may branch
	TR	DG	to other instructions depending upon the con-
	⋮		tents of some word.
DA	—		
	⋮		
DE	—		
	⋮		
DG	—		
	⋮		
DA	WD	—	
ZR	LT	0	
	⋮		Other instructions may follow.

These symbolic instructions will become clearer when we construct a few small programs.

Suppose we wish to count up to ten by ones, starting at zero. If register **R** is less than ten, we add again; if equal to ten we transfer to where we square ten; if greater than ten, we transfer to some error notation which we shall not show.

N	O	A	
	⋮		
	CA	ZR	Clear register **R** and add contents of word **ZR**.
AM	AD	ON	Add to register **R** the contents of word **ON**.
	CM	TN	Compare register **R** to word **TN**.
	TR	AM	If contents of **R** is less than contents of word **TN**, transfer to **AM**.
	TR	SQ	If contents of register **R** is equal to word **TN**, transfer to **SQ**.
	TR	ER	If contents of register **R** is greater than word **TN**, transfer to some error branch.

417

N	O	A	
SQ	ML	TN	The contents of register **R** is multiplied by the contents of word **TN**.
	⋮		Other instructions may follow.
ER	—		The error routine.
	⋮		
ZR	LT	0	The integer zero.
ON	LT	1	The integer one.
TN	LT	10	The integer ten.

Let's find the average† of four words that have been computed.

N	O	A
	⋮	
	CA	A1
	AD	A2
	AD	A3
	AD	A4
	DV	FR
	ST	AN
	⋮	
A1	WD	—
A2	WD	—
A3	WD	—
A4	WD	—
AN	WD	—
FR	LT	4

Exercise 2

1. Write the instructions to add words **XX** and **KK**. Then multiply their sum by 2 and store in word **XK**.

2. Write the instructions that will subtract word **CI** from word **CB**, store the answer in word **ZB**. If **ZB** is less than or equal to zero, go on with the program. If **ZB** is greater than zero, add 1 to word **CT**.

† See Chapter 8, Section 8.

3. Write the instructions needed to add three words and divide the sum by a fourth word. You may pick the word names.

4. Select your own word names and write the instructions needed to multiply two words, divide the product by a third word and store the answer.

5. Explain what is being done by the following instructions:

N	O	A
	⋮	
	CA	ON
MO	AD	ON
	CM	TN
	TR	MO
	TR	OK
OK	DV	NM
	ST	XX
	⋮	
ON	LT	1
TN	LT	10
NM	WD	—
XX	WD	—
	⋮	

6. Explain what the following instructions do.

N	O	A
	⋮	
	CA	ZR
	ST	CT
MR	CA	WW
	SB	TO
	ST	WW
	CM	ZR
	TR	DN
	TR	DN
	CA	CT
	AD	ON
	ST	CT
	TR	MR
DN	—	—
	⋮	
ZR	LT	0
TO	LT	2
ON	LT	1
WW	WD	—
CT	WD	—

7.
NONINTEGRAL NUMBERS

In the previous sections, the numbers we have studied with regard to the digital computer's memory were all **integers.** Integers are elements of the set

$$\{\ldots -3, -2, -1, 0, 1, 2, 3, \ldots\}$$

The integer 19 (base ten) converts to 010011 (base two). This binary numeral when stored in a six-bit computer word would appear thus.

0	1	0	0	1	1

There is no fractional part and the binary point is assumed to be located to the right of the least significant digit thus

$$0\,1\,0\,0\,1\,1\,.$$

If integers were the only type numbers a computer was capable of processing, the machine would be virtually useless. As you can see, if we wish to represent the binary equivalent of a fractional number called a **rational number** (a rational number is one that may be expressed by some fraction $\frac{a}{b}$ where a and b are integers and $b \neq 0$), in the computer's memory, there is no way to indicate the location of the binary point. For instance, if we convert the decimal number 2.375 to its binary equivalent, we have

$$2.375 = 0\,1\,0.\,0\,1\,1_{(two)}$$

If we try to store 010.011 in our six-bit word, we have a problem. There is no way to indicate where the binary point is and no symbol to represent it (our two states are used to represent the binary digits 0 and 1).

The problem of locating the binary point of a fractional number in a computer's storage has a clever solution. In the computer field, rational numbers are referred to as **floating point numbers.** As you will see, the design of the machine actually lets the binary point "float."

A floating point number is composed of two parts; the **characteristic** and the **mantissa.** The characteristic gives the location of the point within the mantissa. The mantissa is the number itself. Suppose we have some decimal integers with six digits. The first digit (most significant) will locate the decimal point within the other five. To locate the point, we will count from left to right. The floating point number:

(a)	263429	means	63.429
(b)	344407	means	444.07
(c)	563789	means	63789.
(d)	011363	means	.11363
(e)	198769	means	9.8769

In (a), 2 is the characteristic which locates the decimal point two places to the right in the mantissa 63429.

The number of binary digits in the characteristic and mantissa of floating point computer words varies with machine design. (The CDC series 6000 with a 60-bit word allocates 12 bits to the characteristic and 48 bits to the mantissa, whereas IBM System/360 has a 32-bit word with a 12-bit characteristic and a 20-bit mantissa.)

For our examples, we will use a 36-bit word with a 9-bit characteristic and a 27-bit mantissa. Since 36 binary digits will become clumsy, where possible we will use the octal equivalent† as would appear in a core dump.†† The octal equivalent of the binary numeral will have a 3-digit characteristic (to locate the octal point) and a 9-digit mantissa, called a fraction.

We have two rules to establish: (1) the most significant bit in the binary fraction must always be a 1—this is called "normalizing"; (2) a characteristic of 010000000 (binary), 200 (octal) will denote the point is located just to the left of the most significant digit of the fraction.

Suppose in a core dump we see the octal word

200100000000

Let's separate this into its characteristic and fraction

200 10000000

The 200 characteristic tells us that the binary point is located to the left of the most significant digit of the fraction. This gives us the octal number

.1

---

† Chapter 12, Section 5 defines the binary octal relationship.
†† See Chapter 12, Section 1 for a definition of a "core dump."

Using the conversion techniques discussed in Chapter 12, we see that .1 base eight is equal to .125 base ten. Thus,

$$20010000000 \quad \text{means} \quad .125$$

In the following examples, where the base being used is not obvious, we will denote the correct base by a subscript, $77_{(eight)}$. Remember, the following floating point numbers are in octal.

Example 1: What decimal number is represented by the floating point computer word 201100000000?

Solution: The characteristic is 201. This means the binary point is moved one binary position to the right ($201 - 200 = 1$). We convert the octal fraction to binary

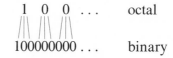

$$1 \quad 0 \quad 0 \quad \ldots \qquad \text{octal}$$

$$100000000 \ldots \qquad \text{binary}$$

and place the binary point one digit to the right

$$.1.00\ 000\ 000 \ldots$$

This gives us the binary number 1.000 which is the same as the decimal number 1.0. Thus,

$$201100000000 \quad \text{means} \quad 1.0_{(ten)}$$

Example 2: What decimal number does 205265000000 represent?

Solution: The characteristic is 205 which indicates the binary point is to be moved 5 places to the right ($205 - 200 = 5$). Let's convert the fraction to binary and locate the binary point.

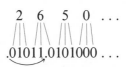

$$2 \quad 6 \quad 5 \quad 0 \quad \ldots$$

$$.01011.0101000 \ldots$$

This gives us the numeral $1011.0101_{(two)}$. Using the binary octal relationship

$$1011.010100_{(two)} = 13.24_{(eight)}$$

The final step is to convert the octal numeral to base ten, thus

205265000000   means
$$1011.0101_{(two)} = 13.24_{(eight)} = 11.3125_{(ten)}$$

Example 3: Find the decimal number represented by the floating point word 212420200000.

Solution: The characteristic is $212_{(eight)}$; $12_{(eight)}$ equals $10_{(ten)}$ which tells us the binary point is ten places to the right $[212 - 200 = 12_{(eight)} = 10_{(ten)}]$. Converting the fraction to binary and inserting the binary point, we have

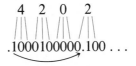

If we separate the binary numeral into groups of three digits, it helps us to convert back to base eight.

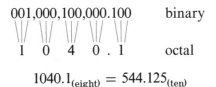

$$1040.1_{(eight)} = 544.125_{(ten)}$$

Therefore,

212420200000   means
$$1000100000.1_{(two)} = 1040.1_{(eight)} = 544.125_{(ten)}$$

Example 4: Convert the floating point word 176477000000 to its decimal equivalent.

Solution: In the previous examples, our characteristics have been greater than 200, indicating the number of places to the right to place the binary point. Our characteristic this time is 176 (remember this is base eight). Characteristics less than 200 indicate the binary point moves to the left. The number of places the point moves to the left is found by subtracting the characteristic from 200.

$$200 - 177 = 1 \text{ place to the left}$$
$$200 - 176 = 2 \text{ places to the left}$$
$$200 - 175 = 3 \text{ places to the left}$$
$$200 - 174 = 4 \text{ places to the left}$$
$$\vdots$$

The characteristic 176 denotes the binary point is moved two places to the left. Let's convert the fraction to binary and position the binary point.

$$.00.100111111000 \ldots$$

Now separate the binary numeral into groups of three digits, using commas, and convert it to base eight.

$$.001,001,111,110, \ldots$$
$$. \quad 1 \quad 1 \quad 7 \quad 6 \quad \ldots$$

Then $.1176_{(eight)} \doteq .15576_{(ten)}$

Therefore,

176477000000 means

$$.001001111110_{(two)} = .1176_{(eight)} \doteq 15576_{(ten)}$$

The following steps are taken to convert a decimal numeral into its floating point representation.

(1) Convert the decimal numeral to base eight.
(2) Convert the base eight numeral to base two using the binary-octal relation.
(3) Determine the number of places the binary point must be moved to place it just to the left of the most significant "1" binary digit (this is called **normalizing**).
(4) Convert the number of places the binary point was moved to octal.
(5) If the point was moved to the left, add the number of places

moved to 200 to form the characteristic. If the point was moved to the right, subtract it from 200 to form the characteristic.

(6) Convert the normalized binary numeral to base eight and attach the characteristic.

Study the following examples.

Example 5: What is the floating point number equal to 10.2?

Solution: Convert 10.2 to base eight then to base two.

$$10.2_{(ten)} \doteq 12.2_{(eight)} = 001010.010_{(two)}$$

Move the binary point so that it is just to the left of the most significant binary digit equal to 1.

$$00.\underset{\frown}{1}010.010$$

The binary point was moved four places to the left, therefore we add 4 to 200 giving the characteristic 204. Separate the normalized binary numeral into groups of three digits and convert it back to base eight.

$$.101,001,000 \ldots$$
$$.\ \ 5\ \ \ \ 1\ \ \ \ 0$$

Attaching the characteristic to the octal fraction gives

$$204510000000$$

Therefore,

$$10.2_{(ten)} \quad \text{means} \quad 204510000000$$

Example 6: What is the floating point representation of .0625?

Solution:
$$.0625_{(ten)} = .04_{(eight)} = .000100_{(two)}$$

Move the binary point so that it is just left of the most significant "1" binary digit.

$$.000.\underset{\frown}{1}00$$

The binary point was moved three places to the right so we subtract 3 from 200, giving a characteristic of 175. Convert the binary fraction to base eight

$$.100,000 \ldots$$

$$. 4 \quad 0 \ldots$$

Attach the characteristic to form the floating point number.

$$.0625_{(ten)} \qquad \text{means} \qquad 175400000000$$

Exercise 3

1. Convert the following floating point numbers to base ten.
(a) 202674000000
(b) 210776600000
(c) 177420000000
(d) 174400000000
(e) 204702000000
(f) 207555000000

2. Convert the following floating point numbers to base ten.
(a) 205400200000
(b) 212660000000
(c) 203411000000
(d) 204722000000
(e) 176400000000
(f) 177700000000

3. Convert the following base ten numerals to their floating point equivalent.
(a) 10.0
(b) 124.0
(c) .25
(d) 200.0
(e) 77.50
(f) 21.375

4. Convert the following base ten numerals to their floating point equivalent.
(a) .28125
(b) .03125
(c) 100.1
(d) 69.25
(e) 879.5
(f) 11.75

John W. Backus (1925–   ) conceived the idea of FOR-
TRAN (formula translation) computer language in 1954
while working on programming research at IBM. His effort
to construct FORTRAN began with the recognition that the
cost of programmers was as great as the cost of computers. Three years later, in 1957, the programming language FORTRAN was developed. Although only fifteen
years old, FORTRAN is solidly established as the language
of thousands of scientists and engineers who communicate
with computers.

A COMPUTER
LANGUAGE

14

1.
INTRODUCTION

A computer's circuitry is designed to perform specific operations such as add, subtract, shift, etc. Since a computer does nothing on its own, there must be some way to tell it what operations to do and when to do them. A sequence of statements, called a **program,** is prepared to instruct the computer's operations. This sequence of instructions must be constructed following specific rules that are determined by the **language** being used. Just as the people in France speak French, and the people in Germany speak German, each type of computer has its own unique language called the **machine language.**

The inefficiency of a different language for each type computer became evident as the number of these machines increased. Although much time, effort, and money was expended developing a program to solve a certain problem on a particular machine, it was impossible to use this program on any other type computer. Therefore, when a decision to change machines was made, it usually meant each program had to be rewritten in a new language. The cost of this conversion could be as high as the cost of the original program itself. To remedy this, some universal languages were developed. These languages tend to be used for specific applications. For example, COBOL, meaning *common business oriented language,* is used by the business community, whereas FORTRAN, a contraction of *formula translation,* is used in scientific areas. These universal languages, called **compiler languages,** not only allow programs written in their vernacular to operate on many different computers, they also permit the program statements to be meaningful with respect to the problem being solved. These statements take on the form of equations or English written commands that allow one to determine in a glance what is being solved, which is certainly more efficient than unraveling a long string of assembly language instructions.

A compiler language is a set of operations (mathematical, logical, etc.) combined with rigid rules governing their usage. These operations are combined to produce program statements that solve some specific problem. Even though statements written in a compiler language may be processed by many different computers, there has to be some way to translate these instructions into each computer's unique machine language. To do this, each manufacturer supplies a program written in his own computer's language that breaks down the compiler language statements and translates them piece by piece into machine instructions. The translating program is called a **compiler.**

FORTRAN, which we shall study here, is widely used throughout the computer world predominantly in science and engineering. The language is extensive, containing details and many many rules that are impossible to include in this chapter. We have chosen the basic framework of the language composed of its most powerful and frequently used operations. The authors do not claim that reading this chapter will make a programmer of the reader. It is hoped that the material selected will shed some light on one way a computer is directed to solve problems.

2.
THE FORTRAN STATEMENT

One feature of FORTRAN is that the instructions contained within the program can be written to resemble closely the equations and formulas to be used. Each separate instruction is called a **statement.** After all statements necessary to solve a particular problem have been written in proper order, the statements are punched on cards. This deck of cards is submitted to the computer where another program, the **compiler,** translates the FORTRAN statements into the language of the machine being used.

FORTRAN statements are written on **coding paper** (Figure 14.1). A typical card on which statements are punched in shown in Figure 14.2. Notice that the paper and card contain eighty columns. The statements themselves are written and punched between columns seven and seventy-two inclusive. Columns one through five are used for statement numbers and column six is used for indicating the continuation of a statement started on the previous card.

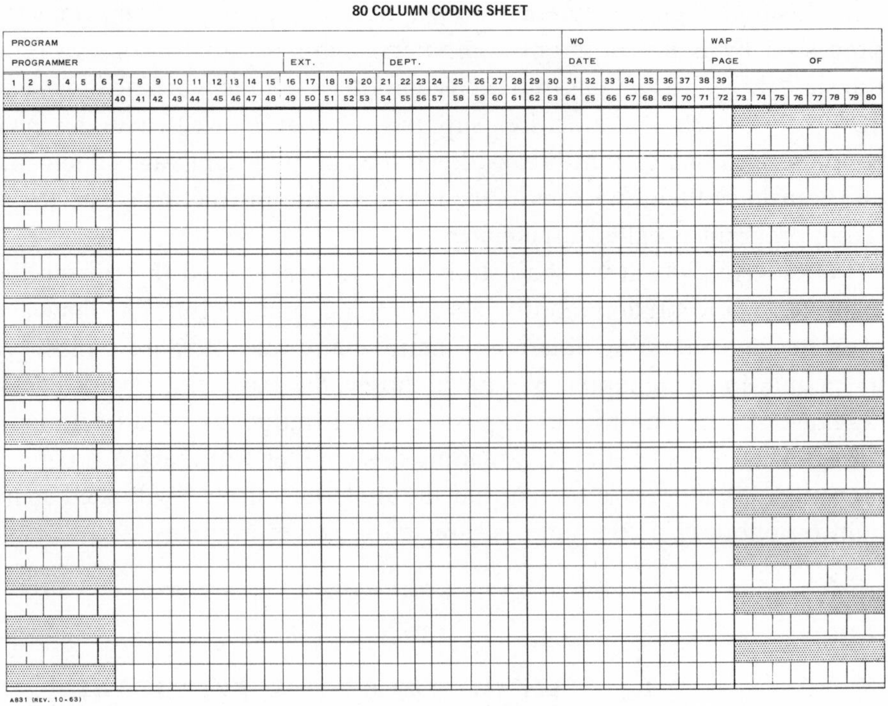

A COMPUTER LANGUAGE

80 COLUMN CODING SHEET

FIGURE 14.1

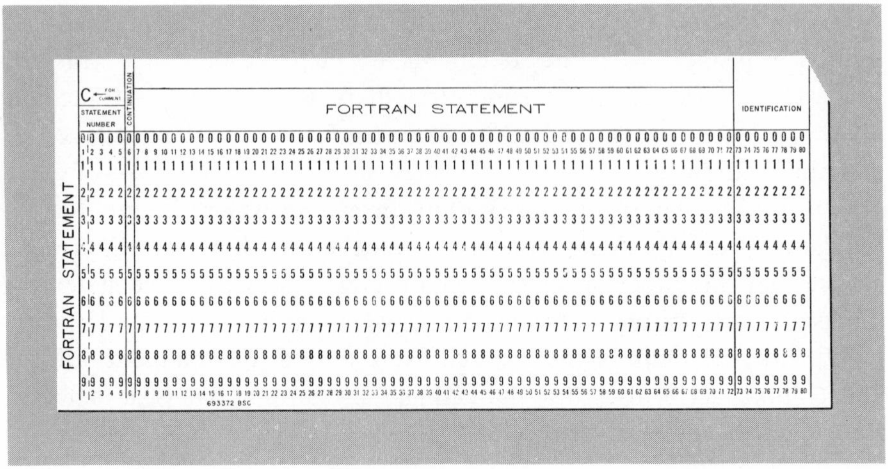

FIGURE 14.2

3.
CONSTANTS, VARIABLES, AND NAMES

The following are four FORTRAN statements representing four algebraic equations:

(1) I = 6 + J
(2) A = 6.5 − C
(3) M = K + 13
(4) X = Y + Z + 4.0

A **constant** is a specific numerical value that appears in a FORTRAN statement. In the statements above; 6, 6.5, 13, and 4.0 are all constants. Two types of constants are used in FORTRAN statements, **integer** and **real. Integer constants** are elements from the set I of integers,

$$I = \{\ldots, -2, -1, 0, 1, 2, \ldots\}.$$

In a FORTRAN statement, an integer constant may *never* contain a decimal point. In statements 1 and 3 above, the numerals 6 and 13 are integer constants. **Real constants** are elements from the set of real numbers. When appearing in FORTRAN statements, real constants must *always* contain a decimal point. In statements 2 and 4 above, 6.5 and 4.0 are real constants.

Although we have been taught to separate large numerals with commas, this *cannot* be done to FORTRAN constants either integer or real. For example, 20,000 must be written 20000 and 13,456.86 is written 13456.86.

Examples of integer and real constants are given in Table 14.1.

In FORTRAN, the term **variable** is used to denote something that is referred to by a name rather than an explicit numerical value, such as a constant. In the statement

$$OVER = CALS - 900.0$$

OVER and CALS are variables and 900.0 is a constant. If this statement were in a program, it might represent: the number of calories over the number specified in our diet (OVER), is equal to the number of calories consumed each day (CALS), minus nine hundred. (This assumes we are

433

TABLE 14.1

Integer constants	Real constants
1	11.2
−74	99.63942
22069	−0.9639
−11	−333.33
−7433	0.0
0	6269.7

on a nine hundred calorie a day diet.) If we were to compute this daily, the number of calories consumed would vary, thus the number over would vary. (If you have willpower, OVER can represent a negative value.) Since OVER and CALS may represent many values, from time to time, they are called **program variables.**

The **naming** of variables used in a FORTRAN program follows certain rules. The name assigned to represent an integer variable must begin with one of the six letters: I, J, K, L, M, N. Names given to represent real variables may begin with any letter of the alphabet *except* those used for integer variables, that is, any letter except I, J, K, L, M, N. Names can be composed of one character (minimum) through the usual maximum of six characters, which we shall use. (The maximum depends upon the computer used, IBM/System 360 uses six whereas CDCs series 6000 uses seven.) With the exception of the starting character, names may contain numerals as well as letters. Usually the names chosen for variables indicate what they represent. Real and integer constants, or variables, should not

TABLE 14.2

Variable	Constant
I	= 3
ALPHA	= 90.0
SUM	= 22.444
NAME	= −25
KEY33	= 33
ZERO	= −74.5
IGUESS	= 6660
A69	= 88.8

be mixed within a single FORTRAN statement. Some examples of variables equated to proper type constants are given in Table 14.2.

There is always a minimum and maximum numerical value that may be defined by a constant or assigned to a variable. The value of this maximum and minimum is determined by the construction of the computer being used but the range is always very large. (The IBM System/360 for example has a range, for real variables and constants, that is roughly $\pm 5.4 \times 10^{-79}$ to $\pm 7.2 \times 10^{75}$).

Exercise 1

1. State the rule for assigning names to FORTRAN variable.
2. Tell whether the following constants are integer or real.

(a) 30.6	(h) -392	(o) -333.3
(b) 6	(i) 63	(p) 0.23
(c) 0.0	(j) 32456	(q) 100
(d) 69.0	(k) 0.666	(r) -6302
(e) 13	(l) 0.723	(s) 19.111
(f) -409	(m) -6969	(t) -45.00
(g) 6834566	(n) 10.00	(u) 1567890

3. Select any appropriate variable name to assign to each of the following constants.

(a) -66.6	(d) -444	(g) 0.0
(b) 432	(e) 100.77	(h) -93
(c) 0	(f) 444269.2	(i) 536.9

4. Each of the following statements contains an error. Identify each error.

(a) INDEX = 22.3	(d) GAMMA = 16
(b) ACAT = 22,369.4	(e) IX = $-1,673$
(c) INTEGER = 2	(f) LAMBDA = -679.

5. Which of the following may name integer variables?

(a) JACK	(e) SQ66	(i) MNS101
(b) MEG	(f) PL401	(j) LPX
(c) SUZIE	(g) MUN86	(k) LZP
(d) KUK	(h) GO69	(l) I93

6. Which columns of FORTRAN coding paper may be used for statements?
7. Which columns on FORTRAN coding paper may be used for statement numbers?
8. Which column on a punched card is used for indicating the continuation of a statement stated on a previous card?
9. Even though FORTRAN is a universal language, this section pointed out two possible areas of machine dependency (the type of computer used imposes some limitation concerning programming aspects). What are these two areas?

10. What is incorrect in the following statements?
(a) PAY = 3.75 ∗ 40
(b) KAY = EYE + 6
(c) LAM = DEL/222
(d) BIRDS = ROOK + FALCON + IBIS + EAGLE

11. If you were writing a FORTRAN program, what variable names would you select to represent the following subjects?
(a) The total number of dollars to revamp the Queen Mary at Long Beach, California.
(b) The total number of transactions each day on the New York Stock Exchange.
(c) The instantaneous speed of a satellite in orbit.
(d) The overall batting average of the Washington Senators baseball team.
(e) The number of albums recorded by the Beatles.
(f) A person's heartbeat per minute.
(g) The amount of light that passes through various types of material.

4.
ARITHMETIC
OPERATIONS

Five arithmetic operations are used in FORTRAN; namely, addition, subtraction, multiplication, division, and exponentiation (that is, raising a number to a power). These five operations combined with the operation of equating (=) relate variables and constants within FORTRAN statements. The symbols for these operations are given in Table 14.3.

TABLE 14.3

Symbol	Operation
+	addition
−	subtraction
∗	multiplication
/	division
∗∗	exponentiation

Some simple FORTRAN arithmetic statements and their meanings follow:

A = B + C	The values assigned to variables B and C are added and the sum is assigned to variable A.
SUBTOT = SUM1 − SUM2	The value of variable SUM2 is subtracted from the value of variable SUM1 and the difference is assigned to variable SUBTOT.
PROD = U1 * U2	The values of variables U1 and U2 are multiplied and the product is assigned to variable PROD.
ANS = YEAR/12.0	The value of variable YEAR is divided by the constant 12.0 and the quotient is assigned to variable ANS.
AREA = SIDE ** 2	The value of variable SIDE is squared and the answer assigned to variable AREA. We could have written this as AREA = SIDE * SIDE.

A single FORTRAN statement may contain many arithmetic operations. For example, we could have the following statement.

$$X = 7.0 * Y ** 4 - Z + 2.0 * A/6.2$$

By combining these constants and variables in different ways, we could obtain several different answers. For example, combining the terms thus

$$X = [7.0 * Y] ** 4 - [Z + 2.0] * A/6.2$$

would give a different answer than combining them

$$X = [7.0 * Y ** 4] - Z + [2.0 * A/6.2]$$

To eliminate this problem, there is an order in which operations are combined within a statement. The rule is: exponentiation first, multiplication and division second, and addition and subtraction last. Let us examine some FORTRAN statements and their order of operation.

437

TOTAL = T1 + T2 + T3 + T4

Variable TOTAL will be assigned the sum of the values of variables T1 + T2 + T3 + T4.

DEGREE = ANG + RAD * 57.29578

The value represented by variable RAD will be multiplied by the constant 57.29578. To this product will be added the value of variable ANG. The results will be assigned to (stored in) variable DEGREE.

FIX = OLDANS − ANS/DEL ** 5

The value of variable DEL will be raised to the fifth power. The value of variable ANS will be divided by the fifth power of DEL and this quotient will be subtracted from the value of variable OLDANS. The result of the subtraction will be assigned to the variable FIX.

Sequences of the same operation are performed from left to right as they appear in the statement. Some examples follow.

Statement	Interpretation
A = A1 + A2 + A3	A = (A1 + A2) + A3
B = B1 − B2 − 6.0	B = (B1 − B2) − 6.0
C = C1 * 80.0 * C3	C = (C1 * 80.0) * C3
D = 77.0/D1/D2	D = (77.0/D1)/D2

Now let us consider the equation

$$X = \frac{y}{a + b}$$

How shall we translate this into a FORTRAN expression? Our first thought might tempt us to write

$$X = Y/A + B$$

which is *wrong*. The FORTRAN compiler would process the division first and the addition second, yielding the mathematical expression

$$X = \frac{y}{a} + b$$

which is not the original equation. One way to reach the correct answer is to use two statements.

SUM = A + B   The arithmetic value of the sum of A and B is saved in SUM.

X = Y/SUM   The value of Y is divided by the value of SUM and the quotient assigned to X.

In order to prevent mathematical expressions from being broken into a series of simple statements, it is possible to group components with parentheses. The parentheses allow operations of lower hierarchy, within their confines, to be treated as a single variable. This allows equations like

$$X = \frac{y}{a + b}$$

to be written in one FORTRAN statement:

X = Y/(A + B)

The parentheses indicate that variables A and B are added first and their sum divides Y. If more than one set of parentheses is used, the expression contained in the innermost set is computed first, then the second set, and so forth, working from inside out. For example:

Z = A/B + C ** 3 $Z = \dfrac{a}{b} + c^3$

Z = A/(B + C) ** 3 $Z = \dfrac{a}{(b + c)^3}$

Z = (A/(B + C)) ** 3 $Z = \left(\dfrac{a}{b + c}\right)^3$

In all statements, the parentheses must balance, that is there must be as many left hand [(] as there are right hand [)].

The operation symbols of the FORTRAN language (+, −, *, /, **) cannot be used side by side. If we wish to multiply Y by negative X, we

439

TABLE 14.4

FORTRAN statement	Mathematical equation
X = A − B + C	$x = a - b + c$
X = A − (B + C)	$x = a - (b + c)$
Z = A * B / C * D	$z = \dfrac{a \cdot b}{c} \cdot d$
Z = (A * B) / (C * D)	$z = \dfrac{a \cdot b}{c \cdot d}$
Y = A * B * C ** 2	$y = a \cdot b \cdot c^2$
Y = (A * B * C) ** 2	$y = (a \cdot b \cdot c)^2$
Z = X * A + B * X + C	$z = ax + bx + c$
Z = (X * A + B) * X + C	$z = ax^2 + bx + c$
X = A + B * A − (C − D) + B / E + C	$x = a + ba - (c - d) + \dfrac{b}{e} + c$
X = A + B * (A − (C − D) + B / (E + C))	$x = a + b\left[a - (c - d) + \dfrac{b}{e + c}\right]$
Z = B / (A + 32.0 / C + D)	$z = \dfrac{b}{\left(a + \dfrac{32}{c} + d\right)}$
Z = B / (A + 32.0 / (C + D))	$z = \dfrac{b}{a + \dfrac{32}{c + d}}$

cannot write Y * − X. Parentheses must be used to separate the operations, thus: Y * (− X).

Table 14.4 gives several FORTRAN statements and their equivalent algebraic equations. Notice the use of parentheses.

The use of spaces between variables, constants, and operation symbols is a matter of individual preference. These spaces have no effect on the meaning of the statements. The three statements below, shown on coding paper, are equivalent.

	STATEMENT NUMBER	Cont.																														FORTRAN
1	2 3 4 5	6	7 8 9 10 11 12 13 14 15 16 17 18 19 20 21 22 23 24 25 26 27 28 29 30 31 32 33 34 35 36 37 38

C FOR COMMENT

Coding:
FOVA = 6.0 * (A + B) / 40.0
FOVA = 6.0 * (A + B) / 40.0
FOVA = 6.0 * (A + B) / 40.0

Exercise 2

1. Write an equivalent FORTRAN statement for each of the following.

(a) $x = a + b(c - d^2)$

(b) $x = a + \dfrac{a \cdot c}{b \cdot d}$

2. Write an equivalent FORTRAN statement for each of the following.

(a) $y = \dfrac{1.0 - \dfrac{a \cdot b}{c \cdot d}}{1.0 + \dfrac{a \cdot b}{c \cdot d}}$

(b) $y = \left[\dfrac{a - (b + c)}{16.0} \right] a \cdot b^2$

3. Write an equivalent FORTRAN statement for each of the following.

(a) $y = \dfrac{a \cdot b}{2.0} + \dfrac{c \cdot d}{2.0}$

(b) $y = \dfrac{a \cdot b}{1.0 + \dfrac{c}{d}}$

4. Write an equivalent FORTRAN statement for each of the following.

(a) $z = \left[\dfrac{(a \cdot b)^x}{(c \cdot d)^y} \right]^2$

(b) $z = \dfrac{a^2}{1.0 + \dfrac{\dfrac{a}{b}}{\dfrac{c}{d}}}$

5. Write an equivalent mathematical equation for each of the following FORTRAN statements. Letters may be substituted for variable names.
   (a) AREA = 3.1415 * R ** 2
   (b) ANS = (A/GEE + B/GEE)
   (c) X2 = ((A * B)/(C * D)) ** 2
6. Write an equivalent mathematical equation for each of the following FORTRAN statements. Letters may be substituted for variable names.
   (a) ADJUST = A − (A/A ** X)
   (b) ANINC = 1.0/(17.0 + (A/B))
   (c) RES = ALPHA * 2.0/EXP * VAL − ALPHA * 2.0/(EXP * VAL)
   (d) TOT = ((A/B * C) ** 2/((B * C)/A) ** 2) ** 4
7. In the following examples, let A = 2.0, B = 3.0, C = 1.0, and D = 4.0. Evaluate the FORTRAN statements for X.
   (a) X = 2.0 * A − (B + C)   (b) X = (A + C)/(B − D)
8. In the following examples, let A = 4.0, B = 6.0, C = 5.0, and D = 2.0. Evaluate the FORTRAN statements for X.
   (a) X = B + D − A ** 2   (b) X = A + C/D − B
9. In the following examples, let A = 10.0, B = 5.0, C = 15.0, and D = 5.0. Evaluate the FORTRAN statements for X.
   (a) X = D/(A/(B + C))   (b) X = ((A + D)/B) ** 3

10. In the following examples, let A = −2.0, B = −7.0, C = −2.0, and D = 3.0. Evaluate the FORTRAN statements for X.

(a) X = ((A + D)/(A − D))/(−A)

(b) X = (D + 5.0 * A)/((C ** 2 + 2.0)/2.0 * B)

5.
INTEGER ARITHMETIC

The arithmetic operations of addition, subtraction, and multiplication performed on integers pose no problem when used in FORTRAN expressions. Table 14.5 contains statements where integers are combined by these three operations.

TABLE 14.5

Statement	Value of I
I = 6 + 9	15
I = 4 + (−7)	−3
I = 32 − 2	30
I = −6 − (−20)	14
I = 7 * 5	35
I = −3 * 6	−18
I = 3 * (6 − 8)	−6

Division of integers, on the other hand, may create a special problem. The division of one integer by another does not always result in an integral

TABLE 14.6

Statement	Value of I
I = 5 / 3 * 6	6
I = 6 * 5 / 3	10
I = 3 / 4 * 2	0
I = 8 / 4 * 12	24
I = 12 / 8 * 4	4
I = 24 / 7 / 2 * 5	5
I = 6 * 9 / 7 * 3	21
I = 3 * 4 / 7 / 2	0

quotient. *When two integers are divided in FORTRAN, the fractional part (the remainder) is truncated (that is, discarded).* For instance, if we divide 6 by 4 the result is 1.5. In FORTRAN, 6/4 would result in a quotient of 1 with the remainder 0.5 discarded. Study the examples of FORTRAN division given in Table 14.6, remembering statements are processed from left to right.

Exercise 3

Determine the value of I in each of the following.

1. $I = 16/4 * 2$
2. $I = 4 * 3/7$
3. $I = 7 + 9/4$
4. $I = (9 + 6)/5$
5. $I = 2 * 6/3 * 8$
6. $I = 2 + 3/4 + 5$
7. $I = (2 * 6)/(3 * 8)$
8. $I = (2 + 3)/(4 + 5)$
9. $I = 8/3 * 5/9$
10. $I = 17/(8 * 2)$
11. $I = 6 * 7/(5 * 9)$
12. $I = 45/8 * (17/(2 * 8))$
13. $I = (16 + 4 * 6 * 3)/20$
14. $I = ((3 + 5) * 7/5)/2 * 3$
15. $I = (6 * 9)/2/2$
16. $I = ((13 - 3) * 6 - 3/2)/2$
17. $I = 7/4 * 6 - 2/4$
18. $I = 2 * 5/6 * 4 * (2 + 3)/7$
19. $I = 9 * 8/3/3$
20. $I = ((9 * 3)/2 - (3 * 2))/2$

6.
TRANSFER STATEMENTS

In the previous sections of this chapter, we studied the composition of FORTRAN statements and the arithmetic operations they represent. We now discuss **transfer statements.** Much of the power in programming is the result of the transfer statements since they allow certain parts of the program to be used and other parts to be skipped. The transfer statements we shall study can be classified into three groups: **unconditional transfers, arithmetic transfers,** and **logical transfers.**

The transfer statements we shall use all refer to a specific statement number. In Section 14.2, we indicated that statement numbers are written and punched in columns 1 through 6 of the coding forms and cards. A statement number is a name for that particular statement and, therefore, is not to be duplicated within the program (if two statements had the same name and you indicated a transfer to that name, the compiling program would indicate an error).

The unconditional transfer:

$$\text{GO TO n}$$

means exactly what it says: go next to execute the statement whose name is n. For example:

> GO TO 6—go next to the statement whose name
> is 6.
>
> GO TO 100—go next to the statement whose name
> is 100.

The arithmetic and logical statements we shall discuss are in the form of **IF statements.** IF statements are **conditional transfers,** that is, they may or may not skip parts of the program. The decision to skip or not to skip depends upon the values or relations of specified variables and/or constants. In arithmetic transfers, decisions are based on an arithmetic quantity being either less than zero, zero, or greater than zero; with logical transfers, decisions are based on a logical quantity being true or false.

The form of the **arithmetic transfer** is

$$\text{IF (a) } n_1, n_2, n_3$$

a an arithmetic expression. It may be a single variable or a series of arithmetic operations that will compute to a single result within the parentheses. The expressions n_1, n_2, and n_3 are statement numbers and must be separated by commas in the IF statement.

The arithmetic IF statement

$$\text{IF (a) } n_1, n_2, n_3$$

means: if the value of the arithmetic expression a is less than 0, execute next the statement named n_1; if the value of a is equal to 0, execute next the statement named n_2; if the value of a is greater than 0, execute next the statement named n_3.

Some examples of arithmetic IF statements and their meanings follow.

Arithmetic IF Statements	Meanings
IF (Z) 10, 20, 30	If Z is less than 0, go to statement 10; if Z is 0, go to statement 20; if Z is greater than 0, go to statement 30.

IF (BAL − WTHDRW) 1000, 500, 20

If BAL − WTHDRW is less than 0, go to statement 1000; if BAL − WTHDRW is 0, go to statement 500; if BAL − WTHDRW is greater than 0, go to statement 20.

IF (INC − IX/5) 100, 200, 300

If INC − IX/5 is less than 0, go to statement 100; if INC − IX/5 is 0, go to statement 200; if INC − IX/5 is greater than 0, go to statement 300.

Either integer or real variables and constants may appear within the parentheses of an arithmetic IF, but the two types should not be mixed.

A logical transfer is in the form

IF (exp) ST

where *exp* is a logical expression. A **logical expression** is an assertion such as, *a* is greater than *b*. It is characterized by the fact that at any given time it can have only two possible values: *true* or *false*. ST will either be an unconditional "GO TO n" or an arithmetic statement.

The logical transfer

IF (exp) ST

means: if the logical expression *exp* is true, then execute the statement ST and proceed from there; if the logical expression *exp* is false, do not execute ST, simply proceed to the next statement.

TABLE 14.7

Relational operator	Mathematical symbol	Definition
.GT.	$>$	Greater than
.GE.	$\geq$	Greater than or equal to
.EQ.	$=$	Equal to
.NE.	$\neq$	Not equal to
.LE.	$\leq$	Less than or equal to
.LT.	$<$	Less than

The expression *exp* can relate two values by using one of six relational operators. The six **relational operators** used in logical transfer statements are given in Table 14.7.

Periods on either side of the relational operators are used to separate them from the variable names or constants. Study the following examples. Notice that care must be taken not to mix real and integer expressions within parentheses.

Logical IF Statements   Meanings

```
C FOR COMMENT
STATEMENT NUMBER | Cont. | FORTRAN
1  2 3 4 5 | 6 | 7 8 9 10 11 12 13 14 15 16 17 18 19 20 21 22 23 24 25 26 27 28 29 30 31 32 33 34 35 36 37 38
           |   | IF(A.GT.12.0)B=90.0-A
```

If A is greater than 12.0, then execute
B = 90.0 − A; If A is less than or equal to 12.0
then go to the next statement.

```
C FOR COMMENT
STATEMENT NUMBER | Cont. | FORTRAN
1  2 3 4 5 | 6 | 7 8 9 10 11 12 13 14 15 16 17 18 19 20 21 22 23 24 25 26 27 28 29 30 31 32 33 34 35 36 37 38
           |   | IF(I.EQ.20)GOTO300
```

If I is equal to 20, then execute the statement
named 300; if I is not equal to 20, then go to
the next statement.

```
C FOR COMMENT
STATEMENT NUMBER | Cont. | FORTRAN
1  2 3 4 5 | 6 | 7 8 9 10 11 12 13 14 15 16 17 18 19 20 21 22 23 24 25 26 27 28 29 30 31 32 33 34 35 36 37 38
           |   | IF(I.NE.1)M=N**2
```

If I is not equal to 1, then execute the statement
M = N ** 2; if I is equal to 1, then go to the
next statement.

C FOR COMMENT

STATEMENT NUMBER	Cont.	FORTRAN
1	2 3 4 5	6	7 8 9 10 11 12 13 14 15 16 17 18 19 20 21 22 23 24 25 26 27 28 29 30 31 32 33 34 35 36 37 38

```
      IF(GEE.GT.32.2)GOTO50
      MACH=0
      GOTO100
   50 MACH=1
```

If GEE is greater than 32.2, then go to statement 50 where MACH is set equal to 1; if GEE is less than or equal to 32.2, then execute the next statement where MACH is set equal to 0 and then go to statement 100.

Exercise 4

In the following examples, determine what will happen under the conditions given (Exercises 1–10).

1.   IF $(2 - 3)$ 100, 200, 300
2.   IF $(7.0 * X)$ 2, 4, 6 $X = 0.0$
3.   IF $(N .GT. M)$ X = 70.0 $N = 2, M = 0$
4.   IF $(ZEE * 3.0 .GE. 35.0)$ GO TO 20 $ZEE = 15.0$
5.   IF $(ICOUNT .LE. 20)$ GO TO 40 $ICOUNT = 21$
     ICOUNT = ICOUNT $- 1$
6.   IF (IND) 10, 20, 30 $IND = 2$
  10 X = Y $Y = 10.0$
     GO TO 40
  20 X = 2.0 * Y
     GO TO 40
  30 X = Y ** 2
7.   IF $(W .LE. 0.0)$ GO TO 20 $W = -32.0$
     IPOS = 1
     GO TO 30
  20 INEG = -1
8.   IF $(X * Y ** 2 - 72.0)$ 15, 40, 25 $X = 2.0$
  15 X = X $+ 1.0$ $Y = 6.0$
     GO TO 40
  20 X = X $- 1.0$
     GO TO 45
9.   IF $(X - Z * (16.0 - Y) .LT. 25.0)$ X = X ** 2 $X = 15.2$
                                        $Y = 5.5$
                                        $Z = 1.5$

10.	IF (7 .GT. IX) GO TO 40		IX = 7
	CALC = FER ** 3		FER = 5.0
	GO TO 30		Z = 2
40	CALC = FER ** Z		
30	. . .		

Write statements without logical IF's (use arithmetic IF's) to duplicate the following (Exercises 11–15).

11. IF (SUDS .LE. BIG) SOAP = X

12. IF (NIGHT .EQ. 28) GO TO 10
     MOON = IOFF
     GO TO 20
  10 MOON = ION
  20 . . .

13. X = 65.0
     IF (IF .GE. −5) GO TO 5
   3 X = X − 2.0
     GO TO 7
   5 X = X + 2.0
   7 SUM = SUM + X

14. IF (ME .LT. 0) M = 0
     M = M + 1

15. IF (HRS .GT. 40.0) GO TO 75
     PAY = RATE * 40.0
     GO TO 100
  75 PAY = RATE * 40.0 + (HRS − 40.0) * RATE * 1.5
 100 . . .

7.
WRITING PROGRAMS

With the addition of four more statements, READ, WRITE, STOP, and END, we have enough tools to write programs. The **READ** and **WRITE** statements are for input and output, respectively. **STOP** signifies the termination of execution for this particular program—a kind of "stop the computer, I want to get off." **END** is physically the last statement of all FORTRAN programs. A complete explanation of the READ and WRITE statements are out of the scope of this book. We introduce them only to give the reader the idea that data must be brought in for processing and answers must be written out. The form we shall use is

READ I, X, Z

meaning: numerical values are read from input and assigned to the variables named I, X, and Z in the program; and

WRITE I, X, Z, Y

meaning the numerical values represented by the names I, X, Z, and Y, in the program, are written on some output device. In both the READ and WRITE statements, variables must be separated by commas.

When beginning a new program it is often helpful to diagram the actual steps to be taken. This diagram is called a **flow chart.** A flow chart is a pictorial representation of the computations and decisions to be made. The results of the decisions determine the flow to the next operation. It is customary to select certain shapes to represent particular functions. We shall use the shapes defined in Figure 14.3.

Suppose an artist mixes red and yellow oils together to produce orange. His actions might be described by the flow chart in Figure 14.4.

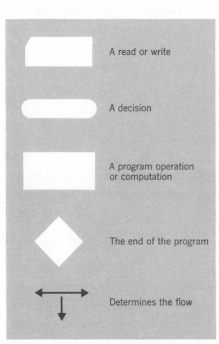

FIGURE 14.3

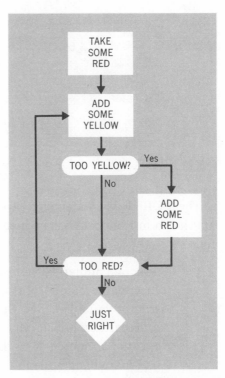

FIGURE 14.4

Figure 14.5 is a flow chart of a program written to square each number in a set of numbers, sum the squares, then print the sum.

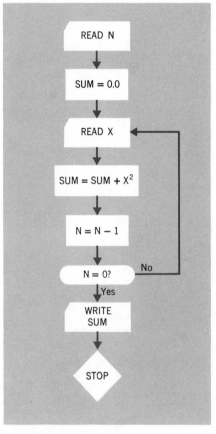

N is the number of numbers to be squared

Set a cell equal to zero

X is a number to be squared

Add to sum the new value of X squared.

Reduce the counter by one

N will equal zero when all numbers have been read and squared

Print the sum of squares

FIGURE 14.5

The FORTRAN statements of this program, written on coding paper are shown in Figure 14.6. The meaning of each statement in the program follows.

1. The actual numerical value of N is read (called **input**) and assigned to a variable named N (the number of numbers to be squared).
2. Assign a variable called SUM the value of zero.
3. An actual numerical value for X is read (input) and assigned to the variable named X in the program.
4. X is squared and added to variable SUM. The new value replaces the old value of SUM.

```
─── C FOR COMMENT
┌─┬─┬────────────────────────────────────────────────────────
│STATEMENT│Cont.│           FORTRAN STATEMENT
│ NUMBER  │  │
│1      5 │6 │7   10      15      20      25      30      35
├─────────┼──┼────────────────────────────────────────────────
│         │  │READ N
├─────────┼──┼────────────────────────────────────────────────
│         │  │SUM = 0.0
├─────────┼──┼────────────────────────────────────────────────
│    10   │  │READ X
├─────────┼──┼────────────────────────────────────────────────
│         │  │SUM = SUM + X**2
├─────────┼──┼────────────────────────────────────────────────
│         │  │N = N-1
├─────────┼──┼────────────────────────────────────────────────
│         │  │IF(N.NE.0)GOTO10
├─────────┼──┼────────────────────────────────────────────────
│         │  │WRITE SUM
├─────────┼──┼────────────────────────────────────────────────
│         │  │STOP
├─────────┼──┼────────────────────────────────────────────────
│         │  │END
└─────────┴──┴────────────────────────────────────────────────
```

FIGURE 14.6

5. One is subtracted from N. This new value replaces the old value of N.
6. If N is not equal to zero, go back to statement 10 to read another value of X.
7. Print the numerical value of SUM (this is called output).
8. STOP—Stops processing.
9. END—Physically the last statement.

N, SUM, and X are names for three variables. Each is assigned to unique core locations where the computer stores their present numerical values. Suppose $N = 5$ and the five X's were 2.0, 4.0, 3.0, 5.0, 10.0. If we were to examine N, SUM and X in core just before the IF test on N was processed, this is what we would find:

		Values of:	
PASS	N	SUM	X
1	4	4.0	2.0
2	3	20.0	4.0
3	2	29.0	3.0
4	1	54.0	5.0
5	0	154.0	10.0

451

In every case, each new X is read into the same core location as the previous X. Note that SUM was set equal to zero before the addition began. The reason for this is that we have no idea what may have been left in that core location by some other program.

Statements such as SUM = SUM + X ** 2 and N = N − 1 cause the value of the word referenced to be modified by the operation indicated. The modified value of the variable replaces the original value in core storage.

The formula for the area of a triangle is $b \cdot h/2$ where b is the measure of the base and h is the measure of the height. Some arbitrary values for b and h follow:

n	b	h
1	7.0	7.0
2	4.0	6.0
3	2.0	5.0
4	6.4	2.3
5	3.5	6.4

Figure 14.7 shows a program to find the area of a triangle. Included are the answers obtained using the above values.

The statements in this program mean:

The actual numerical value of N is read as input and assigned to variable N (the number of areas to be computed).

Variable NUM is given a zero value.

The value of the variable NUM is increased by 1; this is used as a counter.

		NUM	B	H	AREA
	READ N	1	7.0	7.0	24.50
	NUM=0	2	4.0	6.0	12.00
10	NUM=NUM+1	3	2.0	5.0	5.00
	READ B,H	4	6.4	2.3	7.36
	AREA=B*H/2.0	5	3.5	6.4	11.20
	WRITE NUM,B,H,AREA				
	IF(NUM.LT.N)GOTO10				
	STOP				
	END				

FIGURE 14.7

The actual numerical values of B and H are read as input and assigned to variables B and H respectively (the base and height of a triangle).

B is multiplied by H, the product is divided by 2.0, and the result is saved in variable AREA.

Print as output the numerical values of the variables; NUM (the counter), B (the base of the triangle), H (the height of the triangle), and AREA (the area of the triangle).

Is the value of NUM less than N? If it is we have not completed the required number of iterations and we proceed to statement number 10.

If NUM is equal to or greater than N then the process is completed and the program executes STOP.

The arithmetic average, or mean,† of a set of numbers is computed by adding all the numbers of the set and dividing by the number of elements in the set. The following is a FORTRAN program to find the mean of a set on n numbers.

```
        C FOR COMMENT
STATEMENT                                                          FORTRAN
NUMBER
1  2  3  4  5  6  7 8 9 10 11 12 13 14 15 16 17 18 19 20 21 22 23 24 25 26 27 28 29 30 31 32 33 34 35 36 37 38
              READ  N
              SUM=0.0
              I=0
          1   READ  S
              SUM=SUM+S
              I=I+1
              IF (I.NE.N)GOTO1
              RN=N
              AVE=SUM/RN
              WRITE  AVE
              STOP
              END
```

The statements mean:

1. Read N the number of numbers to be averaged.
2. Set SUM, a cell used for totaling, equal to zero.
3. Set I, a counter, equal to zero.

† See Chapter 8, Section 9.

4. Read S, a sample of the set.

5. Add the new member of the set to the total.

6. Increment the counter.

7. If we have not read N samples, go back for another.

8. Convert the integer N† to a real number.

9. Compute the average.

10. Print the answer.

Exercise 5

1. Draw a flow chart for the sample problem in Section 7 which found the areas of triangles.

2. Draw a flow chart for the sample problem to find the average of a set of *n* numbers.

3. Draw a flow chart showing the steps needed to find the greatest number in a set of *n* positive numbers.

4. Write a program for the exercise in problem 3.

5. Draw a flow chart showing the steps needed to compute and compare the mean of a set of *n* positive numbers with the mean of the greatest and least values in the set. All numbers are less than 100.0.

6. Write a program for the exercise in problem 5.

7. What is being computed in the following?

PI = 3.146
READ R
AREA = PI * R ** 2
WRITE AREA
STOP
END

8. What would be the values of A and I in the following cases if N = 4 and B = 23.6?

(a) A = N (b) I = B

9. Draw a flow chart and then write the program to find the mean and standard deviation†† of a set of five numbers. For the square roots needed, use an exponent of 0.5††† that is SQRA = A**.5.

---

† In Section 3 of this chapter we indicated that mixing integer and real variables in a statement could be in error. FORTRAN will automatically convert the expression on the right of the equal sign to that type of variable indicated on the left of the equal sign.
†† See Chapter 8 Section 13 for the formula for the standard deviation.
††† $\sqrt{a} = a^{1/2} = a^{0.5}$.

8.
DIMENSION STATEMENTS AND SUBSCRIPTS

As the compiling program scans the FORTRAN statements, each new and unique variable name is assigned a particular word in core storage. That is, one unique location in memory is reserved for each different name. Each of the constants and variables we have studied thus far have been assigned to a single word. In the statement

$$SUM = X + 12.0$$

the numerical values which SUM, X and 12.0 represent are each in one separate word of core.

More than one core location can be assigned to a variable name by using a **DIMENSION statement.** These statements are written in the form

DIMENSION declaration

The declaration tells the number of variables to which more than one core location is to be allocated. The variable name is followed by the number of locations to be assigned. The number is always enclosed in parentheses. For example, in the statement

DIMENSION A(20)

variable A is assigned twenty words. The twenty words are contiguous in core and are called an **array.** Since variable A is an array of twenty words, we must have a means of referring to each particular element of the array. If we wish to refer to the second element of array A, we would write A(2). The (2) is called the **subscript.** The tenth word of the array is A(10), the first word A(1), and so forth. Subscripts denote a particular element of an array and are always enclosed in parentheses.

If a dimension is assigned to more than one variable, their names must be separated by commas in the DIMENSION statement. For example,

DIMENSION P(3), Q(4)

P(1)	P(2)	P(3)	Q(1)	Q(2)	Q(3)	Q(4)

FIGURE 14.8

means variable P has three words, P(1), P(2), P(3), assigned to it, and variable Q has four words, Q(1), Q(2), Q(3), Q(4), assigned to it. Figure 14.8 is a pictorial representation of the arrays P and Q in core.

The value represented by Q(1) occupies the core location immediately following the value represented by P(3). Care must be taken that subscripts do not refer to elements outside their own array. For example, if we were to refer to P(5) we actually would be referencing Q(2).

Each of the dimensional variables above have one subscript and are called **singularly subscripted variables.** In FORTRAN, variables may have more than one subscript. The maximum number of subscripts that a variable may have depends on the computer used, but it is never less than three. We shall limit our variables to two subscripts, called **doubly subscripted variables.**

The number of core locations to be allotted to doubly subscripted variables is also defined in a DIMENSION statement. For example, the statement

DIMENSION X(2, 2), A(4)

means that X(2, 2) is a $2 \times 2 = 4$ element array. The order in which they appear in core is shown in Figure 14.9.

X(1,1)	X(2,1)	X(1,2)	X(2,2)	A(1)	A(2)	A(3)	A(4)

FIGURE 14.9

The DIMENSION statement, if needed, is the *first* statement of a FORTRAN program. This establishes for the compiling program the sizes and types of arrays so it can alot core storage and check for proper variable use.

Exercise 6

1. Write the DIMENSION statements for the following.
   (a) Array A has six elements.
       Array B has four elements.
       Array C has ten elements.

(b) Array X has seventeen elements.

Array I has twenty elements.

(c) Array P has ten elements.

Array R is a 2 × 2 array.

Array S is a 3 × 3 array.

2. Draw a figure representing the core layout of a three-element array Y, followed by a 2 × 2 array Z.

3. How many core locations would be needed for the doubly subscripted array TEM (10, 20)?

4. Draw a figure representing the arrangement of the elements of M(3, 2) in core.

5. Using DIMENSION AF(30), ZEE(10, 5), M(4, 6).

(a) What is the total number of cells used by the three arrays?

(b) What variable name immediately follows AF(30)?

(c) What variable name immediately follows ZEE(6, 2)?

(d) What variable name immediately precedes ZEE(1, 4)?

(e) What variable name is contained in the tenth cell after ZEE(5, 5)?

6. Write the names of the first, fifth, and eleventh elements of array Z(60).

7. Write the names of the third, twenty-fifth, and thirty-second elements of array HIPPY(20, 3).

8. Name the subscripts of the following variables.

(a) SUZIE(7)   (b) GAI(2, 5)   (c) MEG(1)   (d) ATLAS(13)

9. How many core locations are needed for the statement.

DIMENSION JOHN(7, 11), DOGG(102), ETT(60)

9.
THE DO STATEMENT

Probably the statement that saves the most work in all of FORTRAN is the **DO statement.** DO statements cause iterations to be performed. The form of the DO statement is

$$\text{DO ST IX} = I_1, I_2$$

ST is a statement number (columns 1 through 5). The DO statement causes all statements between it and statement ST, including ST, to be executed repeatedly. The number of times each statement is repeated is controlled by IX, I_1, and I_2. IX is a counter or **index.** Starting at its initial value of I_1, IX is increased by one each iteration until it reaches the value I_2, where the repetition is terminated. IX must be a name of a nonsubscripted integer variable. I_1 and I_2 can be either names for nonsubscripted integer variables or integer constants. Neither I_1 nor I_2 can assume a zero or negative value. Notice a comma separates the limiting names.

One DO statement is

$$\text{DO } 25 \text{ I} = 1, 3$$

This DO statement causes all the subsequent statements up to and including statement 25 to be repeated three times. In the three repetitions I is assigned the values 1, 2, and 3 respectively.

In the statement

$$\text{DO } 60 \text{ N} = 1, \text{M}$$

all statements up to and including 60 will be repeated M times, M having been equated previously to a positive integer greater than zero. In each iteration N is assigned the number indicating the iteration.

In the statement

$$\text{DO } 100 \text{ I} = \text{J}, \text{K}$$

all statements up to and including 100 will be executed $K - J + 1$ times. Both J and K have been assigned positive integer values greater than zero in previous statements. In the first iteration I is assigned the value of J. I is incremented by 1 on each subsequent iteration until the iteration in which $I = K$ is completed. Then the repetition is terminated and the program continues by executing the statement immediately after 100. In this case $J \leq K$.

We shall always use the statement CONTINUE as the last statement in the DO range. The continue statement will have the statement number ST. An example is:

$$\text{DO } 5 \text{ I} = 1, 3$$
$$\vdots$$
$$5 \text{ CONTINUE}$$

All statements between and including the DO and CONTINUE statements comprise what is called a **DO loop.** The word *loop* is used because the program loops through the indicated statements until the value of the index has reached the maximum value.

The index (IX) of a DO loop is used to refer to elements in an array when the array is referenced in statements within the DO's range. Suppose we wish to set all elements of an array named MEG equal to zero (MEG is an array of fifteen words). This is done in the following manner.

```
DIMENSION MEG (15)
DO 10 I = 1, 15
MEG (I) = 0
10 CONTINUE
```

The dimension statement causes fifteen core locations to be assigned to the name MEG. In the first iteration, $I = 1$, and $MEG(1) = 0$; in the second iteration, $I = 2$, and $MEG(2) = 0$; and so forth. This process continues until the last iteration, where $I = 15$, is reached and $MEG(15) = 0$.

Suppose we wish to store the first one hundred positive integers in an array INTGER(200). At the same time we wish to store in array INTSUM(200) the first one hundred sums of the positive integers—that is, cell INTSUM(5) equals the sum of $1 + 2 + 3 + 4 + 5$, cell INTSUM(N) equals the sum of $1 + 2 + 3 + \cdots + N$, and so on.

This may be written:

```
DIMENSION INTGER(200), INTSUM(200)
ISUM = 0
DO 20 N = 1, 100
INTGER(N) = N
ISUM = ISUM + N
INTSUM(N) = ISUM
20 CONTINUE
```

In the example above, the value of the index is used as a subscript as well as integer variable.

A statement that transfers program control to a statement within a DO loop is illegal. The following is an example of an illegal transfer into the range of a DO loop.

```
DIMENSION A (30), I (2)
IF (I (1)) 5, 5, 10
5 DO 20 J = 1, 25
FJ = J
10 A (J) = FJ
20 CONTINUE
```

In this example, the transfer to statement 10 by the positive branch of the arithmetic IF statement is illegal. It transfers into the DO loop. This error would be caught by the compiling program.

An example of a program that finds the arithmetic mean and standard deviation† of *n* numbers stored in an array named S is given below. Array S has been given the dimension 200 which imposes the condition $n \leq 200$. A cell named N contains the value of *n*.

```
DIMENSION S(200)
SUM = 0.0
DO 20 L = 1, N
SUM = SUM + S(L)
20 CONTINUE
FN = N
AM = SUM/FN
SUMSQ = 0.0
DO 40 L = 1, N
SUMSQ = SUMSQ + (S(L) − AM) ** 2
40 CONTINUE
SD = (SUMSQ/FN) ** .5††
```

The variables AM and SD contain the arithmetic mean and standard deviation respectively.

Exercise 7

1. Give the meaning of each FORTRAN statement in the program for finding the arithmetic mean and standard deviation of a set of *n* numbers.

In the following problems (2–8), choose a compatible dimension size and variable names if these are not given.

2. Write the statements needed to sum the first sixteen numbers in an array named FP. Assign a total of fifty cells to array FP. Assign the sum to variable SUM.

3. Write the statements needed to sum the fifth to the ninth numbers in array NUM.

4. Write the statements necessary to find the largest number of a set of forty numbers.

† See Chapter 8, Section 13 for the formula for the standard deviation.
†† $\sqrt{a} = a^{1/2} = a^{0.5}$.

5. Write the statements needed to find the arithmetic mean of a set of numbers.

6. Write the statements necessary to find how many numbers in array FIX(30) have a numerical value less than 100.0.

7. Write the statements necessary to find the maximum and minimum values in an array of n numbers.

8. Write the statements needed to find the arithmetic means of both the negative and positive integers contains in an array N(300). No value is equal to zero.

9. Determine what problem is being solved in the following.

    (a)   DIMENSION FAT(250)

```
          FMIN = 0.0
          DO 20 I = 1, 250
          IF (FAT(I) .GE. FMIN) GO TO 20
          FMIN = FAT(I)
       20 CONTINUE
```

    (b)   DIMENSION SAMP(55)

```
          SUM = 0.0
          SUMSQ = 0.0
          DO 40 K = 1, 55
          SUM = SUM + SAMP(K)
          SUMSQ = SUMSQ + SAMP(K) ** 2
       40 CONTINUE
```

ANSWERS TO ODD-NUMBERED PROBLEMS

Chapter 1

Exercise 1

1. (a) True; (b) Not a statement; (c) False; (d) False; (e) True; (f) True; (g) Not a statement; (h) False; (i) Not a statement.
3. (a) Some girls are not popular; (b) $\sqrt{2}$ is a real number; (c) All career women are beautiful.
5. (a) Meg has two ears; (b) No freshmen are handsome men; (c) Some college professors are not bores; or Some college professors are not excellent talkers.
7. Invalid.
9. (a) Invalid; (b) Invalid.

Exercise 2

1. (a) Conjunction; (b) Implication; (c) Conjunction; (d) Equivalence; (e) Disjunction; (f) Implication; (g) Equivalence; (h) Implication; (i) Implication; (j) Implication.
3. (a) If Mexico is south of the U.S.A., then Canada is a democracy; (b) Mexico is south of the U.S.A. if and only if Canada is a democracy; (c) Mexico is not south of the U.S.A.; (d) If Mexico is south of the U.S.A., then Canada is not a democracy; (e) If Mexico is not south of the U.S.A., then Canada is not a democracy; (f) If Canada is not a democracy, then Mexico is south of the U.S.A.
5. (a) It is not true that it is foggy today, or It is not foggy today; (b) It is false that the waves are six feet high, or The waves are not six feet high; (c) It is false that a circle has a radius; (d) It is not true that an even number is divisible by 2; (e) It is not true that people work hard for a living; (f) It is false that Yorkshire terriers are intelligent dogs.
7. (a) False; (b) True; (c) True; (d) True.
9. (a); (c); (d); (e).
11. (a) Property values are high and taxes are rising; (b) Property values are high and it is not true that taxes are rising; (c) If property values are high then taxes are rising; (d) It is false that property values are high and taxes are rising; (e) It is false that property values are high or taxes are rising; (f) It is false that property values are not high and taxes are rising.
13. (a) Nellie is lucky and Joan is not intelligent; (b) Nellie is not lucky or Joan is intelligent and Betty is stuffy; (c) Nellie is lucky or Joan is not intelligent and Betty is stuffy; (d) It is false that Nellie is lucky or Joan is intelligent and Betty is stuffy.

ANSWERS TO ODD-NUMBERED PROBLEMS

Exercise 3

1.

p	q	$\sim(p \vee q)$	$q \vee p$	$\sim(p \vee q) \vee [(q \vee p)]$
T	T	F	T	T
T	F	F	T	T
F	T	F	T	T
F	F	T	F	T

3.

p	$\sim p$	$p \wedge (\sim p)$
T	F	F
F	T	F

5.

p	q	$\sim p$	$p \rightarrow q$	$p \rightarrow (\sim p)$	$(p \rightarrow q) \vee [p \rightarrow (\sim p)]$
T	T	F	T	F	T
T	F	F	F	F	F
F	T	T	T	T	T
F	F	T	T	T	T

7.

p	q	$p \rightarrow q$	$\sim[p \wedge (\sim q)]$	$(p \rightarrow q) \rightarrow \sim[p \wedge (\sim q)]$
T	T	T	T	T
T	F	F	F	T
F	T	T	T	T
F	F	T	T	T

9.

p	q	$p \wedge q$	$(p \wedge q) \rightarrow p$
T	T	T	T
T	F	F	T
F	T	F	T
F	F	F	T

11.

p	q	$(\sim p)$	$p \wedge (\sim p)$	$[p \wedge (\sim p)] \rightarrow q$
T	T	F	F	T
T	F	F	F	T
F	T	T	F	T
F	F	T	F	T

13.

p	q	$\sim(p \wedge q)$	$(\sim p) \vee (\sim q)$
T	T	F	F
T	F	T	T
F	T	T	T
F	F	T	T

15.

p	q	$p \rightarrow (\sim q)$	$(\sim p) \vee (\sim q)$
T	T	F	F
T	F	T	T
F	T	T	T
F	F	T	T

17.

p	q	$p \rightarrow q$	$(\sim p) \vee q$
T	T	T	T
T	F	F	F
F	T	T	T
F	F	T	T

19. True

21. p $p \wedge p$
  T   T
  F   F

23. p $\sim p$ $\sim(\sim p)$
  T   F   T
  F   T   F

Exercise 4

1. John does not play in the orchestra or he does not play in the marching band.
3. Nellie picks out the date for the picnic and it is not true that the sun will surely shine.
5. My raise will not come through and the company receives a large defense contract; or my raise will come through and it is false that the company receives a large defense contract.
7. He wins this lawsuit and he will not be an excellent candidate for attorney general.
9. You have the winning number and can produce your ticket and you cannot collect the prize money.

Exercise 5

1. Valid.
3. Valid.
5. Invalid.
7. Valid.
9. Invalid.
11. Valid.
13. Invalid.
15. Invalid.
17. (a) X was at the scene of the crime; (b) ABCD is a rectangle.
19. (a) n is an integer; (b) $x > z$.
21. (a) If I have a 3.5 grade-point average, then I shall graduate with honors; (b) All mumbos are jumbos.

Exercise 6

1. Since p and q are both true $p \vee q$ and $p \wedge q$ are true. Since $(p \vee q) \rightarrow (r \wedge q)$ and $(p \vee q)$ are both true, $(r \wedge q)$ is true by the law of detachment. Since $r \wedge q$

is true r is true by the definition of a conjunction.

3. Since p and q are both true, $p \lor q$ is true. Since $(p \lor q)$ is true and $(p \lor q) \to r$ is true, r is true by the law of detachment.

5. Use the following representation:

p: This is a good course.
q: This course is worth taking.
r: Math is easy.

The theorem states:

$$p \to q$$
$$r \lor (\sim q)$$
$$\sim r$$
$$\overline{\qquad\qquad}$$
$$\therefore \sim p$$

Proof: Since $\sim r$ is true, r is false. Since r is false and $r \lor (\sim q)$ is true, $(\sim q)$ is true by the definition of a disjunction. Since $p \to q$ is true, its contrapositive $(\sim q) \to (\sim p)$ is true. Since $(\sim q) \to (\sim p)$ and $(\sim q)$ are true, $(\sim p)$ is true by the law of detachment.

7. Use the following representation:

p: Carl is elected class president.
q: Bill is elected vice-president.
r: Betty is elected secretary.

The theorem states:

$$p \to q$$
$$q \to (\sim r)$$
$$\overline{\qquad\qquad}$$
$$\therefore r \to (\sim p)$$

Proof: Since $q \to (\sim r)$ is true, its contrapositive $r \to (\sim q)$ is true. Since $p \to q$ is true, its contrapositive $(\sim q) \to (\sim p)$ is true. Since $r \to (\sim q)$ and $(\sim q) \to (\sim p)$ are true, $r \to (\sim p)$ is true by the law of syllogisms.

9. Use the following representation:

p: Vincent water skiis.
r: Blanche is in town.
q: It is summer.

The theorem states:

$$p \to q$$
$$r \to p$$
$$\sim q$$
$$\overline{\qquad\qquad}$$
$$\therefore \sim r$$

Proof: Since $r \rightarrow p$ and $p \rightarrow q$ are true, $r \rightarrow q$ is true by the law of syllogisms. Since $r \rightarrow q$ is true, its contrapositive $(\sim q) \rightarrow (\sim r)$ is true. Since $(\sim q) \rightarrow (\sim r)$ and $\sim q$ are true, $\sim r$ is true by the law of detachment.

11. Use the following representation:

p: John is a thief.
q: Newton is a shoplifter.
r: Carl is guilty of car theft.

The theorem states:

$$p$$
$$q$$
$$(\sim p) \rightarrow r$$
$$\underline{q \rightarrow (\sim r)}$$
$$\therefore \sim r$$

Proof: Since $q \rightarrow (\sim r)$ and q are true, $\sim r$ is true by the law of detachment.

13. Use the following representation:

p: This class is a bore.
q: The instructor is interesting.
r: The subject is worthwhile.

The theorem states:

$$p \rightarrow (\sim q)$$
$$(\sim p) \rightarrow r$$
$$\underline{q}$$
$$\therefore r$$

Proof: Since $p \rightarrow (\sim q)$ is true, its contrapositive $q \rightarrow (\sim p)$ is true. Since $q \rightarrow (\sim p)$ and $(\sim p) \rightarrow r$ are true, $q \rightarrow r$ is true by the law of syllogism. Since $q \rightarrow r$ and q are true, r is true by the law of detachment.

Chapter 2

Exercise 1

1. (a) $(2)(24)$; (b) $(7)(8)$; (c) $(-3)(-27)$; (d) $(-5)(-24)$; (e) $(4)(81)$; (f) $(-25)(-29)$; (g) $(16)(43)$; (h) $(-32)(-48)$.

3. (a) 1; 2; 4; 3; 6; 12; (b) 1; 2; 4; 7; 8; 14; 28; 56; (c) 1; 2; 3; 4; 6; 8; 12; 16; 24; 32; 48; 96; (d) 1; 2; 3; 4; 6; 8; 9; 12; 16; 18; 24; 24; 36; 48; 72; 144; (e) 1; 2; 3; 4; 5; 6; 9; 10; 12; 15; 18; 20; 30; 36; 45; 60; 90; 180; (f) 1; 2; 5; 4; 10; 11; 20; 22; 44; 55; 110; 220; (g) 1; 2; 3; 5; 6; 9; 10; 15; 18; 27; 30; 45; 54; 90; 135; 270; (h) 1; 2; 4; 5; 8; 10; 16; 20; 32; 40; 80; 160.

5. (a) ± 1; ± 2; (b) ± 1; ± 3; (c) ± 11; ± 1; (d) ± 1; ± 23; (e) ± 1; ± 41;
(f) ± 1; ± 53; (g) ± 1; ± 83; (h) ± 1; ± 101.
7. (1)(144); (2)(72); (3)(48); (4)(36); (6)(24); (8)(18); (9)(16); (12)(12).
9. (a) True; (b) True; (c) True; (d) False; (e) True; (f) False; (g) True;
(h) True.
11. (a) (3)(2); (b) (3)(10) + 2; (c) (3)(4) + 2; (d) (3)(9); (e) (3)(6) + 2;
(f) (3)(11) + 2; (g) (3)(−20) + 2; (h) (3)(13).
13. (a) Commutative property of addition; (b) Additive identity axiom;
(c) Distributive property; (d) Associative property of multiplication; (e) Multiplicative identity axiom; (f) Commutative property of addition; (g) Multiplication property of zero.
15. (a) -7; (b) 3; (c) -12; (d) 302; (e) $-x$; (f) y or $-(-y)$; (g) $-(x + y)$
or $-x - y$; (h) $-[-(x + y)]$ or $x + y$.

Exercise 2

1. (a) $26 = (8)(3) + 2$; (b) $39 = (7)(5) = 4$; (c) $126 = (15)(8) + 6$; (d) $256 = (27)(9) + 13$; (e) $369 = (21)(17) + 12$; (f) $1274 = (97)(13) + 13$; (g) $8 = (12)(0) + 8$.

3. Let $2k$ and $2h$ represent the even numbers. Then

$$(2k)(2h) = 2(2kh) \qquad \text{Commutative and associative properties}$$
$$\text{of multiplication}$$
$$= 2M \qquad 2kh \text{ is an integer, call it } M$$

Since $2M$ is even, the theorem is proved.
5. Let $2k + 1$ and $2h + 1$ represent the odd numbers. Then

$$(2k + 1) + (2h + 1) = 2k + 2h + 2 \qquad \text{Commutative and associative properties}$$
$$\text{of multiplication and } 1 + 1 = 2$$
$$= 2(k + h + 1) \qquad \text{Distributive property}$$
$$= 2M \qquad k + h + 1 \text{ is an integer, call it } M$$

Since $2M$ is even the theorem is proved.
7. Since 3 divides a and 3 divides b, we can write $a = 3k$ and $b = 3h$, h and k integers, by the definition of divisibility. Then

$$a + b = 3k + 3h$$
$$= 3(k + h) \qquad \text{Distributive property}$$
$$= 3M \qquad k + h \text{ is an integer, call it } M$$

Then 3 divides $a + b$ by the definition of divisibility.
9. (a) $6k + 1$; $6k + 3$; $6k + 5$; (b) $6k$; $6k + 2$; $6k + 4$; (c) $6k$; $6k + 3$.

Exercise 3

1. Since a divides b, we can write $b = ak$, k an integer, by the definition of divisibility. Then

$$
\begin{array}{ll}
a + b = c & \text{Given} \\
a + ak = c & \text{Law of substitution} \\
a(1 + k) = c & \text{Distributive property}
\end{array}
$$

Then a divides c by the definition of divisibility.

3. It is given that d divides 1. The only integral factors of 1 are $+1$ and -1. Since d is greater than zero, $d = 1$.

5. d divides a, hence for some integer k

$$ a = kd $$

But $d = (-1)(-d)$, hence

$$
\begin{array}{ll}
a = [(-1)(-d)]k & \text{Associative and commutative properties} \\
\ = (-d)[(-1)(k)] & \text{of multiplication}
\end{array}
$$

$$ = (-d)(-k) \qquad (-1)(k) = -k $$

and $-d$ divides a by the definition of divisibility.

7. Let n, $n + 1$, and $n + 2$ represent three consecutive integers. Then

$$
\begin{aligned}
n + (n + 1) + (n + 2) &= 3n + 3 \\
&= 3(n + 1) \\
&= 3M
\end{aligned}
$$

and 3 divides $n + (n + 1) + (n + 2)$, the sum of three consecutive integers.

9. Let $2k$ and $2h$ represent the two even integers. Then

$$
\begin{aligned}
2k + 2h &= 2(k + h) \\
&= 2M
\end{aligned}
$$

Since $2M$ is even the theorem is proved.

11. Since a divides b, we write $b = ak$, for some integer k. Since b divides a, we can write $a = bn$ for some integer n. Then

$$
\begin{array}{ll}
b = ak & \\
\ = (bn)k & \text{Law of substitution} \\
\ = b(nk) & \text{Associative property of multiplication}
\end{array}
$$

Since $b \neq 0$, $nk = 1$. Since k and n are both integers $n = 1$ and $k = 1$ or $n = -1$ and $k = -1$. If $n = k = 1$, $a = b(1) = b$. If $n = k = -1$, $a = b(-1) = -b$.

13. Suppose $a_1, a_2, \ldots, a_n$ are integers. If one of them, say a_1, is even, then for some integer k, $a_1 = 2k$, and

$$ a_1 a_2 a_3 \ldots a_n = 2k(a_2 a_3 \ldots a_n) = 2[k(a_2 a_3 \ldots a_n)] = 2M $$

We see that if one of the factors in the product $a_1a_2 \ldots a_n$ is an even integer, the product is even. Now let us assume that every one of the integers in the product is odd. Then a_1a_2 is odd. Since a_1a_2 is odd and a_3 is odd, $a_1a_2a_3$ is odd. Similarly $a_1a_2a_3a_4$ is odd. Continuing in this fashion we find that the product of n odd integers is odd.

15. Every integer can be represented as $4k$, $4k + 1$, $4k + 2$, or $4k + 3$. Of these four forms, $4k = 2(2k)$ and $4k + 2 = 2(2k + 1)$ are even. Integers of the form $4k + 1 = 2(2k) + 1 = 2L + 1$ and $4k + 3 = 4k + 2 + 1 = 2(2k + 1) + 1 = 2M + 1$ are odd.

17. Let $a = 4k + 3$ and $b = 4p + 3$. Then

$$
\begin{aligned}
a - b &= (4k + 3) - (4p + 3) \\
&= 4k + 3 - 4p - 3 \\
&= 4k - 4p + 3 - 3 \\
&= 4(k - p) + 0 \\
&= 4(k - p) = 4L
\end{aligned}
$$

By the definition of divisibility 4 divides $a - b$.

19.
$$
\begin{aligned}
a + b &= (cq + r) + (ct + s) \\
&= (cq + ct) + (r + s) \\
&= c(q + t) + (t + s)
\end{aligned}
$$

Since c divides $a + b$, we can write $a + b = cn$ where n is an integer. Then

$$cn = c(q + t) + (r + s)$$

and

$$cn - c(q + t) = r + s$$
$$c(n - [q + t]) = r + s$$

and c divides $r + s$ by the definition of divisibility.

Exercise 4

1. (a); (b); (d); (e); (g).
3. (c); (d); (e).
5. (g).
7. (a); (c); (d); (f).
9. A number is divisible by 45 if the sum of the digits of its numeral is divisible by 9 and its units digit is 0 or 5.
11. (a); (b); (e).
13. A number divisible by 4 must also be divisible by 2, hence it must be even; no.
15. (a) 0; 3; 6; 9; (b) 2; 5; 8; (c) 1; 4; 7; (d) 0; 3; 6; 9; (e) 2; 5; 8; (f) 1; 4; 7.
17. No; to be divisible by 4 a number must be even.
19. (c); (d).

Chapter 3
Exercise 1

1. The primes less than 200 are: 2; 3; 5; 7; 11; 13; 17; 19; 23; 29; 31; 37; 41; 43; 47; 53; 59; 61; 67; 71; 73; 79; 83; 89; 97; 101; 103; 107; 109; 113; 127; 131; 137; 139; 149; 151; 157; 163; 167; 173; 179; 181; 191; 193; 197; 199.
3. No.
5. 1; 3; 7; 9.
7. (a) 19; 47 (b) 5; 73; (c) 3; 11; 101; (d) 2; 5; 73; (e) 3; 7; 41; (f) 23; 761;
9. (a) 5; 13; 17; 29; (b) 3; 7; 11; 19; (there are infinitely many others).
11. There are no other primes of this form. No; if n were odd $n^2 - 1$ would be even and the only even prime is 2 which is not of this form.
13. $7 = 2^3 - 1; 31 = 2^5 - 1; 127 = 2^7 - 1.$
15. An integer $p > 1$ is a prime if it has only two divisors, 1 and itself.
17. (a); (d); (e).
19. (a) True; (b) True; (c) True; (d) True; (e) True.

Exercise 2

1. 8.
3. 2.
5. 17.
7. 2.
9. 4.
11. $x = 4; y = -7.$
13. $x = 6; y = -17.$
15. $x = 42; y = -55.$
17. (b); (c); (f); (h).
19. $n.$

Exercise 3

1. (a) 2; 3; 13; (b) 5; 73; (c) 2; 101; (d) 2; 3; (e) 2; 3; 5; (f) 2; 211.
3. Every composite number can be factored as the product of prime factors in one and only one way except for the order of the factors.
5. One and itself.
7. 14 divisors; 1; 2; 4; 5; 8; 10; 16; 20; 32; 40; 64; 80; 160; 320.
9. $1 + 2 + 4 + 7 + 14 = 28.$

Exercise 4

1. 137–139; 149–151; 179–181; 191–193; 197–199; 227–229
3. $2^2-2 = 2; 3^2-2 = 7; 5^2-2 = 23; 7^2-2 = 47; 9^2-2 = 79; 13^2-2 = 167; 15^2-2 = 223.$

5. 2; 3; 5; 11; 23.
7. 149–139; 191–181; 251–241; 293–283.
9. 29; 31; 37; 41; 43; 47.

Chapter 4
Exercise 1

1. (a) 5; (b) 0; (c) 3; (d) 5; (e) 4; (f) 5.
3. (a) 1; (b) 5; (c) 3; (d) 1; (e) 5; (f) 6.
5. (a) $3(5 + 4) = (3)(2) = 6$; $(3)(5) + (3)(4) = 1 + 5 = 6$; (b) $6(4 + 6) =$
$(6)(3) = 4$; $(6)(4) + (6)(6) = 3 + 1 = 4$; (c) $3(1 + 5) = (3)(6) = 4$; $(3)(1) +$
$(3)(5) = 3 + 1 = 4$; (d) $6(5 + 2) = (6)(0) = 0$; $(6)(5) + (6)(2) = 2 + 5 = 0$.

7.

+	0	1	2	3	4
0	0	1	2	3	4
1	1	2	3	4	0
2	2	3	4	0	1
3	3	4	0	1	2
4	4	0	1	2	3

9. (a) $(4 + 2) + 3 = 1 + 3 = 4$; $4 + (2 + 3) = 4 + 0 = 4$; (b) $2(4 + 3) =$
$(2)(2) = 4$; $(2)(4) + (2)(3) = 3 + 1 = 4$; (c) The additive inverse of 0 is 0; the
additive inverse of 1 is 4; the additive inverse of 2 is 3; the additive inverse of
3 is 2; the additive inverse of 4 is 1; (d) The multiplicative inverse of 1 is 1;
the multiplicative inverse of 2 is 3; the multiplicative inverse of 3 is 2; the
multiplicative inverse of 4 is 4; (e) $x = 4$; (f) $x = 4$.
11. (a) 6; (b) It has none; (c) No; 0; 2; 4; 6; (c) It has no solution.

Exercise 2

1. 0-class: . . . , $-16, -8, 0, 8, 16, . . .$; 1-class: . . . , $-15, -7, 1, 9, 17, . . .$;
2-class: . . . , $-14, -6, 2, 10, 18, . . .$; 3-class: . . . , $-13, -5, 3, 11, 19, . . .$;
4-class: . . . , $-12, -4, 4, 12, 20, . . .$; 5-class: . . . , $-11, -3, 5, 13, 21, . . .$;
6-class: . . . , $-10, -2, 6, 14, 22, . . .$; 7-class: . . . , $-9, -1, 7, 15, 23, ...$
3. (a); (e).
5. 16 classes; (a) 11; (b) 7; (c) 0; (d) 0; (e) 14; (f) 13.
7. 3.
9. 8.

Exercise 3

1. (b); (c); (d).
3. (a) 6; (b) 5; (c) 8; (d) 6; (e) 5; 7; (f) 4; 8.

5. (a) 3; (b) 2; (c) 0; 2; 4; 6; (d) 1; 3; 5; 7.
7. (a) 8; (b) 5; (c) 2; 6; 10; 14.
9. 3; 11; 19; 27; 35; . . . ; -5; -13; -21; -29; . . . .

Exercise 4

1. $a \equiv a \pmod{m}$ means that $a - a = 0$ is divisible by m. But $0 = (0)(m)$ and hence by the definition of divisibility m divides 0 and $a \equiv a \pmod{m}$.
3. Since $a \equiv b \pmod{m}$ and $b \equiv c \pmod{m}$, we have $a = b + mk$ and $b = c + mh$. Then

$$
\begin{aligned}
a &= (c + mh) + mk \\
&= c + (mh + mk) \\
&= c + m(h + k)
\end{aligned}
$$

and, by the definition of congruence, $a \equiv c \pmod{m}$.
5. (a) 1; (b) 5; (c) 6; (d) 6.
7. (a) 6; 17; 28; 39; 50; (b) -5; 12; (c) -33; -15; 3; 21; 39.
9. 0.
11. Let $n + 0, n + 1, n + 2, n + 3, \ldots, n + (m - 1)$ represent m consecutive integers. No two of these belong to the same residue class modulo m. If there were two of these numbers, say $n + i$ and $n + j$, $i \neq j$, belonging to the same residue class then

$$
\begin{aligned}
n + i &\equiv n + j \pmod{m} \\
i &\equiv j \pmod{m}
\end{aligned}
$$

which is impossible since $i \neq j$ and both i and j are less than m.
13. Since m is odd, $(2, m) = 1$, hence we have m given numbers of the form

$$
0, a, 2a, \ldots, (m - 1)a
$$

where $a = 2$. This is a complete set of residues modulo m.
15. $b^2 - 4ac \equiv b^2 \pmod{4}$. If b is even, b^2 is of the form $4K$ which is congruent to $0 \pmod{4}$. If b is odd, b^2 is of the form $4K + 1$, which is congruent to $1 \pmod{4}$.

Exercise 5

1. (a) 1; (b) 1; 2; (c) No solutions; (d) No solutions; (e) 0; 2; 4; 6.
3. (a) No solutions; (b) 4; (c) 2; 5; (d) 0; 3; 6; (e) No solutions.
5. $1 + 6t$; $2 + 6t$; $4 + 6t$; $5 + 6t$.
7. No solutions.
9. No solutions.

Exercise 6

1. (a) 1; (b) 2; (c) 3; (d) 9; (e) 1.

3. $x \equiv 4 \pmod 5$.

5. $x \equiv 10, 27, 44, 61, 78, 95, 112, 129, 146, 163, 180, 197, 214, 231, 248, 265, 282, 299 \pmod{306}$.

7. No solutions.

9. $x \equiv 0, 4, 8, 12, 16, 20, 24, 28, 32, 36, 40, 44, 48, 52, 56, 60, 64, 68, 72, 76, 80, 84, 88, 92 \pmod{96}$.

11. $x \equiv 13, 27, 41 \pmod{42}$.

13. $x \equiv 620 \pmod{729}$.

15. $x \equiv 22 \pmod{157}$.

Exercise 7

1. If $n = a_0 10^k + a_1 10^{k-1} + \cdots + a_{k-1} 10 + a_k$ is divisible by 2, then

$$a_0 10^k + a_1 10^{k-1} + \cdots + a_{k-1} 10 + a_k \equiv 0 \pmod 2$$
$$a_0 \cdot 0 + a_{k-1} \cdot 0 + \cdots + a_{k-1} \cdot 0 + a_k \equiv 0 \pmod 2$$
$$a_k \equiv 0 \pmod 2$$

3. If $n = a_0 10^k + a_1 10^{k-1} + \cdots + a_{k-1} 10 + a_k$ is divisible by 10, then

$$a_0 10^k + a_1 10^{k-1} + \cdots + a_{k-1} 10 + a_k \equiv 0 \pmod{10}$$
$$a_0 \cdot 0 + a_1 \cdot 0 + \cdots + a_{k-1} \cdot 0 + a_k \equiv 0 \pmod{10}$$
$$a_k \equiv 0 \pmod{10}$$

5. If $n = a_0 10^k + a_1 10^{k-1} + \cdots + a_{k-1} 10 + a_k$ is divisible by 4, then

$$a_0 10^k + a_1 10^{k-1} + \cdots + a_{k-1} 10 + a_k \equiv 0 \pmod 4$$
$$a_0 \cdot 0 + a_1 \cdot 0 + \cdots + a_{k-2} \cdot 0 + a_{k-1} 10 + a_k \equiv 0 \pmod 4$$
$$a_{k-1} 10 + a_k \equiv 0 \pmod 4$$

But $a_{k-1} 10 + a_k$ is the number named by the tens and units digits of the given number.

7. Notice that
$$1 \equiv 10^2 \equiv 10^4 \equiv 10^6 \equiv \cdots \equiv 10^{2n} \pmod{11}$$
$$-1 \equiv 10 \equiv 10^3 \equiv 10^5 \equiv \cdots \equiv 10^{2n+1} \pmod{11}$$

If $n = a_0 10^k + a_1 10^{k-1} + \cdots + a_{k-1} 10 + a_k$ is divisible by 11, then

$$a_0 10^k + a_1 10^{k-1} + \cdots + a_{k-1} 10 + a_k \equiv 0 \pmod{11}$$
$$a_k + a_{k-1} 10 + a_{k-2} 10^2 + \cdots + a_1 10^{k-1} + a_0 10^k \equiv 0 \pmod{11}$$
$$a_k - a_{k-1} + a_{k-2} - \cdots + (-1)^k a_0 \equiv 0 \pmod{11}$$

Chapter 5
Exercise 1

1. (a) {Sunday, Monday, Tuesday, Wednesday, Thursday, Friday, Saturday};
(b) {April, June, September, November}; (c) {Summer, Winter, Fall, Spring};
(d) {Tuesday, Thursday}; (e) {Virginia, Vermont}.
3. (a) $\{x \mid x$ is a natural number and x is less than 8$\}$; (b) $\{x \mid x$ is a month of
the calendar year$\}$; (c) $\{x \mid x$ is one of the first eight letters of the English alphabet$\}$;
(d) $\{x \mid x$ is an even integer$\}$.
5. (a) $\{p, a, s, t, e, r, m, o, n, \ell\}$ (b) $\{\ell, e, r\}$; (c) $\{p, \ell, a, s, t, e, r, m, o, n,$
$b, u, i, d\}$; (d) $\{\ell, e, r, m, a, s, o, n\}$; (e) $\{p, \ell, a, s, t, e, r\} = P$; (f) $\{\ell, a, s,$
$e, r\}$.
7. (a) U; (b) A; (c) $\emptyset$; (d) A; (e) $\emptyset$; (f) U.
11. (a) $\{9\}$; (b) U.
13. (a) $\{3, 5, 7\}$; (b) $\{1, 2, 3, 4, 5, 7\}$.
15. (a) $\emptyset$; (b) U.
17. $\{x \mid x \in R$ and $x \neq \sqrt{10}\}$
19. $\{x \mid x \in R$ and $-\frac{1}{3} \leq x \leq \frac{5}{8}\}$

Exercise 2

1. (a) $\{(-2, -2), (-2, 0), (-2, 2), (0, -2), (0, 0), (0, 2), (2, -2), (2, 0), (2, 2)\}$;
(b) $\{(-2, 1), (-2, 2), (-2, 3), (0, 1), (0, 2), (0, 3), (2, 1), (2, 2), (2, 3)\}$; (c) $\{(3,$
$-2), (3, 0), (3, 2), (2, -2), (2, 0), (2, 2), (1, -2), (1, 0), (1, 2)\}$; (d) $\{(1, 1), (1, 2),$
$(1, 3), (2, 1), (2, 2), (2, 3), (3, 1), (3, 2), (3, 3)\}$.
3. $\{(-1, 0), (-1, -1), (-1, 1), (0, -1), (0, 0), (0, 1), (1, 0), (1, -1), (1, 1)\}$.
5. (a) 24; (b) 36; (c) 16; (d) 24.
7. See graph.

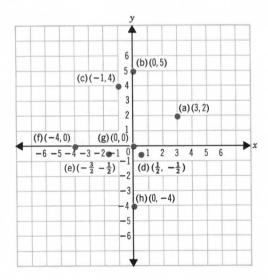

9. (8, 0).

11. (0, −2).

13. (5, 4).

15. (a) {(−3, −3), (−3, 2), (−3, 4), (2, −3), (2, 2), (2, 4), (4, −3), (4, 2), (4, 4)}, (b) See graph.

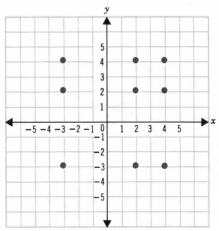

17. See graph.

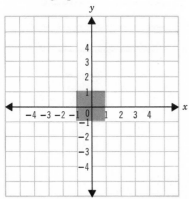

19. (a) {(1, 1), (1, 2), (1, 3), (1, 4), (2, 1), (2, 2), (2, 3), (2, 4), (3, 1), (3, 2), (3, 3), (3, 4), (4, 1), (4, 2), (4, 3), (4, 4)}; (b) See graph; (c) {(1, 4), (2, 3), (4, 1), (3, 2)}.

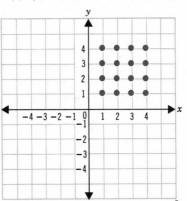

Exercise 3

1. $\{(2, 2), (1, 4), (4, 1)\}$

3. See figures.

(a)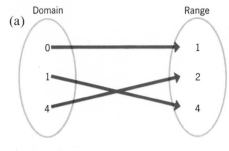

(b)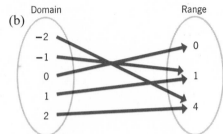

(c)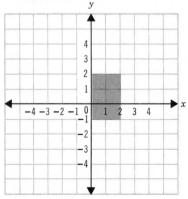

5. (a) Domain: $\{0, 1, 2, 3, 4\}$; Range: $\{6, 7, 8, 9, 10\}$; (b) Domain: $\{-2, 3, 4\}$; Range: $\{3, 4, 5, 7\}$; (c) Domain: $\{-2, -1, 0, 1, 2\}$; Range: $\{-8, -1, 0, 1, 8\}$; (d) Domain: $\{-2, -1, 1, 2, 4\}$; Range: $\{1, 4, 16\}$.

7. See graph.

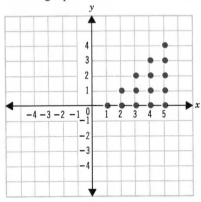

9. See graph.

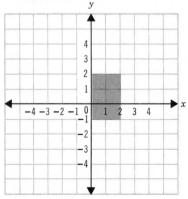

Exercise 4

1. (a) -1; (b) 0; (c) -27; (d) 8; (e) 64; (f) -8.

3. (a) 4; (b) 21; (c) $\frac{23}{18}$; (d) $\frac{17}{2}$; (e) 8; (f) $\frac{1}{8}$.

5. $\frac{3}{2}$.

7. (a) $x^2 + 4x + 6$;   (b) $x^2 - 6x + 11$;   (c) $x^2 + 2xh + h^2 - 2x - 2h + 3$;
(d) $2x - 2 + h$.

9. See graph.

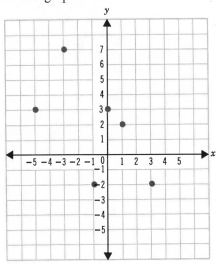

11. See graph.

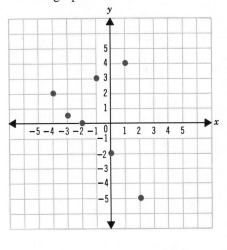

13. See graph.

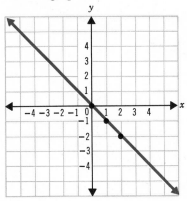

15. See graph.

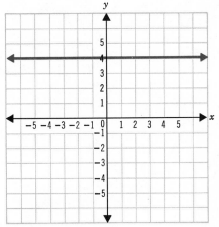

Exercise 5

1. See graph.

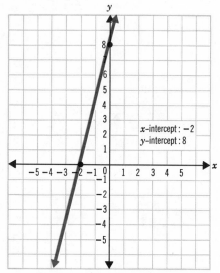

5. See graph.

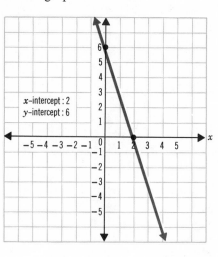

3. See graph.

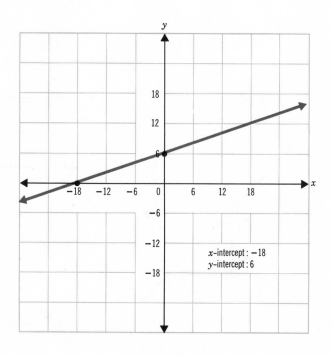

7. See graph.

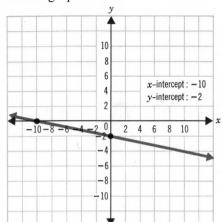

9. See graph.

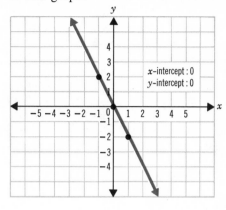

11. 3.

13. $\frac{3}{5}$.

15. $-\frac{2}{3}$.

17. -3

19. $\frac{5}{2}$

21. -6

23. $-\frac{5}{3}$

25. 7

Exercise 6

1. See graph.

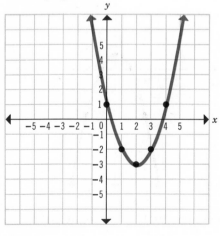

3. See graph.

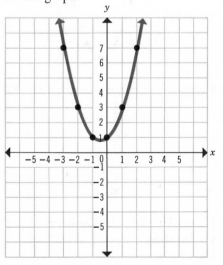

5. See graph.

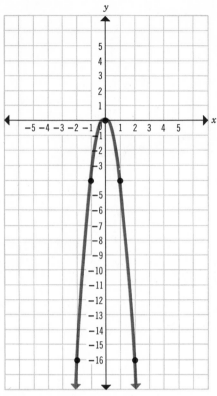

7. See graph.

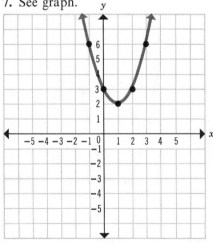

9. See graph.

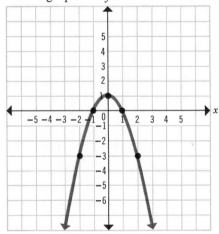

11. (a) Upward; (b) Downward; (c) Downward; (d) Upward; (e) Upward.

13. Parabola.

15. Upward.

Chapter 6
Exercise 1

1. (a) (1) 465; (2) 40,484; (b) (1) 1042; (2) 71,632; (c) (1) 280; (2) 9399; (d) (1) 143; (2) 4872.

3. (c)

5. (a) Transitive; (b) Transitive; (c) Reflexive, symmetric, and transitive.

7. (a) 21; (b) 1; (c) -7; (d) -14; (e) 17; (f) -10.

9. (a) Symmetric; (b) Addition; (c) Transitive; (d) Multiplication.

Exercise 2

1. (a) $\dfrac{5}{3}$; (b) $\dfrac{4}{3}$; (c) $\dfrac{-6}{7}$; (d) $\dfrac{7}{-6}$; (e) $\dfrac{15}{-7}$; (f) $\dfrac{12}{19}$.

3. (a) Not a field; no additive identity; no additive inverses; (b) Not a field; no multiplicative inverses; no multiplicative identity; no additive inverses; (c) Not a field; no additive identity; no additive inverses; no multiplicative inverses; (d) Not a field; no additive identity; no additive inverses; no multiplicative identity; no multiplicative inverses; not closed under multiplication; (e) Field; (f) Not a field; no multiplicative inverses; (g) Not a field; no multiplicative inverses; (h) Not a field; no multiplicative inverses.

5.

$$\left(\frac{a}{b} + \frac{c}{d}\right) + \frac{e}{f} = \frac{ad + bc}{bd} + \frac{e}{f}$$
Definition of addition of rational numbers

$$= \frac{(ad + bc)f + (bd)e}{(bd)f}$$
Definition of addition of rational numbers

$$= \frac{(ad)f + (bc)f + (bd)e}{(bd)f}$$
Distributive property of integers

$$= \frac{a(df) + b(cf) + b(de)}{b(df)}$$
Associative property of multiplication of integers

$$= \frac{a(df) + b(cf + de)}{b(df)}$$
Distributive property of integers

$$= \frac{a}{b} + \frac{cf + de}{df}$$
Definition of addition of rational numbers

$$= \frac{a}{b} + \left(\frac{c}{d} + \frac{e}{f}\right)$$
Definition of addition of rational numbers

7. $\dfrac{a}{b} \cdot \left(\dfrac{c}{d} \cdot \dfrac{e}{f} \right) = \dfrac{a}{b} \left(\dfrac{ce}{df} \right)$ — Definition of multiplication of rational numbers

$= \dfrac{a(ce)}{b(df)}$ — Definition of multiplication of rational numbers

$= \dfrac{(ac)e}{(bd)f}$ — Associative property of multiplication of integers

$= \dfrac{ac}{bd} \cdot \dfrac{e}{f}$ — Definition of multiplication of rational numbers

$= \left(\dfrac{a}{b} \cdot \dfrac{c}{d} \right) \cdot \dfrac{e}{f}$ — Definition of multiplication of rational numbers

9. (a) even + (odd + odd) = even + even = even;
(even + odd) + odd = odd + odd = even;
(b) even × (odd + odd) = even × even = even;
(even × odd) + (even × odd) = even + even = even.
11. odd × odd = odd; odd × even = even × odd = even.
13. The system is not a field; the multiplicative inverse axiom doesn't hold because 2, 4, and 6 do not have multiplicative inverses.
15. $ty = tz$ Given
    $t \neq 0$ Given
Since $t \neq 0$ it has a multiplicative inverse t^{-1} by the multiplicative inverse axiom.

$t^{-1}(ty) = t^{-1}(tz)$ Multiplication property of equality
$(t^{-1}t)y = (t^{-1}t)z$ Associative property of multiplication
$1 \cdot y = 1 \cdot z$ Multiplicative inverse axiom
$y = z$ Multiplicative identity axiom

17. $y + (-y) = 0$ Additive inverse axiom
$t[y + (-y)] = t \cdot 0$ Multiplication property of equality
$t[y + (-y)] = 0$ Theorem 6.1 (multiplication property of zero)
$t[(-y) + y] = 0$ Commutative property of addition
$t(-y) + t \cdot y = 0$ Distributive property

ty is an element of F because of the closure property of multiplication and has an additive inverse, $-ty$, by the additive inverse axiom. Then

$[t(-y) + ty] + (-ty) = 0 + (-ty)$ Addition property of equality
$t(-y) + [ty + (-ty)] = 0 + (-ty)$ Associative property of addition
$t(-y) + 0 = 0 + (-ty)$ Additive inverse axiom
$t(-y) = -ty$ Additive identity axiom

19. Suppose that there are two multiplicative identities, 1 and $1'$. Then

$$1 \cdot 1' = 1 \qquad \text{Multiplicative identity axiom}$$
$$1 \cdot 1' = 1' \qquad \text{Multiplicative identity axiom}$$
$$1 = 1' \qquad \text{Substitution property of equality}$$

Since $1 = 1'$, the two elements are one and the same.

21.

$y + (-y) = 0$	Additive inverse axiom
$(-x)[y + (-y)] = -x \cdot 0$	Multiplication property of equality
$(-x)y + (-x)(-y) = -x \cdot 0$	Distributive property
$(-x)y + (-x)(-y) = 0$	Theorem 6.1
$xy + [(-x)y + (-x)(-y)] = xy + 0$	Addition property of equality
$[xy + (-x)y] + (-x)(-y) = xy + 0$	Associative property of addition
$[x + (-x)]y + (-x)(-y) = xy + 0$	Distributive property
$0 \cdot y + (-x)(-y) = xy + 0$	Additive inverse axiom
$0 + (-x)(-y) = xy + 0$	Theorem 6.1
$(-x)(-y) = xy$	Additive identity axiom

Exercise 3

1. (a) 4; (b) 3; (c) 2; (d) 1; (e) 6; (f) 5.

3. (a) 3; (b) 0; (c) 1; (d) 0; (e) 4; (f) 0.

5. $6(-3) = 6(4) = 3;\ -(6 \cdot 3) = -4 = 3.$

7. $\dfrac{2}{6} \cdot \dfrac{6}{2} = [(2)(6)][(6)(4)] = (5)(3) = 1.$

9. $\dfrac{2}{5} - \dfrac{4}{2} = 2(3) - (4)(4) = 6 - 2 = 6 + 5 = 4;\ \dfrac{2 \cdot 2 - 5 \cdot 4}{5 \cdot 2} = \dfrac{4 - 6}{3} = \dfrac{5}{3} =$

$5 \cdot 5 = 4.$

11. $\dfrac{\frac{3}{4}}{\frac{4}{5}} = \left(\dfrac{3}{4}\right)\left(\dfrac{4}{5}\right)^{-1} = (3 \cdot 2)(4 \cdot 3)^{-1} = 6(5)^{-1} = 6 \cdot 3 = 4.$

13. No; 2, 4, and 6 do not have multiplicative inverses.

15. (a) 1; (b) 3; (c) 2; (d) 3; (e) 3; (f) 1.

Exercise 4

1. (a) $\{6\}$; (b) $\{5\}$; (c) $\{5\}$; (d) $\{2\}$; (e) $\{0\}$; (f) $\{2\}$.

3. (a) 1; (b) 6; (c) 4; (d) 1.

5. 1.

7. (a) $\{3\}$; (b) $\{2\}$; (c) $\{1\}$; (d) $\{2\}$; (e) $\{2\}$; (f) $\{0\}$.

9. (a) $2 \cdot 0 + 1 = 1 \neq 6;\quad 2 \cdot 1 + 1 = 3 \neq 6;\quad 2 \cdot 2 + 1 = 5 \neq 6;\quad 2 \cdot 3 + 1 = 7 \neq 6;\ 2 \cdot 4 + 1 = 1 \neq 6;\ 2 \cdot 5 + 1 = 3 \neq 6;\ 2 \cdot 6 + 1 = 5 \neq 6;\ 2 \cdot 7 + 1 = 7 \neq 6;$ (b) $2 \cdot 0 + 3 = 3 \neq 1; 2 \cdot 1 + 3 = 5 \neq 1; 2 \cdot 2 + 3 = 7 \neq 1; 2 \cdot 3 + 3 = 1; 2 \cdot 4 + 3 = 3 \neq 1; 2 \cdot 5 + 3 = 5 \neq 1; 2 \cdot 6 + 3 = 7 \neq 1; 2 \cdot 7 + 3 = 1;$ the solutions are 3 and 7.

Exercise 5

1. (a) 6; (b) 2; (c) 2; (d) 4; (e) 5; (f) 1.
3. 0; 1; 6
5. {2, 5}
7. {1, 6}
9. Ø
11. {3, 6}
13. {1, 3}
15. Ø.
17. {2, 3, 4, 5, 6}.
19. {5}.

Exercise 6

1. See graph.

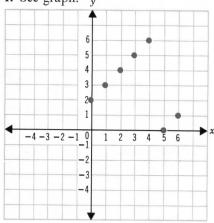

3. See graph.

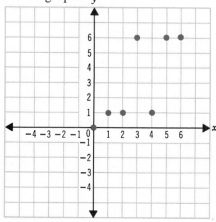

5. See graph.

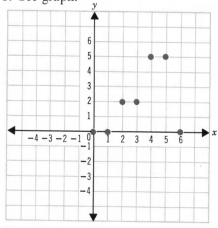

7. See graph.

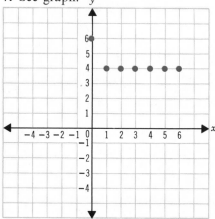

9. See graph.

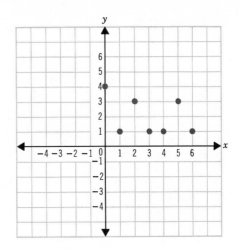

Exercise 7

1. No; no inverses.
3. (e); (f); (g).

5.

*	1	3	5	7
1	1	3	5	7
3	3	1	7	5
5	5	7	1	3
7	7	5	3	1

The identity is 1; the associative property is satisfied; every element has an inverse (every element is its own inverse).

7. Zero is the identity element. The operation is associative since each of the elements is an integer and addition of integers is associative. The inverse of any element km of the set is $-km$. Since addition of integers is commutative, this is a commutative group.
9. yes

Exercise 8

1. (a) V; (b) V; (c) D'; (d) R_3; (e) H; (f) R_0.

3.

*	R_0	R_1	R_2	V_1	V_2	V_3
R_0	R_0	R_1	R_2	V_1	V_2	V_3
R_1	R_1	R_2	R_0	V_2	V_3	V_1
R_2	R_2	R_0	R_1	V_3	V_1	V_2
V_1	V_1	V_3	V_2	R_0	R_2	R_1
V_2	V_2	V_1	V_3	R_1	R_0	R_2
V_3	V_3	V_2	V_1	R_2	R_1	R_0

R_0, R_1, R_2, V_1, V_2, and V_3 are defined as shown below:

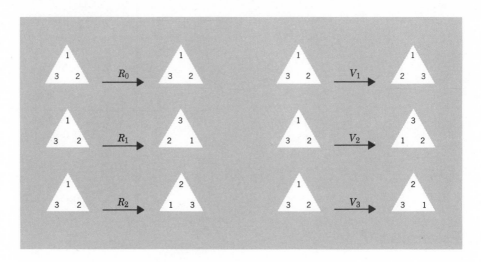

This system is a group, but not a commutative group.

5. (a) 16; (b) 20; (c) 24; (d) $2n$.

Exercise 9

1.

$$b * a = c * a$$ Given

$$a^{-1} \in G$$ Inverse axiom

$$(b * a) * a^{-1} = (c * a) * a^{-1}$$ Theorem 6.6a

$$b * (a * a^{-1}) = c * (a * a^{-1})$$ Associative property

$$b * e = c * e$$ Inverse axiom

$$b = c$$ Identity axiom

3. Since every element of a group G has an inverse, a^{-1} exists and is an element of G. Then

$$a * x = a$$ Given

$$a^{-1} * (a * x) = a^{-1} * a$$ Theorem 6.6

$$(a^{-1} * a) * x = a^{-1} * a$$ Associative property

$$e * x = e$$ Inverse axiom

$$x = e$$ Identity axiom

5.

$$e * e = e$$ Identity axiom

$$e * e^{-1} = e$$ Inverse axiom

$$e * e = e * e^{-1}$$ Law of substitution

$$e = e^{-1}$$ Problem 1, Exercise 9

7.

$a * a^{-1} = e$	Inverse axiom
$b * b^{-1} = e$	Inverse axiom
$a * a^{-1} = b * b^{-1}$	Law of substitution
$a = b$	Given
$a * a^{-1} = a * b^{-1}$	Law of substitution
$a^{-1} = b^{-1}$	Problem 1, Exercise 9

9.

$y * a = b$	Given
$(y * a) * a^{-1} = b * a^{-1}$	Theorem 6.6a
$y * (a * a^{-1}) = b * a^{-1}$	Associative property
$y * e = b * a^{-1}$	Inverse axiom
$y = b * a^{-1}$	Identity axiom

We have shown that there is a solution. Now we must show that there is only one solution. Suppose that there are two solutions y_1 and y_2. Then

$y_1 * a = b$	Definition of a solution of an equation
$y_2 * a = b$	Definition of a solution of an equation
$y_1 * a = y_2 * a$	Law of substitution
$(y_1 * a) * a^{-1} = (y_2 * a) * a^{-1}$	Theorem 6.6a
$y_1 * (a * a^{-1}) = y_2 * (a * a^{-1})$	Associative property
$y_1 * e = y_2 * e$	Inverse axiom
$y_1 = y_2$	Identity axiom

Exercise 10

1.

*	e	p	q	r	s	t
e	e	p	q	r	s	t
p	p	q	e	s	t	r
q	q	e	p	t	r	s
r	r	t	s	e	q	p
s	s	r	t	p	e	q
t	t	s	r	q	p	e

3. (e).

5. (a) 2; (b) 120; (c) 24.

7. $ABCD$; $ABDC$; $ACBD$; $ACDB$; $ADBC$; $ADCB$; $BCDA$; $BCAD$; $BADC$; $BACD$; $CABD$; $CADB$; $CBDA$; $CBAD$; $CDAB$; $CDBA$; $DABC$; $DACB$; $DBAC$; $DBCA$; $DCAB$; $DCBA$; $BDAC$; $BDCA$.

9. $x = D$; $y = A$; $z = B$; $w = C$

ANSWERS TO ODD-NUMBERED PROBLEMS

11. (a) (i) $\begin{pmatrix} A & B & C \\ C & A & B \end{pmatrix}$ (ii) $\begin{pmatrix} A & B & C \\ B & C & A \end{pmatrix}$ (iii) $\begin{pmatrix} A & B & C \\ A & B & C \end{pmatrix}$

(iv) $\begin{pmatrix} A & B & C \\ A & C & B \end{pmatrix}$ (v) $\begin{pmatrix} A & B & C \\ C & B & A \end{pmatrix}$ (vi) $\begin{pmatrix} A & B & C \\ B & A & C \end{pmatrix}$

(b) (i) R_1; (ii) R_2; (iii) R_0; (iv) V_3; (v) V_1; (vi) V_2.

Chapter 7
Exercise 1

1. (a); (c); (d); (e); (f).
3. 3
5. $5\sqrt{5} \doteq 11.18$
7. $\sqrt{468}$ ft. $\doteq 21.6$ ft.
9. 17
11. 3
15. $\sqrt{52}$ ft. $\doteq 7.2$ ft.

Exercise 2

1. (3, 4, 5); (5, 12, 13); (15, 8, 17); (7, 24, 25); (21, 20, 29); (35, 12, 37); (9, 40, 41); (45, 28, 53); (63, 16, 65); (11, 60, 61); (33, 56, 65); (55, 48, 73); (77, 36, 85); (13, 84, 85); (39, 80, 89); (65, 72, 97).
3. (a) (40, 9, 41); (b) (84, 85, 13); (c) (176, 57, 185); (d) (800, 881, 369); (e) (252, 275, 373); (f) (156, 133, 205); (g) (120, 209, 241); (h) (132, 85, 157).
5. Suppose $(y, z) = d \neq 1$. Then for integers y_1 and z_1

$$y = dy_1 \text{ and } z = dz_1$$

From $x^2 + y^2 = z^2$ we have

$$x^2 = z^2 - y^2$$
$$= (dz_1)^2 - (dy_1)^2$$
$$= d^2(z_1^2 - y_1^2)$$

Hence d^2 divides x^2 and hence d divides x. But for a primitive solution $x, y,$ and z must not have a common divisor $d \neq 1$, therefore $(y, z) = 1$.
7. Let the three consecutive positive integers be $n - 1$, n, and $n + 1$. Then

$$(n - 1)^2 + n^2 = (n + 1)^2$$
$$n^2 - 2n + 1 + n^2 = n^2 + 2n - 1$$
$$n^2 - 4n = 0$$
$$n(n - 4) = 0$$
$$n = 0 \quad \text{or} \quad n = 4$$

Exercise 2

1. (a) $2x - 7y + 32 = 0$; (b) $2x + 9y - 39 = 0$; (c) $4x + 3y = 0$; (d) $x - 5y - 33 = 0$; (e) $3x + 2y - 14 = 0$; (f) $14x + 19y - 73 = 0$.

3. (c); (d).

5. $x - y + 1 = 0$; $3x - y - 15 = 0$; $x - 3y + 19 = 0$.

7. The equations of the sides of the figure are (1) $x + 2y - 7 = 0$; (2) $x + 2y + 6 = 0$; (3) $5x - 3y - 9 = 0$; and (4) $5x - 3y - 22 = 0$. Since lines (1) and (2) are parallel and (3) and (4) are parallel, the figure is a parallelogram.

9. (a) Yes; (b) Yes; yes.

11. $(15, -1)$

Exercise 3

1. The coordinates of D are (a, c); the coordinates of E are $(-a, c)$.

$$AD = \sqrt{(-2a - a)^2 + (0 - c)^2} = \sqrt{9a^2 + c^2}$$
$$BE = \sqrt{(2a + a)^2 + (0 - c)^2} = \sqrt{9a^2 + c^2}$$

Since $AD = BE$, the medians are of equal length.

3. $AC = BD = \sqrt{a^2 + b^2}$.

5. Using the distance formula we find

$$AD = \sqrt{(b - 0)^2 + (c - 0)^2} = \sqrt{b^2 + c^2}$$
$$AB = \sqrt{(a - 0)^2 + (0 - 0)^2} = a$$

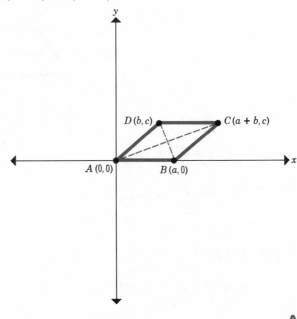

5.

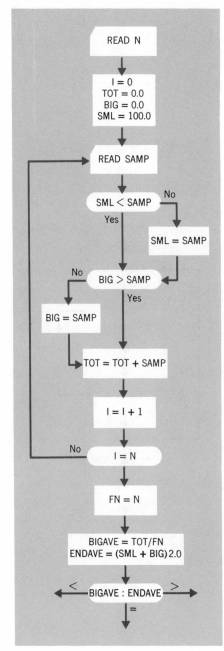

7. The area of a circle with radius R.

9.

$$AV = \frac{W1 + W2 + W3 + W4 + W5}{5.0}$$

$$
\begin{aligned}
S1 &= (W1 - AV)^2 \\
S2 &= (W2 - AV)^2 \\
S3 &= (W3 - AV)^2 \\
S4 &= (W4 - AV)^2 \\
S5 &= (W5 - AV)^2
\end{aligned}
$$

$$SD = \sqrt{\frac{S1 + S2 + S3 + S4 + S5}{5.0}}$$

```
READ W1, W2, W3, W4, W5
AV = (W1 + W2 + W3 + W4 + W5)/5.0
S1 = (W1 - AV) ** 2
S2 = (W2 - AV) ** 2
S3 = (W3 - AV) ** 2
S4 = (W4 - AV) ** 2
S5 = (W5 - AV) ** 2
SD = ((S1 + S2 + S3 + S4 + S5)/5.0) ** .5
```

INDEX

INDEX

We see that if one of the factors in the product $a_1 a_2 \ldots a_n$ is an even integer, the product is even. Now let us assume that every one of the integers in the product is odd. Then $a_1 a_2$ is odd. Since $a_1 a_2$ is odd and a_3 is odd, $a_1 a_2 a_3$ is odd. Similarly $a_1 a_2 a_3 a_4$ is odd. Continuing in this fashion we find that the product of n odd integers is odd.

15. Every integer can be represented as $4k$, $4k + 1$, $4k + 2$, or $4k + 3$. Of these four forms, $4k = 2(2k)$ and $4k + 2 = 2(2k + 1)$ are even. Integers of the form $4k + 1 = 2(2k) + 1 = 2L + 1$ and $4k + 3 = 4k + 2 + 1 = 2(2k + 1) + 1 = 2M + 1$ are odd.

17. Let $a = 4k + 3$ and $b = 4p + 3$. Then

$$
\begin{aligned}
a - b &= (4k + 3) - (4p + 3) \\
&= 4k + 3 - 4p - 3 \\
&= 4k - 4p + 3 - 3 \\
&= 4(k - p) + 0 \\
&= 4(k - p) = 4L
\end{aligned}
$$

By the definition of divisibility 4 divides $a - b$.

19.
$$
\begin{aligned}
a + b &= (cq + r) + (ct + s) \\
&= (cq + ct) + (r + s) \\
&= c(q + t) + (t + s)
\end{aligned}
$$

Since c divides $a + b$, we can write $a + b = cn$ where n is an integer. Then

$$cn = c(q + t) + (r + s)$$

and

$$
\begin{aligned}
cn - c(q + t) &= r + s \\
c(n - [q + t]) &= r + s
\end{aligned}
$$

and c divides $r + s$ by the definition of divisibility.

Exercise 4

1. (a); (b); (d); (e); (g).
3. (c); (d); (e).
5. (g).
7. (a); (c); (d); (f).
9. A number is divisible by 45 if the sum of the digits of its numeral is divisible by 9 and its units digit is 0 or 5.
11. (a); (b); (e).
13. A number divisible by 4 must also be divisible by 2, hence it must be even; no.
15. (a) 0; 3; 6; 9; (b) 2; 5; 8; (c) 1; 4; 7; (d) 0; 3; 6; 9; (e) 2; 5; 8; (f) 1; 4; 7.
17. No; to be divisible by 4 a number must be even.
19. (c); (d).

Chapter 3
Exercise 1

1. The primes less than 200 are: 2; 3; 5; 7; 11; 13; 17; 19; 23; 29; 31; 37; 41; 43; 47; 53; 59; 61; 67; 71; 73; 79; 83; 89; 97; 101; 103; 107; 109; 113; 127; 131; 137; 139; 149; 151; 157; 163; 167; 173; 179; 181; 191; 193; 197; 199.
3. No.
5. 1; 3; 7; 9.
7. (a) 19; 47 (b) 5; 73; (c) 3; 11; 101; (d) 2; 5; 73; (e) 3; 7; 41; (f) 23; 761;
9. (a) 5; 13; 17; 29; (b) 3; 7; 11; 19; (there are infinitely many others).
11. There are no other primes of this form. No; if n were odd $n^2 - 1$ would be even and the only even prime is 2 which is not of this form.
13. $7 = 2^3 - 1$; $31 = 2^5 - 1$; $127 = 2^7 - 1$.
15. An integer $p > 1$ is a prime if it has only two divisors, 1 and itself.
17. (a); (d); (e).
19. (a) True; (b) True; (c) True; (d) True; (e) True.

Exercise 2

1. 8.
3. 2.
5. 17.
7. 2.
9. 4.
11. $x = 4$; $y = -7$.
13. $x = 6$; $y = -17$.
15. $x = 42$; $y = -55$.
17. (b); (c); (f); (h).
19. n.

Exercise 3

1. (a) 2; 3; 13; (b) 5; 73; (c) 2; 101; (d) 2; 3; (e) 2; 3; 5; (f) 2; 211.
3. Every composite number can be factored as the product of prime factors in one and only one way except for the order of the factors.
5. One and itself.
7. 14 divisors; 1; 2; 4; 5; 8; 10; 16; 20; 32; 40; 64; 80; 160; 320.
9. $1 + 2 + 4 + 7 + 14 = 28$.

Exercise 4

1. 137–139; 149–151; 179–181; 191–193; 197–199; 227–229
3. $2^2–2 = 2$; $3^2–2 = 7$; $5^2–2 = 23$; $7^2–2 = 47$; $9^2–2 = 79$; $13^2–2 = 167$; $15^2–2 = 223$.

Since n is a positive integer it cannot be zero, hence $n = 4$. Then $n - 1 = 3$ and $n + 1 = 5$ and the three integers are 3, 4, and 5.

9. (3, 4, 5); (5, 12, 13); (11, 60, 61).

Exercise 3

1. (a) (28, 45, 53); (28, 195, 197); (b) (8, 17, 15); (c) (12, 35, 37); (12, 13, 5); (d) (16, 65, 63); (e) (20, 99, 101); (20, 29, 21); (f) (32, 257, 255); (g) (36, 325, 323); (36, 85, 77); (h) (40, 399, 401); (40, 41, 9).

3. (a); (c); (e); (f).

5. (30, 40, 50); (24, 45, 51); (20, 48, 52); (28, 45, 53).

7. (56, 785, 783); (56, 65, 33).

9. (68, 1155, 1157); (68, 285, 293).

Exercise 4

1. $\sqrt{n^2 + 1}$

5. $\sqrt{168.75}$

7.
$$
\begin{aligned}
(nr)^2 + (ns)^2 &= n^2 r^2 + n^2 s^2 \\
&= n^2(r^2 + s^2) \\
&= n^2 t^2 \qquad \text{Since } r^2 + s^2 = t^2 \\
&= (nt)^2
\end{aligned}
$$

Exercise 5

1. (a) 5; (b) 5; (c) 5; (d) 10; (e) $\sqrt{52} = 2\sqrt{13}$; (f) $\sqrt{89}$.

3. The lengths of the three sides of the triangle are $\sqrt{26}$, $\sqrt{104}$, and $\sqrt{130}$. Since $(\sqrt{26})^2 + (\sqrt{104})^2 = (\sqrt{130})^2$ the triangle is a right triangle.

5. 5; $9\sqrt{2}$; 13

7. The lengths of the sides of the triangle are $\sqrt{5}$, $\sqrt{37}$, and $\sqrt{20}$. The perimeter is $\sqrt{5} + \sqrt{37} + \sqrt{20}$.

9. Let A, B, C, and D have coordinates $(2, -2)$, $(4, 0)$, $(2, 2)$ and $(0, 0)$ respectively. Then $AB = \sqrt{8}$; $BC = \sqrt{8}$; $CD = \sqrt{8}$; and $DA = \sqrt{8}$. We see that the four sides of the figure are equal in length. We now show that the four angles of the figure are right angles. The length of diagonal $\overline{AC}$ is 4. Now $(AB)^2 + (BC)^2 = (AC)^2$, hence triangle ABC is a right triangle and angle ABC is a right angle. In a similar manner we can show that the other three angles of the figure are right angles.

11. 7; -1

13. x-coordinate of midpoint: 3; y-coordinate of midpoint: -1.

15. $d_1 = \sqrt{(4-2)^2 + (4-1)^2} = \sqrt{13}; \quad d_2 = \sqrt{(5-2)^2 + (3-1)^2} = \sqrt{13};$
$d_1 = d_2.$
17. The lengths of the sides of the triangle are 6, 10, and 8. Since $6^2 + 8^2 = 10^2$ the triangle is a right triangle; area: 24.
19. No.

Chapter 8
Exercise 1

1. 18.
3. 10,920.
5. 840.
7. 8.
9. 8.
11. 192.

Exercise 2

1. (a) 12; (b) 180; (c) $\frac{1}{6}$; (d) $\frac{145}{12}$.
3. (a) $6! = 720$; (b) 30; (c) 720; (d) 120.
5. $8! = 40,320.$
7. 5040.
9. 11,880.

Exercise 3

1. (a) 6; (b) 35; (c) 28; (d) 252; (e) 220; (f) 125,970.
3. 220.
5. 15.
7. 166,320.
9. 8,400.
11. Row 9: 1, 9, 36, 84, 126, 126, 84, 96, 9, 1; Row 10: 1, 10, 45, 120, 210, 252, 210, 120, 45, 10, 1.

Exercise 4

1. {H, T}.
3. (a) {(1, 1), (1, 2), (1, 3), (1, 4), (1, 5), (1, 6), (2, 1), (2, 2), (2, 3), (2, 4), (2, 5), (2, 6), (3, 1), (3, 2), (3, 3), (3, 4), (3, 5), (3, 6), (4, 1), (4, 2), (4, 3), (4, 4), (4, 5), (4, 6), (5, 1), (5, 2), (5, 3), (5, 4), (5, 5), (5, 6), (6, 1), (6, 2), (6, 3), (6, 4), (6, 5), (6, 6)}; (b) 36; (c) {(1, 6), (2, 5), (3, 4), (4, 3), (5, 2), (6, 1)}}; (d) {(1, 6), (2, 6), (3, 6), (4, 6), (5, 6), (6, 6), (6, 1), (6, 2), (6, 3), (6, 4), (6, 5)}.

5. (a)

1¢	5¢	10¢	Outcome

(b) {HHT, HTH, THH}: (c) {HTT, THT, TTH}; (d) {HHT, HTH, HTT, THH, THT, TTH, TTT}.

7. (a) {HHHH}; (b) {HHTT, HTHT, THTH, TTHH, HTTH, THHT}:
(c) {HHHH, HHTT, HTHT, HTTH, THHT, THTH, TTTT, TTHH}.

9. (a) $\frac{1}{6}$; (b) $\frac{1}{6}$; (c) $\frac{11}{36}$; (d) $\frac{1}{4}$; (e) $\frac{13}{18}$.

11. (a) $\frac{1}{2}$; (b) $\frac{4}{5}$; (c) $\frac{1}{2}$.

13. (a) $\frac{8}{15}$; (b) $\frac{8}{15}$; (c) $\frac{1}{5}$; (d) $\frac{4}{15}$.

15. 0.99.

17. (a) $\frac{7}{64}$; (b) $\frac{1}{64}$; (c) $\frac{3}{32}$; (d) $\frac{3}{32}$.

19. (a) Yes; (b) $\frac{1}{10}$; (c) $\frac{3}{20}$; (d) 1.

Exercise 5

1. (a) $\frac{1}{12}$; (b) $\frac{5}{18}$; (c) $\frac{5}{12}$; (d) 0; (e) $\frac{1}{36}$.

3. (a) $\frac{4}{29}$; (b) $\frac{12}{29}$; (c) $\frac{13}{29}$; (d) $\frac{16}{29}$.

5. $\frac{3}{4}$.

7. $\frac{3}{5}$.

9. (a) $\frac{1}{20}$; (b) $\frac{1}{50}$; (c) $\frac{7}{100}$.

Exercise 6

1. $\frac{2}{11}$.

3. $\frac{2}{3}$.

5. $\frac{7}{24}$.

7. $\frac{9}{98}$.

9. $\frac{1}{6}$.

Exercise 7

1. (a) Mean: 5; median: 4; mode: 3; (b) Mean: 23; median: $22\frac{1}{2}$; mode: 21;
(c) Mean: 3; median: 3; mode: 3.

3. 1464.

5. 24,475 barrels.

7. (a) \$7500; (b) \$6000.
9. Mode.
11. (a) \$7123.08; (b) \$5000; (c) \$5000; (d) Mode; (e) Mean; (f) Mean.
13. (a) Mean; (b) Mode; (c) Median; (d) Mode.
15. 70.
17. (a) 86; (b) 80.
19. Maisie was thinking of the mode; the customer of the mean.

Exercise 8

1. (a) 4.3; (b) 2.
3. Mean: 3.01; standard deviation: 0.04.
5. 2.

Chapter 9
Exercise 1

1. (a) -6; (b) -7; (c) 4; (d) -2.
3. (a) $x = 1$; $y = -2$; (b) $x = 6$; $y = -4$; (c) $x = -6$; $y = -12$; $z = -6$;
(d) $x = 3\sqrt{3}$; $y = -5\sqrt{6}$; $z = -4$.

5. (a) $\begin{pmatrix} 2 & 3 & 4 \\ 3 & 4 & 5 \\ 4 & 5 & 6 \end{pmatrix}$ (b) $\begin{pmatrix} 0 & -3 & -8 \\ 3 & 0 & -5 \\ 8 & 5 & 0 \end{pmatrix}$ (c) $\begin{pmatrix} 1 & \sqrt{2} & \sqrt{3} \\ \sqrt{2} & 2 & \sqrt{6} \\ \sqrt{3} & \sqrt{6} & 3 \end{pmatrix}$

(d) $\begin{pmatrix} 1 & 2 & 3 \\ 2 & 4 & 6 \\ 3 & 6 & 9 \end{pmatrix}$ (e) $\begin{pmatrix} 2 & 9 & 28 \\ 9 & 16 & 35 \\ 28 & 35 & 54 \end{pmatrix}$ (f) $\begin{pmatrix} -1 & -4 & -7 \\ 1 & -2 & -5 \\ 3 & 0 & -3 \end{pmatrix}$

7. (a) 10; (b) 0; (c) 14; (d) B; (e) 35; (f) 5; (g) 37; (h) 14.
9. (a) 0.3; (b) 0.6; (c) 0.3.

Exercise 2

1. (a) $\begin{pmatrix} 1 & -1 \\ 6 & -4 \end{pmatrix}$ (b) $\begin{pmatrix} 40 & 122 \\ -18 & -108 \end{pmatrix}$

3. (a) $\begin{pmatrix} \frac{4}{3} & \frac{11}{10} & \frac{7}{6} \\ \frac{9}{8} & \frac{3}{2} & 1 \\ \frac{11}{8} & \frac{11}{12} & \frac{13}{10} \end{pmatrix}$ (b) $\begin{pmatrix} 0.2 & -0.8 \\ 0.5 & 12.6 \\ 8.4 & -12.1 \end{pmatrix}$

5. (a) $\begin{pmatrix} 2\sqrt{5} & 3\sqrt{2} \\ -7\sqrt{5} & 8\sqrt{2} \end{pmatrix}$ **(b)** $\begin{pmatrix} 5\sqrt{3} & -\sqrt{2} & -3\sqrt{5} \\ 9\sqrt{2} & 5\sqrt{5} & 2\sqrt{3} \\ 2\sqrt{5} & 0 & -2 \end{pmatrix}$

7. (a) $\begin{pmatrix} -3 & 2 \\ 4 & -1 \end{pmatrix}$ **(b)** $\begin{pmatrix} 1.8 & 3.2 \\ -6.7 & -1.5 \end{pmatrix}$ **(c)** $\begin{pmatrix} 0 & 0 & -3 \\ 2 & -7 & 9 \end{pmatrix}$

(d) $\begin{pmatrix} -\frac{1}{2} & -\frac{2}{3} & \frac{3}{4} \\ 2 & 0 & -1 \end{pmatrix}$ **(e)** $\begin{pmatrix} 1 \\ -3 \\ -2 \end{pmatrix}$ **(f)** $(1 \quad -\frac{1}{2} \quad -5 \quad 2)$

9. (a) $\begin{pmatrix} 8 & 2 \\ -3 & 5 \end{pmatrix}$ **(b)** $\begin{pmatrix} 6 & 4 & 4 \\ 11 & -5 & 1 \\ -6 & 7 & 1 \end{pmatrix}$ **(c)** $\begin{pmatrix} \frac{1}{4} & \frac{11}{6} \\ \frac{11}{8} & -\frac{1}{4} \\ \frac{1}{6} & -\frac{1}{4} \end{pmatrix}$

11. (a) $\begin{pmatrix} 850 & 550 \\ 1050 & 750 \end{pmatrix}$; **(b)** 1900; **(c)** 1400; **(d)** 1300; **(e)** 1800

13. (a) $A + B = B + A = \begin{pmatrix} 4 & 3 & 2 & -5 \\ 3 & 0 & -2 & 9 \end{pmatrix}$

(b) $C + (-C) = \begin{pmatrix} 0 & 0 & 0 & 0 \\ 0 & 0 & 0 & 0 \end{pmatrix} = D$

(c) $(A + B) + C = A + (B + C) = \begin{pmatrix} 2 & -2 & -5 & -4 \\ 2 & 0 & -3 & 9 \end{pmatrix}$

(d) $B + D = B$

Exercise 3

1. (a) (1) $\begin{pmatrix} -19 & -8 \\ 7 & -6 \end{pmatrix}$ **(2)** $\begin{pmatrix} -15 & -20 \\ 1 & -10 \end{pmatrix}$

(b) (1) $\begin{pmatrix} 0 & 31 \\ 8 & 98 \end{pmatrix}$ **(2)** $\begin{pmatrix} 56 & 40 \\ 65 & 42 \end{pmatrix}$

3. (a) (1) $\begin{pmatrix} -8 & -4 \\ -4 & -2 \end{pmatrix}$ **(2)** $\begin{pmatrix} -12 & -2 \\ 12 & 2 \end{pmatrix}$

(3) $\begin{pmatrix} 9 & 2 \\ 0 & 1 \end{pmatrix}$ **(4)** $\begin{pmatrix} 8 & 4 \\ -8 & -4 \end{pmatrix}$

(b) (1) $\begin{pmatrix} -18 & 19 \\ 20 & -18 \end{pmatrix}$ **(2)** $\begin{pmatrix} -14 & -13 \\ -28 & -22 \end{pmatrix}$

(3) $\begin{pmatrix} -8 & 0 \\ 0 & -8 \end{pmatrix}$ **(4)** $\begin{pmatrix} 1 & 4 \\ -8 & 17 \end{pmatrix}$

5. (a) $\begin{pmatrix} 23 & 27 \\ 16 & 34 \end{pmatrix}$ (b) $\begin{pmatrix} -5 & -15 \\ 103 & 94 \end{pmatrix}$ (c) $\begin{pmatrix} 12 & -2 \\ 0 & 10 \end{pmatrix}$

(d) $\begin{pmatrix} 22 & 38 \\ -16 & -14 \end{pmatrix}$ (e) $\begin{pmatrix} -12 & 34 \\ 22 & 38 \end{pmatrix}$ (f) $\begin{pmatrix} -50 & -52 \\ 50 & 46 \end{pmatrix}$

(g) $\begin{pmatrix} 16 & 14 \\ 12 & 18 \end{pmatrix}$ (h) $\begin{pmatrix} 16 & 20 \\ 36 & 24 \end{pmatrix}$ (i) $\begin{pmatrix} 38 & 37 \\ 12 & -20 \end{pmatrix}$

7. $(A \cdot B) \cdot C$

$$= \begin{pmatrix} a_{11}b_{11}+a_{12}b_{21} & a_{11}b_{12}+a_{12}b_{22} \\ a_{21}b_{11}+a_{22}b_{21} & a_{21}b_{12}+a_{22}b_{22} \end{pmatrix} \cdot \begin{pmatrix} c_{11} & c_{12} \\ c_{21} & c_{22} \end{pmatrix}$$

$$= \begin{pmatrix} (a_{11}b_{11}+a_{12}b_{21})c_{11}+(a_{11}b_{12}+a_{12}b_{22})c_{21} & (a_{11}b_{11}+a_{12}b_{21})c_{12}+(a_{11}b_{12}+a_{12}b_{22})c_{22} \\ (a_{21}b_{11}+a_{22}b_{21})c_{11}+(a_{21}b_{12}+a_{22}b_{22})c_{21} & (a_{21}b_{11}+a_{22}b_{21})c_{12}+(a_{21}b_{12}+a_{22}b_{22})c_{22} \end{pmatrix}$$

$$= \begin{pmatrix} a_{11}b_{11}c_{11}+a_{12}b_{21}c_{11}+a_{11}b_{12}c_{21}+a_{12}b_{22}c_{21} & a_{11}b_{11}c_{12}+a_{12}b_{21}c_{12}+a_{11}b_{12}c_{22}+a_{12}b_{22}c_{22} \\ a_{21}b_{11}c_{11}+a_{22}b_{21}c_{11}+a_{21}b_{12}c_{21}+a_{22}b_{22}c_{21} & a_{21}b_{11}c_{12}+a_{22}b_{21}c_{12}+a_{21}b_{12}c_{22}+a_{22}b_{22}c_{22} \end{pmatrix}$$

$$= \begin{pmatrix} a_{11}(b_{11}c_{11}+b_{12}c_{21})+a_{21}(b_{21}c_{11}+b_{22}c_{21}) & a_{11}(b_{11}c_{12}+b_{12}c_{22})+a_{12}(b_{21}c_{12}+b_{22}c_{22}) \\ a_{21}(b_{11}c_{11}+b_{12}c_{21})+a_{22}(b_{21}c_{11}+b_{22}c_{21}) & a_{21}(b_{11}c_{12}+b_{12}c_{22})+a_{22}(b_{21}c_{12}+b_{22}c_{22}) \end{pmatrix}$$

$$= \begin{pmatrix} a_{11} & a_{12} \\ a_{21} & a_{22} \end{pmatrix} \cdot \begin{pmatrix} b_{11}c_{11}+b_{12}c_{21} & b_{11}c_{12}+b_{12}c_{22} \\ b_{11}c_{11}+b_{22}c_{21} & b_{21}c_{12}+b_{22}c_{22} \end{pmatrix}$$

$$= A \cdot (B \cdot C)$$

9.
$$A \cdot (B + C) = \begin{pmatrix} 1 & 1 \\ 0 & 1 \end{pmatrix}; \quad A \cdot B + A \cdot C = \begin{pmatrix} 1 & 1 \\ 0 & 1 \end{pmatrix}$$

11. $\begin{pmatrix} -13 & 10 & 2 & -3 & -1 \\ -14 & 15 & 1 & 1 & 2 \\ 18 & -13 & -3 & 5 & 2 \end{pmatrix}$

13. (a) $\begin{pmatrix} 1 & 2 & 2 \\ 2 & 1 & 4 \\ 1 & 2 & 2 \end{pmatrix}$ (b) Impossible   (c) $\begin{pmatrix} 3 & -1 & 3 \\ -3 & -2 & -3 \\ 3 & 1 & 3 \\ 2 & -2 & 2 \end{pmatrix}$

(d) $\begin{pmatrix} 9 & -3 & 6 \\ -3 & 1 & -2 \\ 6 & -2 & 4 \end{pmatrix}$

15. (a) $\begin{pmatrix} 0 & 0 & 2 \\ 0 & 3 & 0 \\ 4 & 0 & 0 \end{pmatrix}$ (b) $\begin{pmatrix} 0 & 0 & 4 \\ 0 & 3 & 0 \\ 2 & 0 & 0 \end{pmatrix}$

Exercise 4

1. (a) $\begin{pmatrix} 1 & -\frac{1}{2} \\ -\frac{1}{2} & \frac{1}{2} \end{pmatrix}$ (b) $\begin{pmatrix} \frac{2}{5} & -\frac{1}{5} \\ -\frac{1}{5} & \frac{3}{5} \end{pmatrix}$

3. (a) $\begin{pmatrix} -\frac{1}{2} & \frac{1}{2} \\ 1 & 0 \end{pmatrix}$ (b) $\begin{pmatrix} \frac{1}{2} & \frac{1}{2} \\ \frac{1}{4} & \frac{3}{4} \end{pmatrix}$

5. (a) $9 \cdot 2 - 6 \cdot 3 = 0$; (b) $2 \cdot 8 - 4 \cdot 4 = 0$; (c) $3 \cdot 8 - 4 \cdot 6 = 0$.
7. Have inverses: a; d; do not have inverses: b; c.
9. When $a_{11}a_{22} - a_{12}a_{21} = 1$.

Exercise 5

1. $x = 2$; $y = 1$.
3. $x = -2$, $y = 4$.
5. $x = 3$; $y = 5$.
7. $x = \frac{1}{2}$; $y = -\frac{1}{2}$.
9. $x = \frac{1}{3}$; $y = -\frac{3}{5}$.

Exercise 6

1. (a) 14; (b) -3; (c) -14; (d) -2.
3. 6.
5. 0.
7. 49.
9. 4.
11. 0
13. -20
15. 0

17. $\begin{vmatrix} a & b \\ c & d \end{vmatrix} = ad - bc = -(bc - ad) = -(cb - ad) = -\begin{vmatrix} c & d \\ a & b \end{vmatrix}$

19. $\begin{vmatrix} a & b & c \\ d & e & f \\ g & h & i \end{vmatrix} = a(ei - hf) - b(di - gf) + c(dh - ge)$

$$= \begin{vmatrix} a & d & g \\ b & e & h \\ c & f & i \end{vmatrix}$$

Chapter 10
Exercise 1

1. Postulate 1: Each pair of wires in S has at least one bead in common.
Postulate 2: Each pair of wires in S has not more than one bead in common.
Postulate 3: Every bead in S is on at least two wires.
Postulate 4: Every bead in S is on not more than two wires.
Postulate 5: The total number of wires in S is four.
3. Postulate 1: Each pair of rows in S has at least one tree in common.
Postulate 2: Each pair of rows in S has not more than one tree in common.
Postulate 3: Every tree in S is in at least two rows.
Postulate 4: Every tree in S is in not more than two rows.
Postulate 5: The total number of rows in S is four.
5. Postulate 1: Each pair of classes in S has at least one student in common.
Postulate 2: Each pair of classes in S has not more than one student in common.
Postulate 3: Every student in S is in at least two classes.
Postulate 4: Every student in S is in not more than two classes.
Postulate 5: The total number of classes in S is four.
7. Flight 1: San Diego, Los Angeles, Oakland.
Flight 2: San Diego, San Francisco, Portland.
Flight 3: Los Angeles, San Francisco, Seattle.
Flight 4: Oakland, Portland, Seattle.

Exercise 2

1. By Postulate 1 each pair of lines have at least one point in common. By Postulate 2 each pair of lines have not more than one point in common. Therefore, each pair of lines has exactly one point in common.
3. Let "point" be interpreted as "member" and "line" be interpreted as "committee." Then the postulates read:
A-1: Every pair of members in S is on at least one committee together.
A-2: Every pair of members in S is on not more than one committee together.
A-3: Every committee in S contains at least two members.
A-4: Every committee in S contains not more than two members.
A-5: The total number of members is four.
Note: The wording in the postulates have been changed to make better English statements.
5. See figure at top of p. 499.
7. New York–Boston; New York–Philadelphia; New York–Washington; Boston–Philadelphia; Boston–Washington; Philadelphia–Washington.
9. By Problem 8 there are exactly six lines in S. These are A-B, A-C, A-D, B-C, B-D, and C-D. We see that through point A we have three lines A-B, A-C, and A-D; through point B we have three lines, A-B, B-C, and B-D; through point

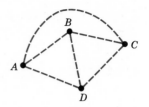

The points are A, B, C, and D
The lines are: AB
$\qquad AC$
$\qquad AD$
$\qquad BC$
$\qquad BD$
$\qquad CD$

C we have three lines, A-C, B-C, and C-D; through point D we have three lines, A-D, B-D, and C-D. Since this list contains all the lines in S there are exactly three lines passing through each point.

Exercise 3

1. To show the independence of Postulate 2, let us interpret S to consist of seven points, A, B, C, D, E, F, and G, distributed among four lines, I, II, III, and IV as follows:

$$\begin{array}{lcccc}
\text{I} & A & B & C & G \\
\text{II} & A & D & E & G \\
\text{III} & B & F & D & \\
\text{IV} & C & F & E &
\end{array}$$

In this interpretation Postulate 2 fails because lines I and II have more than one point in common. Postulate 1 is satisfied because each pair of lines has at least one point in common. Postulate 3 is satisfied because every point is on at least two lines. Postulate 4 is satisfied because every point is on at most two lines. Postulate 5 is satisfied because there are four lines.

3. To show the independence of Postulate 3 let us interpret S to consist of seven points, A, B, C, D, E, F, and G, distributed among four lines, I, II, III, and IV, as follows:

$$\begin{array}{lcccc}
\text{I} & A & B & C & \\
\text{II} & A & D & E & \\
\text{III} & B & E & F & G \\
\text{IV} & C & D & G &
\end{array}$$

In this interpretation Postulate 3 fails because each point is not on two lines. Postulate 1 is satisfied because each pair of lines has at least one point in common. Postulate 2 is satisfied because each pair of lines has not more than one point in common. Postulate 4 is satisfied because every point is on not more than two lines. Postulate 5 is satisfied because there are four lines in S.

Exercise 4

1. See figure.

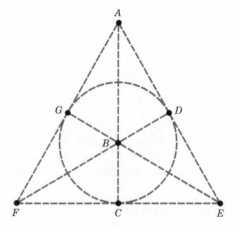

The points are A, B, C, D, E, and F
The lines are: ABC
$\qquad\qquad ADE$
$\qquad\qquad DBF$
$\qquad\qquad AGF$
$\qquad\qquad FCE$
$\qquad\qquad GBE$
$\qquad\qquad CDG$ (the circle)

3. L_1: P_2, P_3, P_4; L_2: P_1, P_3, P_5; L_3: P_1, P_2, P_6; L_4: P_1, P_4, P_7; L_5: P_2, P_5, P_7; L_6: P_3, P_6, P_7; L_7: P_4, P_5, P_6.
5. Let P_1 and P_2 be two points of S. By Postulate 1 at least one line contains P_1 and P_2. By Postulate 2 there is at most one line containing P_1 and P_2. Therefore, P_1 and P_2 are on exactly one line.
7. Let "point" be interpreted as "bead" and "line" be interpreted as "wire."
9. By S-6, there exists at least one point. By S-4 exactly three lines contain each point. Hence the assertion that there is at least one line is certainly true.

Chapter 11
Exercise 1

1. (a) $\sqrt{29}$; (b) $\sqrt{89}$; (c) $\sqrt{97}$; (d) $\sqrt{13}$; (e) 3; (f) $\sqrt{82}$; (g) $5\sqrt{2}$; (h) $5\sqrt{2}$.
3. $(-\frac{5}{2}, -4)$; $(-2, -1)$; $(-\frac{13}{2}, -2)$.
5. $(\frac{11}{2}, \frac{9}{2})$.
7. $(-\frac{7}{5}, 0)$.
9. Vertices of a triangle; $AB + BC \neq AC$.

Exercise 2

1. (a) $2x - 7y + 32 = 0$; (b) $2x + 9y - 39 = 0$; (c) $4x + 3y = 0$; (d) $x - 5y - 33 = 0$; (e) $3x + 2y - 14 = 0$; (f) $14x + 19y - 73 = 0$.
3. (c); (d).
5. $x - y + 1 = 0$; $3x - y - 15 = 0$; $x - 3y + 19 = 0$.
7. The equations of the sides of the figure are (1) $x + 2y - 7 = 0$; (2) $x + 2y + 6 = 0$; (3) $5x - 3y - 9 = 0$; and (4) $5x - 3y - 22 = 0$. Since lines (1) and (2) are parallel and (3) and (4) are parallel, the figure is a parallelogram.
9. (a) Yes; (b) Yes; yes.
11. $(15, -1)$

Exercise 3

1. The coordinates of D are (a, c); the coordinates of E are $(-a, c)$.

$$AD = \sqrt{(-2a - a)^2 + (0 - c)^2} = \sqrt{9a^2 + c^2}$$
$$BE = \sqrt{(2a + a)^2 + (0 - c)^2} = \sqrt{9a^2 + c^2}$$

Since $AD = BE$, the medians are of equal length.
3. $AC = BD = \sqrt{a^2 + b^2}$.
5. Using the distance formula we find

$$AD = \sqrt{(b - 0)^2 + (c - 0)^2} = \sqrt{b^2 + c^2}$$
$$AB = \sqrt{(a - 0)^2 + (0 - 0)^2} = a$$

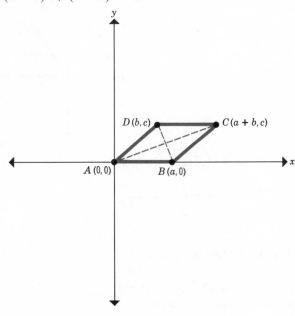

Since $ABCD$ is a rhombus, $AB = AD$, hence $a^2 = b^2 + c^2$ and $c = \sqrt{a^2 - b^2}$
The equation of $\overline{BD}$ is

$$x[\sqrt{a + b}] + y[\sqrt{a - b}] - a[\sqrt{a + b}] = 0$$

The equation of AC is

$$x[\sqrt{a - b}] - y[\sqrt{a + b}] = 0$$

But these two lines are perpendicular, hence the diagonals of a rhombus are perpendicular to each other.

7.

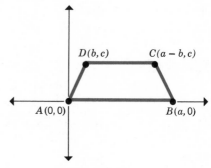

$$AC = BD = \sqrt{a^2 - 2ab + b^2 + c^2}$$

9.

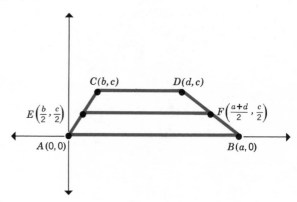

$$EF = \frac{a + d - b}{2}$$

$$\frac{CD + AB}{2} = \frac{d - b + a}{2} = EF$$

Exercise 4

1. $x^2 - 4x + y^2 - 8y + 16 = 0$
3. $16x^2 + 16y^2 - 1 = 0$

5. $x^2 - 2x + y^2 = 0$
7. $x^2 + 10x + y^2 - 24y + 104 = 0$
9. $x^2 + 3x + y^2 - 2y - 13 = 0$
11. Center: $(0, 0)$; Radius: 4.
13. Center: $(0, -3)$; Radius: 3.
15. Center: $\left(\dfrac{1}{2}, -\dfrac{3}{2}\right)$; Radius: $\dfrac{3}{2}$.

17. Center: $\left(\dfrac{1}{2}, -\dfrac{1}{3}\right)$; Radius: $\dfrac{\sqrt{3}}{3}$.

19. Center: $\left(\dfrac{1}{2}, \dfrac{1}{2}\right)$; Radius: $\dfrac{1}{6}$.

Chapter 12
Exercise 1

1. (a) $(7 \cdot 10^1) + (8 \cdot 10^0) + (1 \cdot 10^{-1})$
(b) $(1 \cdot 10^2) + (6 \cdot 10^1) + (3 \cdot 10^0) + (2 \cdot 10^{-1}) + (2 \cdot 10^{-2})$
(c) $(8 \cdot 10^2) + (2 \cdot 10^1) + (7 \cdot 10^0) + (6 \cdot 10^{-1})$
(d) $(7 \cdot 10^3) + (6 \cdot 10^2) + (1 \cdot 10^1) + (9 \cdot 10^0) + (3 \cdot 10^{-1}) + (2 \cdot 10^{-2})$
(e) $(8 \cdot 10^4) + (2 \cdot 10^3)$
(f) $(1 \cdot 10^5) + (3 \cdot 10^4) + (9 \cdot 10^2) + (9 \cdot 10^1) + (9 \cdot 10^0) + (9 \cdot 10^{-1})$.
3. (a) 10^2; (b) 10^6; (c) 10^{-3}; (d) 10^{-2}; (e) 10^0; (f) 10^{-4}; (g) 10^1; (h) 10^1.
5. (a) 1,002,003.05; (b) 70,000,000; (c) 9,100,040.36; (d) 90,000; (e) .004101.

Exercise 2

1. (a) 7; (b) 10; (c) 25; (d) 45; (e) 511; (f) 325.
3. (a) 4.5; (b) 7.125; (c) 10.65625; (d) 8.375; (e) 15.875; (f) 128.
5. (a) 1,001,110,001.11
(b) 1,000,101.0001
(c) 1,100,011.0101
(d) 1,101,010.01001
(e) 1,111,011,010.011
(f) 100,001.1001.
7. When written as a common fraction in simplest form its denominator is a power of 2.
9. (a) When the units digit is 1; (b) when the units digit is 0.

Exercise 3

1. (a) 1063.640625; (b) 343.375; (c) 52.03125; (d) 64.984375; (e) 63.109375;
(f) 29,696.5.

3. (a) 7730.110 (b) 100,000.06314; (c) 1644.1135; (d) 100.7354.
5. 8.
7. (a) 8^2; (b) 8^0; (c) 8^{-3}; (d) 8^{-2}.

Exercise 4

1. 1, 10, 11, 100, 101, 110, 111, 1000, 1001, 1010, 1011, 1100, 1101, 1110, 1111, 10000, 10001, 10010, 10011, 10100.
3. (a) 11; (b) 52; (c) 77; (d) 444; (e) 353; (f) 115; (g) 207; (h) 636; (i) 572.
5. (a) 1,010,011,100
   (b) 111,100,110,010
   (c) 101,111,011,001
   (d) 11,110,010,100,101
   (e) 111,111
   (f) 1,000,000
   (g) 101,110,010,000,000,000,110,011
   (h) 100,100,111,110,110
   (i) 10,011,110,001.

Exercise 5

1. (a) 4660; (b) 3627; (c) 2989; (d) 8173; (e) 13,114; (f) 26,892.
3. (a) 6AC; (b) AAA; (c) 28C; (d) 309; (e) BAD; (f) 1000.
5. (a) 0.875; (b) 0.015625; (c) 0.84375.
7. 0 to 4095.

Chapter 13
Exercise 1

1. (a) 10000; (b) 100101; (c) 10110; (d) 100011.
3. (a) 101; (b) 101; (c) 1111; (d) 1101.
5. (a) 001100; (b) 00000; (c) 01100; (d) 00110; (e) 01110; (f) 1110; (g) 1111; (h) 01011.
7. (a) 001001; (b) 100111; (c) 000010; (d) 011100; (e) 010110; (f) 011110.

Exercise 2

1. N	O	A
	CA	XX
	AD	KK
	ML	TO
	ST	XK
	⋮	
TO	LT	2
3.	CA	W1
	AD	W2
	AD	W3
	DV	W4
	⋮	
W1	WD	—
W2	WD	—
W3	WD	—
W4	WD	—

5. Increment until register becomes 10. At 10 transfer to OK. At OK the 10 is divided by the contents of word NM. The quotient is stored in word XX.

Exercise 3

1. (a) 3.46875; (b) 255.375; (c) .265625; (d) .03125; (e) 14.0625; (f) 91.25.
3. (a) 204500000000; (b) 207760000000; (c) 177400000000; (d) 210620000000;
(e) 207466000000; (f) 205526000000.

Chapter 14
Exercise 1

1. Integer names begin with any of the letters I, J, K, L, M, or N; Real names begin with any letter except I, J, K, L, M, or N.
3. (a) SIXTY; (b) I432; (c) IOH; (d) NEG444; (e) HUN; (f) BIG; (g) ZERO;
(h) NNT; (i) REST.
5. (a); (b); (d); (g); (i); (j); (k); (1).
7. One through five.
9. (1) The minimum and maximum size of a number. (2) The number of characters allowed in naming variables.
11. (a) MONYQM; (b) NUMSTK; (c) SPDSAT; (d) WASHBA; (e) NOBEAT;
(f) HB; (g) TRANS.

Exercise 2

1. (a) $X = A + B * (C - D ** 2)$; (b) $X = A + A * C/(B * D)$.

3. (a) $Y = A * B/2.0 + C * D/2.0$ (b) $Y = A * B/(1.0 + (A/D))$.

5. (a) $a = 3.1415r^2$; (b) $an = \dfrac{a}{g} + \dfrac{b}{g}$; (c) $x = \left(\dfrac{a \cdot b}{c \cdot d}\right)^2$

7. (a) 0.0; (b) -3.0.

9. (a) 10.0; (b) 27.0.

Exercise 3

1. 8.

3. 9.

5. 32.

7. 0.

9. 1.

11. 0.

13. 4.

15. 13.

17. 6.

19. 8.

Exercise 4

1. Transfer to statement 100.

3. $X = 70.0$.

5. ICOUNT is not less than or equal to 20, therefore the next sequential statement is executed where $\text{ICOUNT} = \text{ICOUNT} - 1$.

7. INEG will be set equal to -1.

9. The expression left of the logical operator reduces to -0.55 which is less than 25, therefore X is squared, giving the value $X = 231.04$.

11.　　　If (SUDS $-$ BIG) 10, 10, 20

　　　　10 SOAP $= X$

　　　　20 . . .

13.　　　$X = 65.0$

　　　　IF (IF $+$ 5) 3, 5, 5

　　　　3 $X = X - 2.0$

　　　　GO TO 7

　　　　5 $X = X + 2.0$

　　　　7 SUM $=$ SUM $+ X$

15.   IF (HRS − 40.0) 60, 60, 75
   60 PAY = RATE * 40.0
       GO TO 100
   75 PAY = RATE * 40.0 + (HRS − 40.0) * RATE * 1.5
 100 . . .

Exercise 5

1.

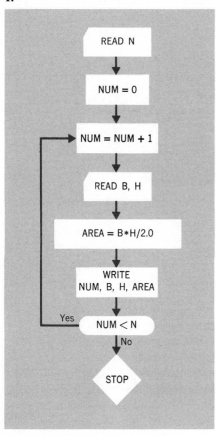

3.

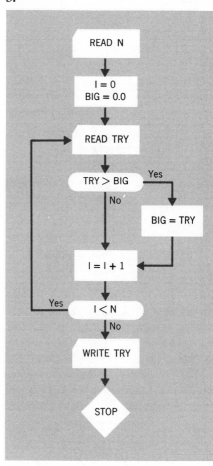

5.

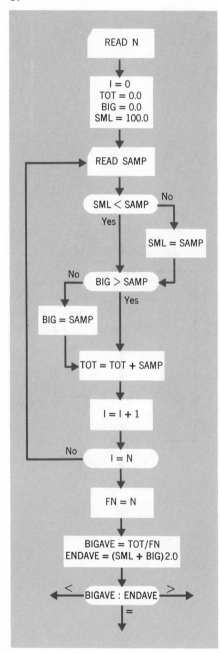

7. The area of a circle with radius R.

9.

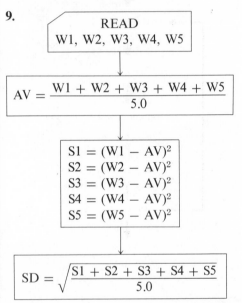

```
READ W1, W2, W3, W4, W5
AV = (W1 + W2 + W3 + W4 + W5)/5.0
S1 = (W1 − AV) ** 2
S2 = (W2 − AV) ** 2
S3 = (W3 − AV) ** 2
S4 = (W4 − AV) ** 2
S5 = (W5 − AV) ** 2
SD = ((S1 + S2 + S3 + S4 + S5)/5.0) ** .5
```

Exercise 6

1. (a) DIMENSION A(6), B(4), C(10); (b) DIMENSION X(17), I(20); (c) DI-MENSION P(10), R(2, 2), S(3, 3).
3. 200.
5. (a) 104; (b) ZEE(1, 1); (c) ZEE(7, 2); (d) ZEE(10, 3); (e) M(1, 2)
7. HIPPY(3, 1) HIPPY(5, 2), HIPPY(12, 2).
9. 239.

Exercise 7

1.

DIMENSION S(200)	Allots array S200 core locations.
SUM = 0.0	Clear word SUM.
DO 20 L = 1,N	Repeat all statements through statement 20 N times.
SUM = SUM + S(L)	Total the elements in array S.
20 CONTINUE	End of the DO LOOP
FN = N	Real value of integer N.
AM = SUM/FN	Mean of elements in array S.
SUMSQ = 0.0	Clear word SUMSQ.
DO 40 L = 1,N	Repeat all statements through 40 N times.
SUMSQ = SUMSQ + (S(L) − AM) ** 2	Sum of elements minus the average sample, squared.
40 CONTINUE	End of DO loop.
SD = (SUMSQ/FV) ** .5	Compute standard deviation.

3.
```
    DIMENSION NUM (50)
    NT = 0
    DO 66 K = 5, 9
    NT = NT + NUM(K)
66 CONTINUE
```
Another answer is:
```
DIMENSION NUM(50)
NT = 0
NT = NUM(5) + NUM(6) + NUM(7) + NUM(8) + NUM(9)
```
5. In this problem let us assume that the number of members in the set was read in and assigned to variable MSET. The set of numbers is in array X which is dimensional 2000.
```
    DIMENSION X(2000)
    TOT = 0.0
    DO 500 K = 1, MSET
    TOT = TOT + X(K)
```

```
500 CONTINUE
    RSET = MSET
    AM = TOT/RSET
```

7. In this problem we include the input and output statements:

```
    DIMENSION X(4000)
    READ N
    DO 100 I = 1, N
    IF (I .NE. 1) GO TO 75
    XMAX = X(1)
    XMIN = X(1)
    GO TO 100
 75 IF (XMAX .GT. X(1)) GO TO 50
    XMAX = X(1)
 50 IF (XMIN .LT. X(1)) GO TO 100
    XMIN = X(1)
100 CONTINUE
    WRITE XMIN, XMAX
    STOP
    END
```

9. (a) The statements find the minimum value of the 250 members of array FAT.

   (b) The statements compute the sum and the sum of the squares of each of the 55 elements of array SAMP.

INDEX

INDEX

INDEX

Randall Library – UNCW

NXWW

QA39.2 .W555 1972

Willerding / Mathematics, the alphabet of science

304900157290/